AF173205

The Diamond Mines of South Africa

Some Account of their Rise and Development

GARDNER F. WILLIAMS

CAMBRIDGE
UNIVERSITY PRESS

CAMBRIDGE UNIVERSITY PRESS

Cambridge, New York, Melbourne, Madrid, Cape Town, Singapore,
São Paolo, Delhi, Dubai, Tokyo, Mexico City

Published in the United States of America by Cambridge University Press, New York

www.cambridge.org
Information on this title: www.cambridge.org/9781108026598

© in this compilation Cambridge University Press 2011

This edition first published 1902
This digitally printed version 2011

ISBN 978-1-108-02659-8 Paperback

This book reproduces the text of the original edition. The content and language reflect
the beliefs, practices and terminology of their time, and have not been updated.

Cambridge University Press wishes to make clear that the book, unless originally published
by Cambridge, is not being republished by, in association or collaboration with, or
with the endorsement or approval of, the original publisher or its successors in title.

CAMBRIDGE LIBRARY COLLECTION

Books of enduring scholarly value

Technology

The focus of this series is engineering, broadly construed. It covers technological innovation from a range of periods and cultures, but centres on the technological achievements of the industrial era in the West, particularly in the nineteenth century, as understood by their contemporaries. Infrastructure is one major focus, covering the building of railways and canals, bridges and tunnels, land drainage, the laying of submarine cables, and the construction of docks and lighthouses. Other key topics include developments in industrial and manufacturing fields such as mining technology, the production of iron and steel, the use of steam power, and chemical processes such as photography and textile dyes.

The Diamond Mines of South Africa

A life-long friend of Cecil Rhodes, Gardner F. Williams (1842–1922) was the first general manager of the De Beers Consolidated Mines, which once produced 95% of the diamond yield of the world. First published in 1902, this work opens with a chapter on notable historical diamonds, and goes on to describe the initial discovery of diamonds in South Africa, the development of mines, the mining methods adopted, the welfare of the miners and a history of the whole diamond-producing region. Williams also includes an appendix, which describes the mechanical workings of winding engines and pumps, and the value of different types of coals, and provides a table listing the yearly yield of the mines since 1888. Nearly 700 pages long, *The Diamond Mines of South Africa* contains over 500 illustrations and maps.

Cambridge University Press has long been a pioneer in the reissuing of out-of-print titles from its own backlist, producing digital reprints of books that are still sought after by scholars and students but could not be reprinted economically using traditional technology. The Cambridge Library Collection extends this activity to a wider range of books which are still of importance to researchers and professionals, either for the source material they contain, or as landmarks in the history of their academic discipline.

Drawing from the world-renowned collections in the Cambridge University Library, and guided by the advice of experts in each subject area, Cambridge University Press is using state-of-the-art scanning machines in its own Printing House to capture the content of each book selected for inclusion. The files are processed to give a consistently clear, crisp image, and the books finished to the high quality standard for which the Press is recognised around the world. The latest print-on-demand technology ensures that the books will remain available indefinitely, and that orders for single or multiple copies can quickly be supplied.

The Cambridge Library Collection will bring back to life books of enduring scholarly value (including out-of-copyright works originally issued by other publishers) across a wide range of disciplines in the humanities and social sciences and in science and technology.

PORTRAIT OF GARDNER F. WILLIAMS

THE DIAMOND MINES OF
SOUTH AFRICA

SOME ACCOUNT OF THEIR RISE
AND DEVELOPMENT

BY

GARDNER F WILLIAMS, M.A.

GENERAL MANAGER OF DE BEERS CONSOLIDATED MINES, LTD.

ILLUSTRATED

New York
THE MACMILLAN COMPANY
LONDON: MACMILLAN & CO., LTD.
1902

All rights reserved

COPYRIGHT, 1902,
BY THE MACMILLAN COMPANY.

Set up and electrotyped October, 1902.

Norwood Press
J. S. Cushing & Co.—Berwick & Smith
Norwood, Mass., U.S.A.

Contents

Illustrations

MAPS

PHOTOGRAVURES

THE DIAMOND MINES OF
SOUTH AFRICA

SOME ACCOUNT OF THEIR RISE
AND DEVELOPMENT

The Diamond Mines of South Africa

CHAPTER I

THE ANCIENT ADAMAS

T the beginning of the last century, when the blinded Shah-Shuja sought refuge in the lair of the "Lion of Punjaub," Runjeet Singh, his chief treasure was the crystal pebble which Nadir Shah had snatched from the head of the last of the Great Moguls.

For the sake of the pebble, Runjeet starved the wife and children of his friend until he was driven to lay the Koh-i-nûr at the feet of his host. "At what price do you value it?" said the Lion, showing his teeth in a grim smile.

The Koh-i-nûr. (Old Cutting.)

"At good luck," replied the blind Shah, "for it has ever been the bosom companion of him who has triumphed over his enemies."

It may have been the traditional talisman of Carna, Rajah of Anga, fighting in legendary wars, hundreds of years before the great Achilles stormed and sulked under the walls of Troy.[1] From its earliest known appearance it had been so coveted that agas and sultans and

[1] "Tales from Indian History," J. Talboys Wheeler, assistant secretary of the government of India in the Foreign Department, Calcutta, 1881; "The Great Diamonds of the World," Edwin W. Streeter.

rajahs and shahs had snatched it in the first spoils of victory, or tried to extort it by starvation or blinding or boiling oil or some other device of torture; and the adventurous and blood-stained career of this famous diamond is only one of many like

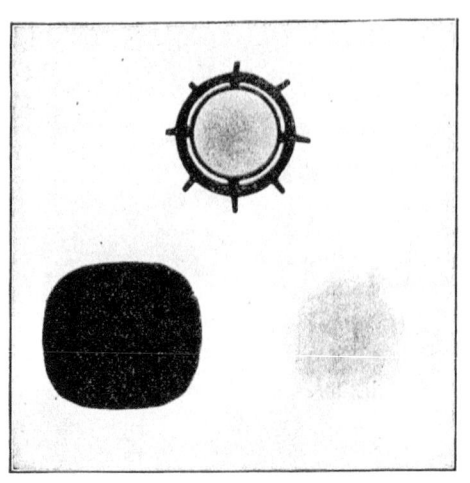

passages, for every precious stone of renown has a trail like a meteor. Some have gleamed weirdly in the eye-sockets of idols in Indian temples or flashed from the splendid thrones of emperors, or glittered in golden basins amid gems of every hue heaped up in tribute, or sparkled on the crests of warriors, the tur-bans of rajahs, the breasts of begums, and the san-dals of courtesans. To

1. A Black Diamond in Gold Setting.   2. Ordinary Window Glass.   3. A Pink Diamond.   (Photo-graphed with the Roentgen Rays.)

win them temples have been profaned, palaces looted, thrones torn to fragments, princes tortured, women strangled, guests poisoned by their hosts, and slaves disembowelled. Some have fallen on battlefields, to be picked up by ignorant freebooters and sold for a few silver coins, and others have been cast into ditches by thieves or swallowed by guards, or sunk in ship-wrecks, or broken to powder in moments of frenzy. No strain of fancy in an Arabian tale has outstripped the marvels of fact in the diamond's history.

Among all the stones that our world's fancy holds precious, the diamond stands preëminent. It is pure crystallized carbon. It crystallizes in almost all the forms of the isometric system, commonly the octahedral or dodecahedral, and frequently with curved faces.[1] Two pyramids with triangular sides and a

---

[1] Dana and others mention that diamonds in the form of cubes have been found. While one might expect to find a diamond in cubic form, as this is the fundamental form of the isometric system, still no specimen of this form has come under the

common base make up the octahedron. The dodecahedron has twelve rhombs or natural facets of lozenge shape.

It is the most impenetrable of all known substances, for the edge of one of its facets will scratch the face of any other stone or the hardest steel. It is the most perfect reflector of light. It refracts entering rays more than any other translucent substance except crocolite, the chromate of lead.[1] Chrysolite alone exceeds its dispersive power to dissolve white light into rainbow tints, but its combined powers of reflection, refraction, and dispersion are unmatched.[2] Hence appears the play of color in its crystalline heart and the resplendent flashing of its radiant fire. It may be as purely transparent and colorless as a drop of dew, or it may display all the primary colors, such as red, orange, yellow, blue, and violet; so that, as John Mandeville quaintly observed, " It seems to take pleasure in assuming in turn the colors proper to other gems." [3] It is highly phosphorescent. Even the blackest of diamonds are transparent to the X-rays. No acid will mar it, no solvent will dissolve it. Its brilliance is undecaying, and ages might roll by without rubbing the minutest particle from its adamantine face. The diamond that gleamed with such strange fire in an idol's eye before the rising of the Star of Bethlehem may be sparkling to-day with more dazzling radiance in the crown of an emperor. Koh-i-nûr and Darya-i-nûr and Taj-e-mah and Regent and Orloff and Sancy and Shah will shine no less resplendent when the sovereigns that now treasure them shall be dust.

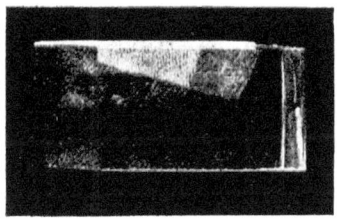

The Shah.

observation of those whose duty it is to look over every stone that comes from the South African mines. The South African diamonds differ in appearance from those found in India or Brazil. They are bright and without any incrustation, and the imperfections, if any, are visible in their natural state.

[1] " Table of Indices of Refraction," Dufrenoy, p. 87. " Treatise on Gems," Feuchtwanger, New York, 1867.

[2] " Table of the Distinguishing Characteristics of Gems," Feuchtwanger, pp. 494–499. " Optical Properties of the Diamond," Sir David Brewster, Phil. Trans., VIII, 157, 1817. [3] " Le Grand Lapidaire," Paris, 1561.

"With the point of a diamond," Jeremiah (B.C. 600) says,[1] records were graven when stones were writing-tablets; but, unfortunately for our knowledge, the diamond did not tell its own story; and it is, at best, a groping effort that would search out the rising of this gem through the mists of tradition.

"Thou hast been in Eden, the garden of God; every precious stone was thy covering, the sardius, topaz, and the diamond, the beryl, the onyx, and the jasper, the sapphire, the emerald, and the carbuncle. Thou wast upon the holy mountain of God; thou hast walked up and down in the midst of the stones of fire."[2]

How glowing are the words of the Prophet of the Captivity, declaring the vainglory forerunning the doom of Tyre's princes and people (588 B.C.). Did the three rivers of Eden flow through sands glittering with stones of fire? Did the eating of a little green apple from the tree of knowledge open the eyes of the first woman of earth to the lure of the gems that are now so tempting to every daughter of Eve? If not, how long was it before the topaz and the diamond, the emerald and the ruby and the sapphire were added to the fig-leaf covering of our first parents?

The Egyptian Pascha.

Multicycles of refining are needed for a clear perception of beauty. The aboriginal Adams and Eves did not have it. The children of the twentieth century will open their eyes to its light more quickly than those of the Stone Age, because the children of to-day inherit the quickened sense of unnumbered generations, and are taught to trace the range of beauty in nature and art. Prehistoric man, a weakling in perception, turned his eyes to the grand orb of the sun, rising above the horizon and flooding the earth with its rays, to the pale bow

[1] Jeremiah xvii.

[2] Ezekiel xviii. 13 and 14 (588 B.C.). Babylonian captivity of the Jews (588–537 B.C.).

of the moon and the sparkling of the firmament of stars, to the ceaseless surge of the ocean and the mountain summits wreathed in clouds, — to all the grander aspects and motions of nature, — before his eyes were drawn to lesser things outside the petty circle of his rambling and the sating of his crude animal wants. Mayhap thousands of years of brutal life rolled by before the savage stooped to pick up any one of the gleaming pebbles which the fierce tiger spurned with bounding foot and the flying deer trampled heedlessly on the river's bank.

Any one may guess, and any one's guess is as good as another's, what little pebble first drew the glance of the barbarian's eye or the stoop of the rover's knee. The first-known precious stones of the world were undoubtedly found on the face of the ground, without any wearisome digging or quarrying, as they lay shining in the gravel, washed from hillsides over the plains, or along the courses of rivers swelled by floods and sweeping the parings of the earth's crust to the sea. Thousands of carnelians, garnets, jasper, amethysts, sapphires, rubies, and diamonds were picked up, maybe by children rummaging in gravel beds or the clefts of rocks, and thrown away as carelessly as splinters of flint, before one was preserved and prized. White and tinted shells were much easier to collect and pierce and link together, and rude armlets and leg-bands of copper and silver and gold were easily forged, and more to the savage taste than any necklace of stones.[1]

When some of the precious stones were lifted and borne away from their beds in drifts of gravel, they were valued first chiefly for the mystic powers attributed to pebbles of such rich hues, phenomenal hardness, and peculiar lustre. One of them would be worn in a pouch next to the bosom as an amulet or charm, averting peril, inspiring courage, healing diseases, repelling evil spirits, or winning the love of scornful maidens. Or, if any one of these magic stones was set to gleam in the buckle of a warrior's plume, it was less for a show of ornament than for

[1] "A Treatise on Diamonds and Precious Stones," John Mawe, London, 1813.

its mystic shielding power and redoubling of valor. The tradition of these virtues has passed from generation to generation,
and still finds credence among the masses of Asia. The poor
natives of India believe to this day in the efficacy of sapphires
and rubies in purifying the blood, strengthening the body,
quenching thirst, dispelling melancholy, averting danger, and
assuring honor and fortune. The emerald in their eyes is
potent to dispel bad dreams, give courage, and cure palsies,
colds, and acute dysentery. The turquoise they say will brighten
and heal weak and sore eyes, and serve as an antidote for veno-
mous snake bites.[1] Like the other precious stones, the diamond
was early endowed by fancy with medical virtues, and particularly prized as a safeguard from madness, in its power to "raze
out the written troubles of the brain."[2] It was also believed to
be potent to touch the heart, and there is a pretty conceit that
the darts of Cupid were diamond tipped. Perhaps the passion
of women for gems gave point to this fiction.

As the diverse stones of fire became better known and more
sharply distinguished, special significance was given to each by
some nations of the East, associating them with the planets,
the march of the seasons, or with various divinities. Sometimes
they were of emblematic service. For the representation of the
twelve tribes of Israel, twelve distinct gems were set in gold
plates on the robe of the high-priest.[3] When the rise of letters
and the fine arts brought the devising of symbols and graven
inscriptions, the supposed potency of these stone amulets was
increased by the craft of priests and sorcerers, cutting the face
of the charms themselves or directing the hands of expert work-

---

[1] "Oriental Accounts of Precious Minerals," Journal of Asiatic Society of
Bengal, August, 1832.

[2] "Treatise on Gems," Feuchtwanger, New York, 1867.

[3] Exodus xxviii. "Natural History of the Bible," Thaddeus M. Harris,
Boston, 1820. "Precious Stones in their Scientific and Artistic Relations," A.
H. Church, London, 1883. "De Duodecim Gemmis in Veste Aaronis."
Epiphanius, 1565. John Peter Lange, Professor University of Bonn, in Schaff's
"Critical, Doctrinal, and Homiletical Commentary" on the Bible.

men. The Chaldeans are especially charged with the fomenting of superstitions by the exaggeration of this conceit. These engraved stones served often as distinctive seals, and for convenience in carrying and the gratification of a spreading taste for such ornaments, the talismans were set in rings and clasps. So Solomon's seal, summoning and mastering genii, was the wonder of legends, and so, too, the famous ring of Polycrates and the rival marvels of Oriental romancers familiar in the tales of the " Arabian Nights."

As time and art disclosed more and more of the marvels of the stones of fire in the crust of the earth, the wonder grew and the supernatural potency of the various gems was more deeply impressed. Thus we reach the belief and tribute of the priest Onomacritus (500 B.C.), who declared of the lucent crystal, " Whoso goes into the temple with this in his hand may be sure of having his prayer granted, as the gods cannot withstand its power." Its use to concentrate the sun's rays as a burning glass was highly prized also in priestly ministrations.

Onomacritus says crudely of this use that " when a transparent crystal is laid on wood, so that the sun's rays may shine upon it, there will soon be seen smoke, then fire, then a bright flame." [1] Fire kindled through this agency was holy in the sight of priests and people, and no burnt offering was so pleasing to the gods as one set in these sacred flames.

The precious stones are so greatly dependent upon the advance in the art of polishing and cutting for the revelation of their qualities and beauty that it was doubtless long after their discovery before they came into any considerable use as ornaments. Their hardness defied, at first, any effort to fashion their shape with primitive tools. The most that could be effected was the rude polish that might be obtained by the tedious rubbing of the face of one stone against another. But, as time went on, the lines of natural cleavage were noted, and grinding wheels in the hands of skilful artisans gave a smooth face to the natural contours of the softer stones, and, later, even to the sapphire

[1] " Precious Stones and Gems," Streeter.

and diamond.   With the advance in art the demand for precious stones increased apace, and, to meet the demand, keener and wider ranging searches developed new and greater supplies.

The Polar Star.

There is a certain tracing of the use of precious stones for ornaments to the ancient Babylonian civilization, whose existing ruins extend back to from 6000 to 7000 years B.C.[1] Babylonian lapidaries were cutting and polishing carnelians, sards, onyx, and rock crystals before the Egyptians had advanced beyond the carving of their soft steatite.   Then the Phœnicians drew from all parts of the known earth its treasures.[2] So Ezekiel testifies of Tyre: "Syria was thy merchant by reason of the multitude of wares of thy making: they occupied in thy fairs with emeralds, purple and broidered work and fine linen and coral and agate. The merchants of Sheba and Raamah, they were thy merchants: they occupied in thy fairs with chief of all spices, and with all precious stones, and gold."[3]

Judea had some share of this stream.   The Queen of Sheba bore a "great store of precious stones " to Solomon (B.C. 1015–975) with her tribute of gold,[4] but this was a trivial trickle compared with the flow to Phœnicia and Babylonia.   Long before the days of the Captivity (B.C. 598),[5] the robes of the princes and nobles of these rich realms were glittering with jewels, and their gorgeous array was the marvel of the poor

The Hope Blue.

exiles, crying with the voice of their prophet, Ezekiel : " Every

[1] " Encyclopedia of Religious Knowledge," Schaff-Herzog.   " Archæology of the Past Century," Professor W. M. F. Petrie.

[2] The Story of the Nations, " Phoenicia," George Rawlinson, M.A.

[3] Ezekiel xxvii. 22.

[4] " Old Testament History," William Smith.   " Precious Stones in the Scriptures," R. Hindmarsh, London, 1851.

[5] Date of removal of Jehoiachin, according to Prideaux and to Clinton.   Ewald makes the date 597 B.C.

precious stone was thy covering. Thou hast walked up and down in the midst of the stones of fire." As tradition placed the garden of Eden in the valley of the Euphrates, Ezekiel makes the garden typical of the splendor of Babylon in his fervid outpouring.

How the stones of fire were brought into being in the garden of Eden or elsewhere, Ezekiel was not moved to reveal, and the savants that have sought to tell are but groping seers. When a sprinkling of stones was uncovered by the rains and floods, or dug and washed from the beds of gravel, or traced by rude mining through clay or conglomerate layers or enclosing rocks, there was still no widespread knowledge of the deposits, and even among the most familiar with the search there was ever the hope of finding, some day, some marvellous store. Hence sprung up the romances. Even in the days when the sharp tooth of history had cut into legends, a story was told of the climbing of Zulmat by the great Alexander, to the rim of the inaccessible valley, where, beneath sheer precipices, glittered a coverlet of the stones of fire. There was no way of winning the diamonds that glowed so temptingly except by flinging down masses of flesh and waiting for swooping vultures to bear the lumps up to their perches on the mountain with precious stones sticking in the meat.[1]

Sindbad the sailor had this tale in mind fortunately in his second voyage. It will be remembered that he was stranded by shipwreck on a desert island and carried away by the flight of a gigantic rukh to the top of a distant mountain. From this mountain he descended into a neighboring " valley, exceeding great and wide and deep and bounded by vast mountains that spired high in air." Walking along the wady, he found that "its soil was of diamond, the stone wherewith they pierce minerals and precious stones and porcelain and the onyx, for that it is a dense stone and a stubborn, whereon neither iron or hardhead hath effect, neither can we cut off aught therefrom, nor break it save by means of lead stone."

[1] "Oriental Accounts of Precious Minerals," Journal of Asiatic Society of Bengal, August, 1832.

Luckily for the sailor, his descent was by day, for "the valley swarmed with snakes and vipers, each as big as a palm tree, that would have made but one gulp of an elephant; and they came out by night, hiding during the day, lest the rukhs and eagles pounce on them and tear them to pieces." In view of the horrid prospect of soon dropping through the throat of one of these snakes, Sindbad began to wish that he had not flown away from the island, where he was, at least, out of reach of vast vipers, but he soon bethought himself of the old story of the valley from which diamond-studded meat was "plucked by eagles." So he quickly filled his pockets and shawl girdle and turban with the choicest diamonds. Then he put a piece of raw meat on his breast and lay down on his back. Soon a big eagle swooped into the valley, clutched the meat in his talons, and flew up to a mountain above, "where, dropping the carcass, he fell to rending it," leaving the lucky sailor to scramble off with his booty. He gave a parcel of the diamonds to the disappointed merchant, who had cast down the meat, but he had stuffed his clothes so full of the gems that he went home, after some strange sight-seeing, with a great store of diamonds and money and goods.[1]

This amazing tale is less teeming with interest than it was in the days when it was first told, for, even hundreds of years afterwards, diamond-lined valleys and monstrous rukhs and snakes that could gulp down elephants were not beyond credence. If in valleys there might be a diamond lining, why should there not be a massing of diamonds and rubies in the dwellings of genii in caves, awaiting the entry of some lucky Aladdin? Oriental fancy, teeming with visions, disdained any curbing within the petty confines of crawling experience, and was prolific in marvels far more pleasing to the masses that egged on the story-tellers with craving credulity. Who then could explode these bubbles with any sharp prick of positive contradiction? Even if in all known fields the precious stones were gathered by toilsome searches only rarely rewarded, who had the

[1] "Arabian Nights," Lady Burton's edition, Vol. III, pp. 476–482.

range of knowledge to deny the possible existence of caverns
filled with rubies or mountain summits studded with diamonds?

Seeing that to this day so little can be asserted positively of
the forming of the precious stones scattered in the earth's crust,
it is not surprising that the origin of the stones of fire has been,
from the first, a baffling puzzle and a fountain-head of conflicting surmises.   Some wondering people viewed them as splinters dropping from the stars, and some, as the creations or
transformations of genii.   Some Hindoo miners still believe
that diamonds grow like onions, though much less quickly, and
that their age is marked by the difference in their size and
quality.   Others suppose the common rock crystals to be
immature diamonds, and the distinction is marked by calling
the rock crystal kacha (unripe), while the diamond is pakka
(ripe).[1]

For the ripening of the crystals and the quickening of their
seeming inward fire, the lightning bolts, that sometimes rived
the ground, were thought to be potent.   Others again, observ-
ing the liquid purity and likeness which is marked to this day
in the term " diamonds of the purest water," attributed the
forming of the crystals to the supernormal trickle and hardening
of dewdrops.   It is of this fancy that Dryden makes poetic use
in his likening of the tears of Almahide : —

> " What precious drops are those,
> Which silently each other's track pursue,
> Bright as young diamonds in their infant dew ? " [2]

Bizarre speculation was stretched even to the point of attrib-
uting to these strange crystals animal instincts and reproductive
powers.   Thus Barreto is quoted in the dictionary of Antonio
de Moraes Silva as saying : —

> " Que os diamantes se unem, amam e procream." [3]

[1] " Oriental Accounts of Precious Minerals."   Translation by Rajah Kalikis-
ken, Asiatic Society of Bengal.

[2] " The Conquest of Granada," Second Part, Act III, Scene 1, Dryden.

[3] " Commonplace Book," Second Series, p. 668, Southey.

The tradition of the generative power of this marvellous crystal originates with the Hindoos, and to this day the natives of Pharrah will affirm that the diamond beds yield fresh supplies of well-grown stones at intervals of from fifteen to twenty years.

It is seemingly hopeless to attempt to fix with any certainty the time when the diamond was first singled out from the pebbles in which it lay, and was prized by any one, or even when it entered the list of gems known to the chief nations of Asia. Traditions coming down through the mists of legendary ages are conflicting and uncertain reliances at best. The ancient writers add to this perplexity by loose or erroneous descriptions when the advance of the science had not marked precise distinctions of structure and composition. Thus the Carbunculus of Pliny was probably stretched to cover the spinal or Balas ruby, the garnet and other red stones, besides embracing the Anthrax of Theophrastus or our modern ruby. Many ancient writers confounded also under the general term Smaragdus various distinct minerals of green color, not only the true emerald, but green jasper, malachite, chryscolla, and fluor spar.[1] Among the common people, pretending to no mineralogical knowledge, there was less thought of distinction, and, in days approaching our own, Tavernier observes in his travels, A.D. 1669, after describing the true ruby of Pegu, in Ceylon, "the fatherland of rubies," that "all other stones in this country are called by the name Ruby, and are only distinguished by color, thus, in the language of Pegu, the sapphire is a Blue Ruby," etc.[2] This confusion is not surprising, and a much more discreditable one occurred within the last thirty years in the sensational touting of the discovery of rubies in the garnets of the Macdonnell Ranges in South Australia. It seems highly probable that the stone of exquisite blue, now particularly distinguished as the typical sapphire, was the ancient Hyacinthus ; and the Sapphirus of the ancients certainly included the lapis lazuli and

[1] "Precious Stones and Gems," Streeter, London, 1892.
[2] "Voyages en Turquie, en Perse et aux Indes," Paris, 1676.

covered the range of corundums of every tint except red. Thus green sapphires ar noted, although very rarely, and yellow and gray, as well as pure white or colorless, and this stone is presumed by Streeter and other investigators to have been the "adamas" first known to the Greeks.[1]

There can be no question that sapphires or corundums of varied hue were much more common than diamonds in the hands of the merchants of the East or any other ancient collectors before the Christian era. The sapphire was, indeed, one of the most widely known of all gems, and how highly it was valued may be surmised from the dignity given to it by the sacred writers. The prophet Ezekiel likens to a "Sapphire stone" the appearance of the throne in the firmament above the cherubim. Job makes it the representative of all gems in his splendid description of the daring of miners.[2]

Like the sapphire, the diamond is repeatedly referred to by the Hebrew writers. It formed one of the typical stones in the high priest's breastplate, and Ezekiel puts it in the first rank of the stones of fire. Jeremiah speaks of the sin of Judah as written with the point of a diamond, "puncto adamantinis" of the Latin Bible, but Streeter holds that this pen point was probably a corundum and not the true diamond.[3]

This is a stretch of assumption largely based upon the lack of any precise description applying to the diamond until close to the beginning of the first century of our reckoning. Adamas, the indomitable, the adamant of the ancients, was the name given to the diamond because of its distinguishing hardness. Pliny was greatly impressed by what he heard of this characteristic, but obviously knew little or nothing of the stone by personal handling or test. For he wrote down soberly: " The most valuable thing on earth is the Diamond, known only to kings, and to them imperfectly. It is only engendered in the finest gold. Six different kinds are known, among these the Indian

---

[1] "Traité de Minéralogie, avec application aux Arts," Brongniart, Paris, 1807.

[2] Job xxviii. 1–11.   [3] "Precious Stones and Gems," Streeter.

and Arabian of such indomitable, unspeakable hardness, that
when laid on the anvil it gives the blow back in such force as to
shiver the hammer and anvil to pieces.[1]

Unfortunately for the aim of identifying the diamond with
the references to the ancient adamas, the term was commonly and
loosely applied to any substance of peculiar hardness. So mountains of iron-stone, like unto that upon which the ship of Sindbad
was dashed, were called adamant, and so too were the arms and
armor of gods and heroes. Addison only transmits a tradition
in the fine lines of his poem —

> "And mighty Mars, for war renowned,
> In adamantine armor frowned."[2]

In Homer, as Streeter notes, adamas occurs only as a personal name, and in Hesiod, Pindar, and other early Greek poets
it is used to signify any hard weapon or metal like steel or an
alloy of the harder metals.[3]   No distinct identification of the
diamond with adamas appears, according to Streeter's view, until
the first century A.D., in the writings of the Latin poet and astron-
omer Manilius, and his contemporary Pliny (A.D. 62–114).   In
the fourth book of Manilius's poem " Astronomicum," occurs
this line, " Sic Adamas, punctum lapidis, pretiosior auro," which,
Streeter says, " is supposed to be the earliest indubitable reference
to the true diamond."   It is difficult to see how this " stone's
point, more precious than gold," is any more distinct and indubitable in its reference to the diamond than the diamond pen point
of Jeremiah hundreds of years before.   But Pliny, with all his
erroneous amplifications, unquestionably describes the true Indian
diamond as " colorless, transparent, with polished facets and six
angles ending either in a pyramid with a sharp point or with two
points like whipping tops joined at the base."[4]

---

[1] " Historia Naturalis," XXXVII, 15.

[2] Poem addressed to Sir Godfrey Kneller, referring to William III. of England.

[3] " αδαμας γενος οιδυηον," Æschylus.   See Stanley's Commentary on Æschylus, " Prometheus Vinctus."

[4] Plinii Secundi (Caii), " Naturalis Historia," XXXVII, 15.

In view of the hardness of the sapphire, so great that it will scratch every other precious stone except the diamond, it is therefore contended that this was the stone known to the earliest Greek writers as adamas.[1] This may be so, and it cannot be doubted that, even at a much later day, a white corundum or a pale yellow topaz or a good rock crystal often passed for a diamond in the hands of collectors or in the sharp practice of gem selling. Whatever may have been the blundering of the Greeks or the application of adamas, there is, nevertheless, no sufficient reason in this for questioning the probability that genuine diamonds were found in the gravels of India many centuries before the Christian era. As far back as tradition goes the largest stones were particularly prized by the native princes, and were strictly exacted in tribute from the diamond-bed washers. But the smaller stones were less jealously guarded, and may readily have found their way into the hands of traders with the other peoples of Asia or with Egypt. It seems most probable that the Jews derived their first knowledge of precious stones from the Egyptians chiefly, for the Hebrew names of the stones are of Egyptian derivation.[2] Thus there is no approach to certainty for the assumption that the stones called diamonds in the English version of the Hebrew Scriptures were not rightly named, or that allusions to the diamond in other ancient writings were wholly unreliable or mistaken.

The Empress Eugenie.

The main support for the questioning of the mingling of diamonds with the other gems noted by the ancient writers is the apparent failure to uncover diamonds in the excavations on the site of ancient temples and cities where other precious stones are brought to light. Thus emeralds and other gems in various settings have been exhumed from the volcanic overflow that

[1] "History of Stones," Theophrastus.   Edited by Sir John Hill, London, 1746.   "Elem. de Min.," Lessing, II, 61.   "The Great Diamonds of the World," Streeter.

[2] "Encyclopedia of Religious Knowledge," Schaff-Herzog.

buried Herculaneum and Pompeii, from the ruins of old Rome, and the tombs of Egypt.[1]   In the course of explorations on the site of Curium and other ancient towns in Cyprus, scarabs and scaraboids of agate, onyx, jasper, and variously tinted carnelians were found, as well as gold ornaments, relics traced to the days of Eteandros, king of Paphos in the seventh century B.C.; but no diamonds were unearthed in this collection.[2]   Nor is

The Nassak.

there record, as yet, of the discovery of diamonds in the explorations in Babylonia.[3]

But this is, at most, evidence pointing to what is undoubted, — the comparative rarity of the diamond among the gems that served as amulets or ornaments for the people of western Asia, northern Africa, or southern Europe prior to the Christian era and for centuries afterward.   Pliny expressly asserts this rarity in his allusion to the diamond; but the fact that the gem was scarce, outside of India, is entirely compatible with its occasional inclusion in the collections of sovereigns, which the same writer remarks, and the high value set upon it would naturally limit its use as an ornament.

It is impossible to mark with any precision in what district of India a search for diamonds first began.   Rajah Sourindo Mohun Tagore, in his account of the precious stones of India, gives the names of eight localities in which diamonds have been found according to tradition or more certain report.   These are Harma (Himalayas), Mátanga (Kistna), and Godaveri (or Golconda), Saurashtra (Surat), Paunda (probably including the Chutia Nagpur Province), Kalinga (the tract between Orissa and the Godaveri), Kosala (the modern Ajodhya or Berar), Vera Ganga

[1] Clarke's "Travels," Vol. VIII, p. 150.

[2] Story of the Nations, "Phoenicia," George Rawlinson.

[3] "Nineveh and Babylon," pp. 160–161, 602 et seq., Layard; "Archæology of the Past Century," Professor W. M. F. Petrie.

(the Wemganga), and Saubira (the stretch between the Sarhund and Indus rivers).[1]

According to this showing the diamond is scattered over a wide ranging region, but it occurs everywhere in one of two comprehensive formations, — alluvial or otherwise disintegrated surface deposits, and conglomerate rocks of far receding geologic antiquity, belonging to the Vindyhan formation, which borrows its name from the Vindyhan Hills of old geographers.[2] It seems reasonable to presume that the surface wash comes from the disintegration of the seat of the diamond in conglomerate beds, — for even in alluvial gravels there are frequently no diamonds found outside of a conglomerate of rounded pebbles and sandstone breccia. It is likely that the first diamonds were taken from the surface wash and that the more solid breccia was opened later.

In some of the diamond-bearing districts of India to-day the native villagers are searching for diamonds exactly as their fathers did in days of remotest tradition. After a heavy rain that washes away loose soil, a sprinkling of diamonds may be found in exposed sandstone breccia, and sharp-eyed Hindoos scrape the face of the ground for the precious crystals.

The Great Mogul.

Along the banks of the Kistna and Godaveri rivers the Golconda of tradition outstretched, and this diamond-studded ground came later into the hands of the Nizam of Hyderabad, and was included in the bounds of the Madras Presidency. Here, it is claimed, was the bed of the Koh-i-nûr and Regent and Great Mogul, and others of the jewels most renowned in history and romance. Here, of a certainty, was the richest diamond field in India, in

[1] "Mani Mala," Calcutta, 1879.
[2] "Manual of Geological Survey of India," Professor V. Ball, Vol. III.

c

the days of Tavernier's travels (1669 A.D.). Here was the famous mine, " Gani-Coulour," that he saw, where sixty thousand natives were then at work, and "Gani-Parteal," and twenty more of lesser note.[1]   Gani-Coulour has probably been identified with the modern Kolur on the Kistna, Gani being simply a slight change of the Persian " Kan-i" or "mine of," so that Gani-Coulour is the mine of Kolur as Gani-Parteal is the mine of Parteal.[2]   The surface ground of this district along the rivers is a black "cotton soil" washed down by floods, and underlying this at an average depth of twenty feet is a layer of broken sandstone, quartz, jasper, flint, and granite, interspersed with masses of calcareous conglomerate, forming the stratum in which the diamonds were embedded.   When the black soil had been dug up laboriously and carried away, the diamond-bearing layer was exposed, and was removed, piecemeal, to level stretches of ground or prepared floors, where it was scraped and picked over by hand to find the diamonds.

The whole of this rich mining district and a tract stretching for many miles away was loosely called Golconda, or the Kingdom of Golconda, by foreign traders and travellers, because the town of Golconda was its capital and the trading centre where the diamonds from the mines were chiefly bought and sold. The only mark of this old mart to-day is a deserted fort near Hyderabad, but its fame will endure until traditionary Golconda ceases to be a standard of riches.

Next in importance and prestige to the mines of Golconda was the diamond field of Sumbulpur, in the Central Provinces, between the rivers Mahanadi and Brahmini.   The diamonds of this district were remarkable for their purity and beauty, though no very large crystals have been traced to this region, and the few which the washings still yield rank with the finest of the Indian stones.   Here the precious stones were found chiefly along the course of the Mahanadi, in a stratum of tough clay and pebbles stained reddish by iron oxide.   At the opening of

[1] " Voyages en Turquie, en Perse et aux Indes," Tavernier, Paris, 1676.
[2] " Manual of Geological Survey of India," Vol. III.

the dry season, thousands of villagers, men, women, and children, began to search every cleft and cranny in the river beds for diamonds. With ankovas, or light picks, the men broke and scraped out the diamond-bearing bed and piled the broken ground on the river bank. Then the women scooped up ground from the heaps with their daers. These were shovel-shaped boards, about five feet long, with ridged sides and hollowed in the centre. Resting one end of the daer on the ground and tilting the other slightly, they washed away the clay and sand and picked off the rock splinters and larger pebbles. After this rude sorting they spread out the finer gravel on a smaller board, the kootla, and scraped it over very carefully to separate the diamond crystals and grains of gold. When there was a level stretch along a bank, the native workers would sometimes make an enclosure on this flat, with a low wall pierced at several points by small waterways. Then they would dump the diamond-bearing ground into this shallow basin and wash away the clay and dirt with running water. After two or three washings they would pick out the larger stones from the cleaned gravel, and dry the remainder, to be picked over on their kootlas or on any smooth, hard flooring.

Perhaps the most laborious diamond digging in India has been in the pits of Panna and neighboring villages in the Province of Bundelkhund. Here the diamond-bearing conglomerate was buried under a cover of heavy ground, ranging in places over thirty feet in thickness. To reach the diamond strata large pits were dug, with inclines leading to the bottom in or below the conglomerate. There was no drainage, and the diamond diggers were forced to work in the rainy season knee-deep in water, breaking the conglomerate, and filling baskets which were hauled by hand to the top of the pits. In this primitive fashion the diamond beds of India were opened, and diamonds are to-day won by these simple methods or others essentially similar.[1]

[1] "A Treatise on Diamonds and Precious Stones," John Mawe, London, 1813. "A Treatise on Gems," Feuchtwanger, New York, 1867. "Precious Stones and Gems," Streeter, London, 1892.

Color and size were the chief distinction in diamonds, as in the other precious stones, in the early days before the advance of the art of diamond cutting which has added so greatly to the brilliancy and beauty of this gem. Centuries ran by before the ancient lapidaries attempted more than the polishing of the surfaces of the natural facets of the crystal, though the comparative ease with which this hardest of stones may be split by following the natural cleavage lines may have been observed. Size was rated so highly by the Hindoos in valuing a gem that the conception of increasing the worth of a jewel by cutting away the greater part of it would not have been tolerated even if it had been feasible. When cutting to a limited extent began to be practised in India, it was generally unsymmetrical and unscientific, as the oldest known diamonds bear witness, and there was comparatively little advance for many centuries, as every celebrated gem of Indian workmanship plainly shows.[1] But even with imperfect cutting and crude polishing the inherent beauties of the ancient stones were more or less fully disclosed.

In the mines of Panna there were four noted divisions in grading. Clear and brilliant stones were in the class Motichul, Mansk was the class name applied to diamonds of greenish tint, Panna to light yellow, and Bunsput to sepia colored stones.[2] In India at large there was a comprehensive divisional grading corresponding to the main caste distinctions, — the "twice-born," priests, warriors, and merchants, and the "once-born," tillers of the land.[3] The Brahmans were the diamonds of highest range, clear and colorless crystals; the Kshatriyas, clear crystals, amber tinted or of the color of honey; Vaisyas, the cream colored; and the servile Sudras, the grayish white stones. Grades in rank were more minutely marked in the rubies of the famous Badakshan mines in Persia, where the common people believed that the precious stones were deposited in the "rag-i-lal" or parent vein in successive layers. The outside layer contained the small

[1] "A Treatise on Diamonds and Pearls," David Jeffries, London, 1751.
[2] "Precious Stones and Gems," Streeter.
[3] "Annals of India," J. Talboys Wheeler, Calcutta, 1881.

and imperfect stones styled piadehs, foot soldiers; the next, a better class of stones called sawars, horse soldiers; and so on through layers of amirs, bakshis, and vazirs until a single stone was reached, transcending all in size and beauty, which the miners polished dutifully, and took in tribute to their sovereign.[1]

With the expansion of Greek commerce and the entry of Greek mercenaries into the employ of satraps in Asia Minor (about 500 B.C.), the riches of the Orient were made known, and precious stones began to pass into Europe. Herodotus, 484 B.C., was first of the early Greek writers[2] to mark particularly the displays of precious stones in palaces and temples — the signet rings of Darius, the magnificent emerald in the ring of Polycrates, and the marvellous show of the emerald column in the temple of Hercules in Tyre, gleaming like a pillar of green fire at night. This fiery column has a certain likeness to the traditional stone as big as an ostrich egg, to which homage was paid as the "Goddess of Emeralds" by the people of the Manca Valley in Peru. Sceptics would clip the marvel of both by substitution of beryl, or aquamarine, or colored glass; but this trimming of legend does not question the extraction of true emeralds from mines in Upper Egypt, or the superb yield of the deposits in Peru and New Grenada.[3]

The conquests of Alexander the Great (334–323 B.C.) made the Greeks familiar with the precious stones of India as well as of Western and Central Asia. His successors revelled in profuse displays of jewelled rings and bracelets, and wine cups and candelabra, in luxurious banquets. Pliny tells a glowing tale of a statue of Arsinoë, wife of Ptolemy Philadelphus (283 B.C.), four cubits in height, made of topazon.[4] The true topaz was undoubtedly known to the ancient Egyptians, and is still obtained at Risk Allah near the old emerald mines of Jebel Zabara; but the Oriental topaz is presumed to have been the yellow sapphire;

---

[1] "Oriental Accounts of Precious Minerals," Journal of Asiatic Society of Bengal, August, 1832.   [2] Rawlinson's "Herodotus."

[3] Brun's "Travels." Rawlinson's "Herodotus," II. 44. Prescott's "Conquest of Mexico."   [4] "Historia Naturalis," XXXVII, 32.

and the Greek topazios, the yellowish green chrysolite or the peridot, of deeper green tint. The word is derived from τοπαζα, "to seek," because the traditional source was an island in the Red Sea, often difficult to reach through its envelope of fog.[1] The loose use of the term by Pliny and other old writers makes it impracticable to mark with any certainty from what greenish hued stone Arsinoë's statue was cut. Still, in spite of current exaggeration and confusion of distinctions, there can be no doubt of the rising production and circulation of the precious stones.

With the spread of the Roman Empire prodigality in displays ran riot. After Pompey's victory over Mithradates, (B.C. 66) precious stones and pearls poured into Rome and the demand of vanity rose to a passion.[2] Pliny writes: "We drink out of a mass of gems crusting our wine bowls, and our drinking cups are emeralds." To heighten the wonder he tells in his gossiping way how emeralds were set as the eyes of a lion sculptured in marble on the tomb of King Hermias in the island of Cyprus. So great was the size and so piercing the light of these emerald eyes that the tunny fish in the surrounding sea were frightened away until the fishermen of Cyprus put common stones in place of the dazzling gems. Later scepticism would make these emerald eyes of malachite, for copper ores were of common occurrence in Cyprus[3] and the glory of the emerald was scattered by loose usage over green fluor spar, jasper, aquamarine, malachite, and perhaps even green glass. There is also a shaking of the marvel of the cups, holding a pint, that were made out of solid carbuncles; for these are supposed to be cuttings from the common garnets of the Barbary coast, flowing out from Carthage in such profusion that the carbuncle was called "the Carthaginian stone."[4]

Beryl was largely used in the ornamentation of cups and

[1] Diodorus Siculus, Lib. III, c. 38.   Jameson's "Mineralogy," p. 48. Kidd's "Mineralogy," I, 121.   [2] "Historia Naturalis," XXXVII, 6-7.

[3] Cleaveland's "Mineralogy," p. 565.   Theophrastus, "De Lapid.," c. 49.

[4] "The Story of Carthage," p. 121, Alfred J. Church, M.A.   "Story of the Nations."

for cameos;[1] and carnelian was particularly prized as a base for the engraving of seals or cameos, sometimes elaborately pictorial. The great scarab in the Prussian cabinet, representing the five heroes of Thebes, is a recognized masterpiece of old Etruscan art, and a deep-cut carnelian once belonging to Michael Angelo portrays the birthday festival of Dionysius.[2] Amethyst ranked with carnelian as a favorite stone with engravers, and it was of peculiar traditional service in the fashioning of drinking cups, from its supposed checking of drunkenness, whence its Greek name, — α, "not," and μεθυω, "to intoxicate." Opals were placed in the first rank of gems, and Pliny tells of a senator, Nonius, who bore banishment and the loss of all his estate rather than the sacrifice of his opal ring to the greed of Mark Antony.[3]

Pearls were even more highly valued and lavishly displayed than any of the precious stones. Swelling the yield of the Mediterranean shores there flowed into Rome a profusion of still finer pearls from the Persian Gulf and Ceylon, to be set in necklaces, bracelets, earrings, and clasps of all kinds. Rich robes were bespangled with jewels, and it is reported that Lollia Paulina, the wife of Caligula (A.D. 37–41), wore a dress covered with pearls and emeralds. Cleopatra's famous pearls were said to have cost her £80,000. Julius Cæsar (B.C. 102–44) gave Servilia, Cato's sister, a pearl valued at over £50,000, and Nero dropped handfuls of pearls in the laps of his mistresses (A.D. 54–68).

From personal adornments, the decoration of arms and trappings, and the embellishing of banquets, the use of gems spread to the mounting of pictures in frames studded with precious stones, and the ornamentation of statuary. Nero viewed the combats of gladiators in a mirror of jewels,[4] and Constantine

---

[1] "Historia Naturalis," XXXVII, 20.

[2] "A Treatise on Gems," Feuchtwanger.

[3] "Historia Naturalis," XXXVII, 21.

[4] *Ibid.* XXXVII, 16. Beckmann thinks the mirror of Smaragdus in which Nero gazed may have been green obsidian, green jasper, or even green glass. "History of Inventions," III, 177.

challenged the splendor of Oriental monarchs by his entry into Rome in a chariot of gold sparkling with precious stones (A.D. 312–337).

Amid all this profusion, in which millions of sesterces were lavished, the diamond is noted only by rare allusions. This is probably accounted for by the check in the advance of lapidary art on reaching a stone of such indomitable hardness. Even the diamonds set in the clasp of the regal mantle of Charlemagne, after the opening of the ninth century, show only a partial polishing of the natural planes of the crystals. There was no scientific cutting of facets to heighten the brilliancy of the stone until the fifteenth century. When artificial shaping was attempted before that time, it did not go beyond the production of a flat top or table, or a convex surface, with a truncated pyramid as a base. Even when a large number of facets were cut, as was sometimes done by East Indian lapidaries, there was no scientific proportioning, as was signally shown in the instance of the remarkable stone known as the " Beau Sancy," which came into the possession of Charles the Bold, Duke of Burgundy. It was the recutting of this stone in 1465, by the true artist Louis de Berquem of Bruges, that marks the rising of the modern art that has enhanced so immensely the resplendence and beauty of the diamond, and established its place securely as the chief among gems that are prized for adornment.

Then begins the entry of the famous diamonds passing over the face of Europe with meteoric trains of adventure. The Beau Sancy glitters for a moment in the splendid array led by Charles the Bold against the Swiss peasants. On the bloody field of Granson (3d March, A.D. 1476) where the best knights of Burgundy were killed or put to flight by the mountaineers, the jewel that might ransom a king is trampled under foot in the rout. A Swiss soldier picks it up. It is no more in his eye than a bit of glass which he is well pleased to sell for a florin to a priest. Philip de Commines says that the priest knew no more of its value than the soldier, and thought he did well to make a franc by selling the diamond to the burghers of Berne.

There the diamond disappears. One current story makes it reappear one hundred years later in the possession of the king of Portugal, who pledges it with other jewels for a loan from Nicholas Harlai, Seigneur de Sancy, and treasurer of the king of France. M. de Sancy soon buys it outright for one hundred thousand francs and loans it to sparkle for a time on the head of his king, Henry III. (A.D. 1574–1589).

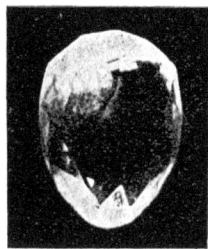

The Sancy.

When Henry of Navarre comes to the throne (1589), M. de Sancy sends the diamond to him by a trusted servant. Thieves waylay and kill the messenger, but the precious stone is seemingly not in his keeping. So his body is thrown into a grave hastily made by his murderers. When the place of burial is later searched out by direction of M. de Sancy, the lost diamond is found in the dead man's stomach.

Undimmed in this ghastly adventure, it rises from the grave to shine on the breast of Elizabeth of England (A.D. 1558–1603). From the last of the Tudors it passes to the Stuarts, and one of the few treasures that James the Second carries off in his flight from his throne (A.D. 1688) is the brilliant Sancy. Louis XIV. buys the gem from the king in exile (A.D. 1695), and it is held as one of the most precious of the crown jewels until the Revolution. In 1792 robbers break open the treasure chamber and bear it off with other plunder. Again it is beyond tracing for years, till it reappears in the hands of a noble Russian family, the Demidoffs, from whom it passes to London merchants, and finally to the Maharajah of Puttiala. It may be that the adventures of two diamonds are fused in this tale, but it is none the less an outline of truth with the marvel of romance.[1]

Even Aladdin's wonderful palace, reared in a night by the hands of obedient genii, scarcely outstripped the glittering show of the court of the Great Moguls, enthroned in Delhi (A.D. 1526) by the arms of the Sultan Baber and his grandson Akbar, of the

[1] " A Treatise on Gems," Feuchtwanger. " Great Diamonds of the World," Streeter.

line of Timour the Tartar. Here embassies passed through the main gate of the palace along a magnificent avenue to the grand central square. Thousands of bodyguards in splendid dress lined the way, and behind the ranks richly caparisoned elephants were massed, waving flags of satin and silver. Dark eyes peered through the crimson hangings of the howdahs and the gilded lattices of the zenana cloisters bordering the square. Beyond the cloisters gardens outspread, with beds of lovely flowers and sheltering arbors and fountains splashing in sculptured basins.

The entrance to the durbar or audience hall was through a pavilion hung with tapestries of purple and gold to a stately marble chamber, whose pillars and walls gleamed with rainbow hues. Under a canopy of flowered tissue on silver poles was set the imperial throne, the matchless triumph of Indian art. There strutted two peacocks fashioned deftly of jewels and gold to depict every plume and hue of the living creature. The out-spread tail seemed to flutter in mimicry of life with the sheen of sapphires and emeralds. The body was of enamelled gold and the eyes two radiant diamonds. Peacocks were emblems of the sun and of the descent of the Great Moguls from the sun through Chenghiz Khan. Ranged beside these splendid figures were stands bearing masses of unfading flowers, for every stem and leaf and petal was counterfeited in precious stones and metals.

When the Great Mogul took his seat on his throne of solid gold studded with jewels, all bent low before his imperial majesty attired in cloth of gold blazing with precious stones in armlets and necklaces and crusted embroidery. Over the entrance to the hall was engraven in letters of gold: " If there be an elysium on earth, it is this." Here was at least a splendor of luxury beyond all rivalry. Never was shown, in vain Babylon, adventurous Tyre, or imperial Rome, any display as dazzling as the jewels of Delhi.[1]

[1] "The Turks in India," Henry George Keene, London, 1879. "History of British India," Sir W. W. Hunter. Hunter's "Indian Empire." "Tales from Indian History," J. Talboys Wheeler. "Travels in the East," Vol. III, Forbes.

Here the Koh-i-nûr, Mountain of Light, sparkled, a price-less trophy. In the great battle of Pariput (April 21st, 1526), when the last emperor of the Afghan-Lodi dynasty, Ibrahim, was beaten by Baber, the Rajah of Gwalior was "sent to hell," as Baber wrote grimly, and his most precious jewel — valued "at half the daily expense of the whole world" — came in tribute to Hûmaiûn, the great sultan's favorite son.[1] Here, too, were the Koh-i-tûr, Mountain of Sinai, and the Darya-i-nûr, Sea of Light, and the Taj-e-mah, Crown of the Moon, and that prodigy of diamonds, the Great Mogul, presented to Shah Jehan by the Emir Jemla.[2]

These precious stones were coveted and hoarded with insane passion when every other lure in the boasted elysium was as Dead Sea fruit to the jaded senses. Shah Jehan, dethroned and impris-oned at Agra, sank to dotage, clasp-ing his casket of jewels, and trickling diamonds and rubies over his head and breast. When his son, Aurung-zeb, sent a messenger to borrow some of this hoard, the resentful old man threatened to break up the gems in a mortar. Shah Rokh, the feeble son

The Koh-i-nûr. (Present Cutting.)

of Nadir Shah, who broke the peacock throne of the Moguls, was blinded by the Aga Mohammed in the vain effort to extort the Koh-i-nûr. Then his head was shaved and circled with a ring of paste to hold boiling oil, but even this intensity of torture only forced the surrender of a ruby plucked from the crown of Aurungzeb. Shah Zamân, blinded by his brother Shuja, hid the Koh-i-nûr defiantly for years in the plaster of his prison cell; and Shuja, blinded by a third brother, Mahmûd, yielded up the priceless stone to Runjeet Singh, only to save his family from agonizing death.

In the sack of Delhi by Nadir Shah in 1739, the wonderful

[1] Memoirs of Sultan Baber.

[2] "Voyages en Turquie, en Perse et aux Indes," Tavernier, Paris, 1676.

store of jewels in the court of the Mogul emperors was borne away
by the plunderers. It is supposed that the Great Mogul was
broken at that time, and other famous diamonds were beyond
tracing for years. The great gems were still more widely scattered upon the assassination of Nadir Shah, and some of the finest
of the crown jewels of Europe have probably come from the
hoards of Delhi. The Darya-i-nûr and Taj-e-mah were set in a
pair of bracelets which Sir John Malcolm saw at the court of
Persia,[1] and they are still the most precious of the jewels of the
Shah. Some have seen in the Orloff or Sceptre diamond of the
Czar, the reappearance of the Great Mogul, but Streeter thinks
that the Great Mogul has never come to light since the loot of
the treasures of Nadir Shah by the Abdalli-Afghans.

When the Koh-i-nûr came into the hands of Runjeet Singh,
he had the stone set in a bracelet which he wore proudly on
every parade day. On his death-bed he sought to propitiate
the gods by presenting this, the chief of his jewels, to the shrine
of Jaga-nath (Juggernaut), but his hand was too weak to sign the
warrant of delivery. So the gem descended to the young rajah

The Orloff.

Dhulip-Singh, and was held until the
Indian mutiny and the seizure of the
Punjaub by the English forces. Then
the state property of the province was
confiscated to pay debts due to the East
India Company, but the Koh-i-nûr was
reserved for the English crown, and on
June 3d, 1850, this jewel, from earliest
tradition the emblem of conquest, was
placed in the hands of Queen Victoria by the messengers of
Lord Dalhousie.

Every precious stone of uncommon size has some adventure
to tell, though its tale may not be a drama of as many acts as the
Koh-i-nûr's career. What a strange story might be drawn from
the Orloff of the sights in the temple of Mysore, when it was
the eye of the Hindoo god, Sri-Ranga.[2] There was no other

[1] "Sketches of Persia," Sir John Malcolm, 1827.   [2] *Ibid.*

witness of the sacrilege of the French grenadier, masquerading as
a devotee on the black and stormy night when he plucked out
the precious stone eye and ran off through the British army lines
to Madras. Here the captain of an English ship gave him
£2,000 for his prize, but it cost Prince Orloff more than fifty
times this sum when he bought it in Amsterdam to win back the
favor of the Empress Catherine.

The Regent lies in state, most lustrous and precious of the
gems of the old French crown. The slave who found it buried
in the bank of the Kistna River, A.D.
1701, cut his leg deeply to pouch
the stone in his flesh, and wrapped the
wound in a thick bandage. At the
first opening he ran away to the seacoast and found refuge on an English
merchant ship. But the lure of the
big diamond was too tempting to
the captain. When his ship was in the
open sea, he flung the slave overboard

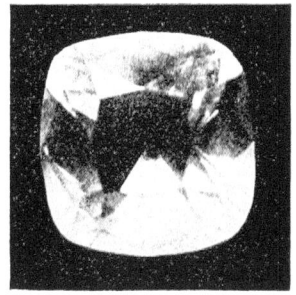

The Regent.

to drown, and took the stolen diamond to sell to an Indian
merchant, from whom it passed to the governor of Fort St.
George, Thomas Pitt, grandfather of the great Earl of Chatham.

It was one of the largest of all known diamonds, the rough
stone weighing 410 carats, and Thomas Pitt would not suffer it
to be out of his sight or touch day or night, though he was
racked by the fear of thieves and murderers. While the alarm-
ful gem was in his keeping, it is said that he never slept twice
under the same roof, and moved from place to place in disguise,
at a moment's caprice, to cover his tracks. Fortunately for his
peace of mind, as well as his purse, he was able to sell his prize
for £135,000 to the Duke of Orleans, Regent of France in the
minority of Louis XV. (A.D. 1715–1723). So the splendid stone
made the fortune of the house of Pitt, and came to glitter in the
most prodigal and luxurious court of Europe. It was held by
the Bourbons until the French Revolution, and in 1792 it was
stolen by the robbers who carried off the Sancy and thrown into

a ditch in the Champs Elysees. Here it was picked up with other plunder which the thieves did not dare to keep or offer for sale.

Then it was uplifted again to the French crown and has held its place through revolutions that have unmade kings and emperors.

So it might be told how " The Florentine " wandered from India through Tuscany to the Austrian crown, — how the " Piggott " saw Clive's conquests (A.D. 1751–1767) and travelled to England with the governor of Madras and

The Florentine.

was crushed to powder by the dying Ali Pasha, — how the " Star of the South " made its way from the sands of Brazil to glitter on the breast of the fantastic Gaikwak of Baroda while he killed disagreeable people with diamond dust, — and how banished convicts won their pardon from the Portuguese crown by the discovery of the Braganza, the largest diamond, if genuine, that the world ever saw.[1]

The Piggott.

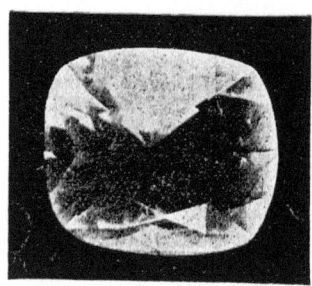

No one can say of a true diamond story, "it is closed"; for diamonds outlast dynasties, and their wanderings may be on the verge of renewal when they seem to be ended. "A jewel may rest on an English lady's arm that saw Alaric sack Rome, and beheld before — what not? The

The Star of the South.

[1] " Great Diamonds of the World," Streeter.   " Diamonds," W. Pole, London Archæological Trans., London, 1861.   " Diamonds and Precious Stones," H. Emanuel, London, 1865.   " Outlines of Mineralogy," J. Kidd, Oxford. " Traité Complet des Pierres précieuses," Charles Barbot, Paris, 1838.   " The People of India," J. Forbes Watson and J. W. Kaye, Editors.   " Gemmarum et Lapidum Historia," etc., Boetius, 1647.

treasures of the palaces of the Pharaohs and of Darius, or the camp of the Ptolemies, come into Europe on the neck of a vulgar pro-consul's wife to glitter at every gladiator's butchery at the amphitheatre; then pass in a Gothic ox-wagon to an Arab seraglio at Seville; and so back to its native India, to figure in the peacock throne of the Great Mogul; to be bought by an Armenian for a few rupees from an English soldier, and so, at last, come hither."

The illustrations of the historic diamonds shown in this chapter have been made from photographs of facsimiles of the stones, and are the exact sizes of the originals.

CHAPTER II

CHILD picking a shining pebble for a plaything from the gravel edging a river — was this sport of blind chance the revelation of the marvellous diamond fields of Africa? In narrow fact, yes; but in a wider, truer range of view, this discovery was the crown that sooner or later must reward the search of daring adventurers and the push of stubborn pioneers into the dark heart of the continent.

There was no chance in the strain of pluck that braved strange perils to reach traditional Ophir and the pits of King Solomon's mines, that wandered far in quest of the golden cities of Monomotapa, that tore the wilderness from the clutch of the lion and vulture, and beat back the frantic impis of Tchaka, Dingaan, and Umsilikazi. The ardor and the toil and the courage and the blood of ten generations of explorers were spent before it was possible for a little child to play pitch and toss with the pebbles of the Orange River and clasp a rough diamond in his heedless hand.

Two dominant motives were fused with the high-spirited zeal for exploration that so signally stamped the fifteenth century, — the opening of an all-sea route to the Indies, and the grasp of the riches of lands behind the veil. In the unknown there is space for any vault of fancy, and in that romantic age her soaring wings were rarely clipped. One may be moved to smile at the fantastic visions of the men who found the southern waterway to the Indies, and added a new world to the old; but there will be no sneer in the smile of any one who can measure his own debt to experience, and put himself back five centuries to stand

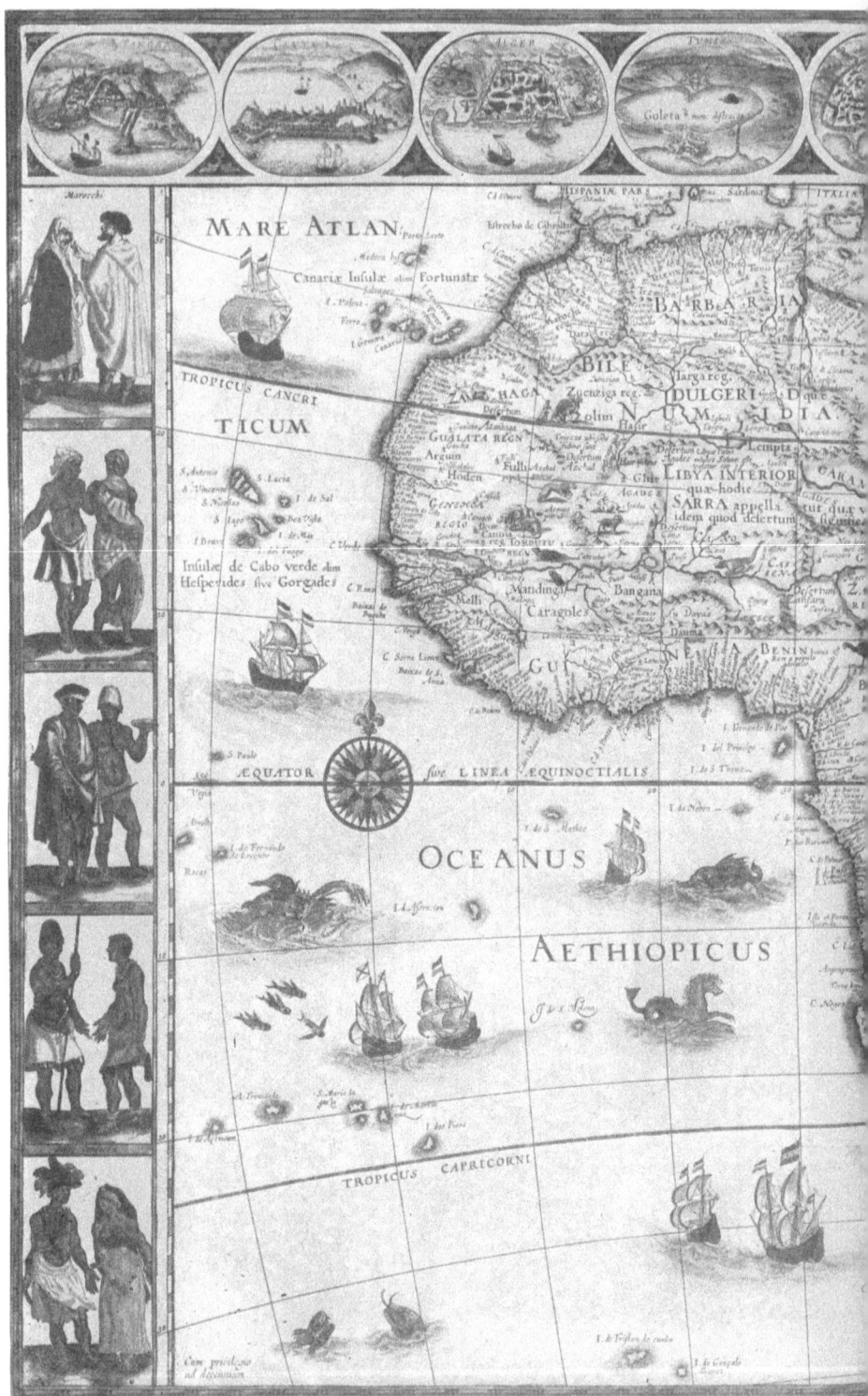

BLAEUW'S MAP OF AFRICA.

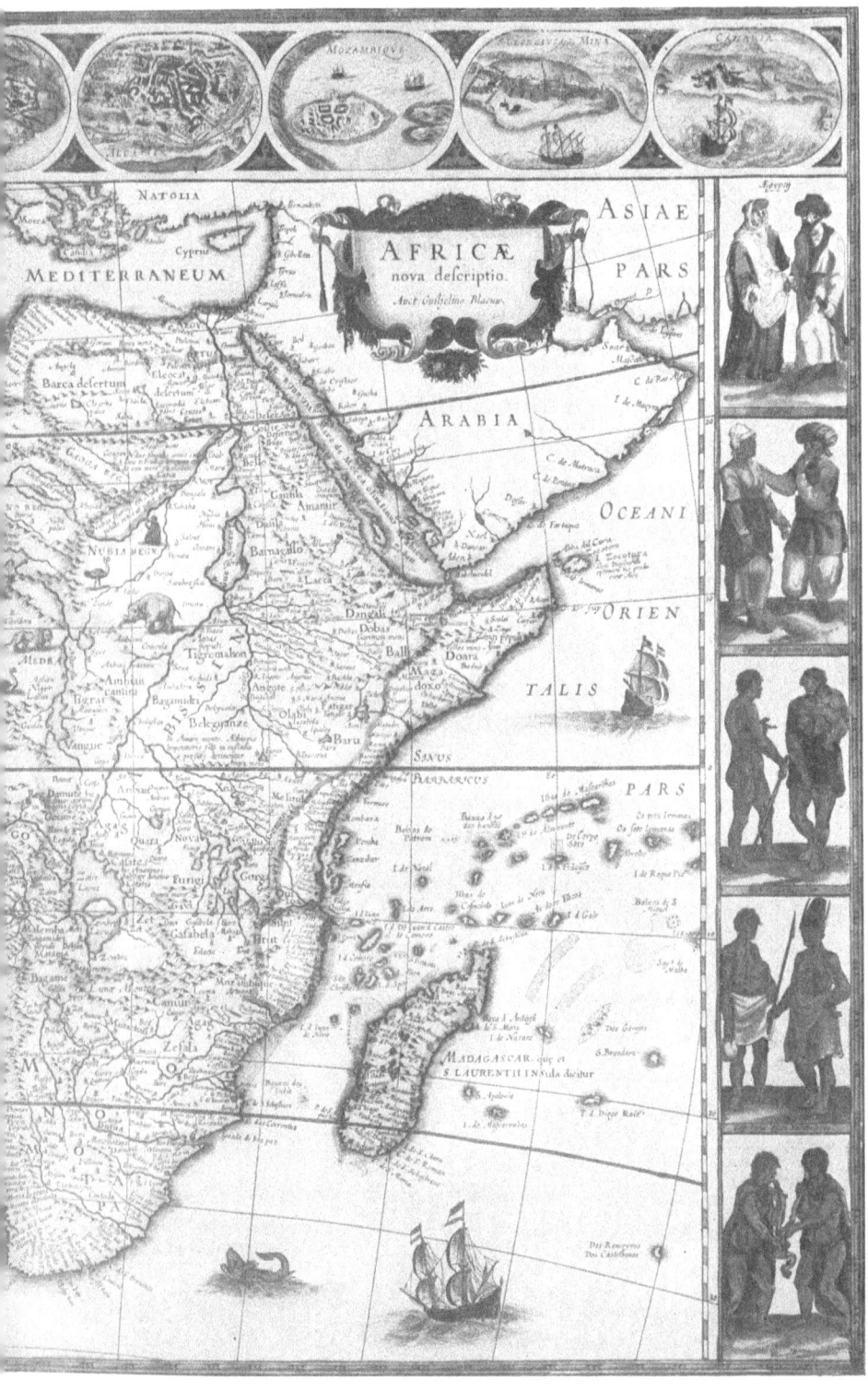

AFRICÆ
nova descriptio.
Auct Guilielmo Blaeuw.

on the deck with Cam, Dias, and Da Gama, or the still more greatly daring Columbus.

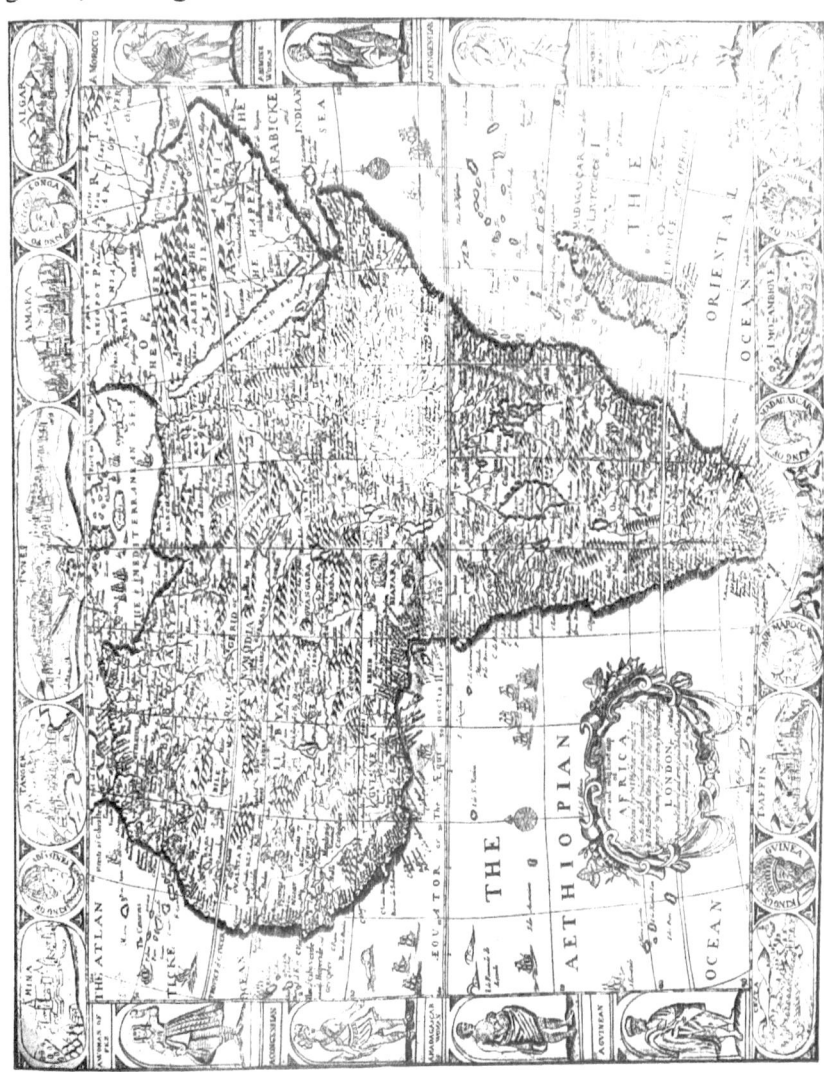

Visscher's Map of Africa, 1662.   (From the original in the British Museum.)

But who can to-day feel the hopes and fears that shook those strong hearts? Who can lay the course for their clumsy caravels over the unknown stretches of ocean? Who can sail on with them day after day and night after night without a chart

D

or buoy or beacon or surf-rocked bell? Who can start from fitful sleep to pierce the night with straining eyes or watch for the glimmer of the dawn on sea-girt horizons? Who can recall their racking fears or the dazzling images ever forming and dis-

Africa, from an early Dutch Map.

solving in the alembic of their fancy? With every daybreak the isles of Atlantis might spring into view, or gardens fairer than the golden Hesperides, or monsters more horrific than dragons, guarding hoards beyond the dreams of avarice, or, perchance, even the realms of some potentate accustomed to make footstools of princes with stiffer necks than haughty Xerxes or the terrible Tamburlane.

Amid the drift of such cloudy conceits there was one more clearly shaped and persistent than the rest. Somewhere below the equator, in the unknown expanse of Africa, tradition placed the home of the Queen of Sheba, King Solomon's mines, and the marvels of Ophir. Every adventurer skirting the South African coast hoped to touch with certainty the shore of this delectable country. The alluring recital in Kings and Chronicles glittered before his eyes.[1] In fancy he saw the gathering of the ships in " Ezion-Geber, which is beside Eloth on the shore of the Red Sea," and how this fleet came to Ophir and fetched from thence gold, four hundred and twenty talents, and brought it to King Solomon. He saw, too, the coming of the Queen of Sheba to the king to prove him with hard questions, and the great train that followed her with camels that bare spices and very much gold and precious stones. Then it was told him how the queen was overcome by Solomon's wisdom and grandeur until " there was no more spirit in her," and she gave the king one hundred and twenty talents of gold, and of spices very great store, and precious stones. Following this tribute came the regular flow, from Ophir to Judea, of gold and gems and almug trees in the transports of Tyre. With such a fountain of supply, it was easy to credit the wonderful tale of the targets and shields of beaten gold, of the throne of ivory overlaid with gold, and of all the other displays of Solomon's splendor. If the king's gold made silver to be in Jerusalem as stones in the eyes of the chronicler, it is not surprising that this vision came down undimmed to the days of Da Gama.

But how to find the source of this flow was the puzzle that faced the explorer. Unfortunately the old chroniclers had omitted to give any landmarks of King Solomon's mines. Surmise strayed down the eastern coast of Africa, and the close commercial connection between southwestern Arabia and the

[1] 1 Kings ix., x ; 2 Chronicles viii., ix.

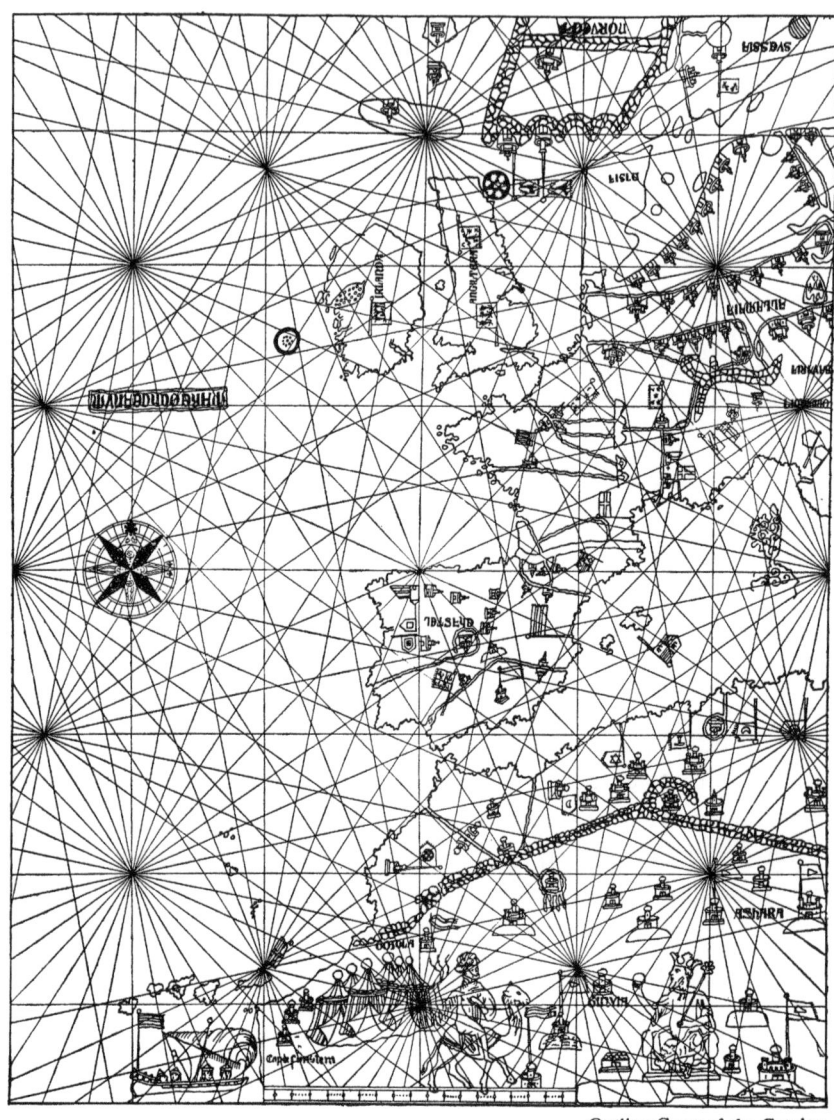

Outline Copy of the Catalan
(In the original the shore line has almost illegible names,

equatorial coast region of East Africa was unquestionable. Herodotus declares that East Africa at its furthest known limits supplied gold in great plenty as well as huge elephants and ebony. The Alexandrian geographers mark rudely the East African coast line to Zanzibar, and attest the relations between

Mappermonde, 1375.
which, for the sake of clearness, have been omitted here.)

this coast and Arabia Felix.   Eratosthenes observes that naviga-
tion extends down East Africa beyond Bab-el-Mandeb, "along
the myrrh country, south and east as far as the Cinnamon country, about five thousand stadia." [1]   Ptolemy, in the second cen-

[1] Strabo, XVI, Chap. IV, 4.

tury A.D., describes quite accurately the east coast of Africa as far as Zanzibar and Ras Mamba Mku.   His information was chiefly derived from Arabian merchants.   But, as Schlechter has closely pointed out in his admirable monograph,[1] there is no trace or hint anywhere during the Greek and Roman periods of antiquity of any colony or emporium south of the Zanzibar

Outline Copy of the Portolano Laurenziano, 1351.

coast, and not long after the time of Herodotus the gold imports of Arabia had shrunk to inconsiderable importance.   With the decline of the Himyaritic Kingdom in Arabia, soon after the second century of our era, there was a falling off of commercial enterprise and intercourse with Africa, so marked that even the

[1] "Periplus of the Erythræan Sea," Henry Schlechter, The Geographical Journal of the Royal Geographical Society, July, 1893.

notable map of the Arabian Edrisi, in 1154 A.D., shows how slight and vague was the advance in the knowledge of the Dark Continent from the days of the Alexandrian geographers. Still this old chart gives some substantial proof of the communication of Arabian traders with the natives on the East African coast. But on this map the African coast appears to curve

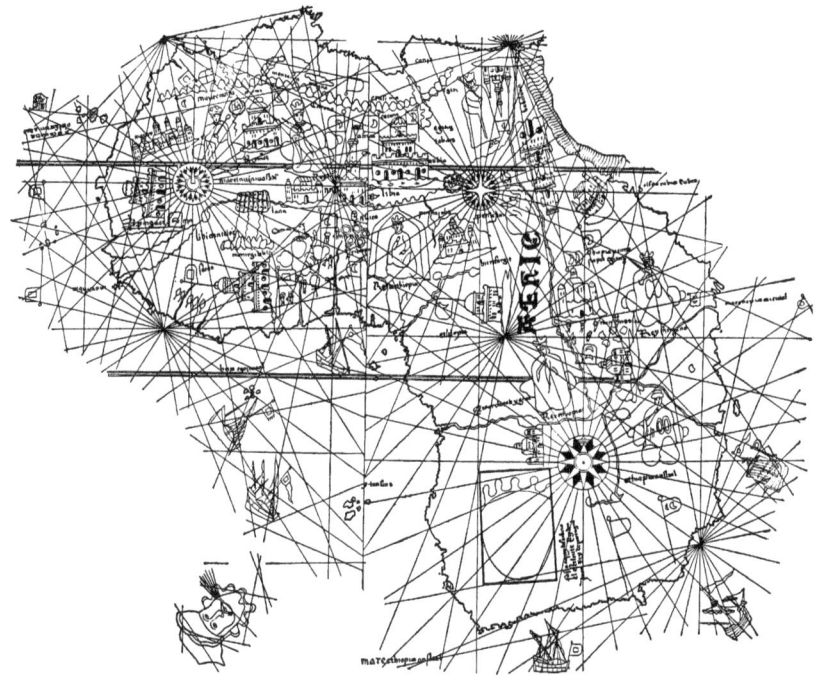

Africa de Mappermonde, Juan de la Cosa, 1500. (This map was made only fourteen years after the discovery of the Cape of Good Hope, and is one of the earliest known maps giving the entire contour of Africa with approximate accuracy.)

east continuously from the mouth of the Red Sea, and Edrisi was plainly ignorant of the abrupt trend to the south from Cape Guar-da-fui. Yet he shows rudely the islands lying off the east coast of Africa, and, south of Sokotra, traces the African mainland in three divisions, Zendj (Zanzibar), Sofala, and Vakvak.

With all its imperfections this Arabian map was in advance of any European portrayal of South Africa. It was the prevailing belief in the Middle Ages, " bequeathed from antiquity," as

Justin Winsor observes, that " owing to the impassable heats of
the torrid zone, it could not be discovered whether this region
were inhabited or whether land existed there." Map makers
plainly made the bounds of land and water beyond the equator
from sheer surmise, and the confession was commonly frank that
the land was terra incognita and the ocean a sea of darkness.
" Most famous of all these early maps " (of the Atlantic Ocean),

Dutch Ship of the XVIIIth Century.

says Winsor,[1] " was the Catalan Mappermonde of 1375." It was
probably the one best known by the sailors sent out by Prince
Henry of Portugal, in the year 1413, to follow down the Atlantic shore line of Africa. On this map, all known Africa is
bounded on the south by a line drawn eastward from Finisterra,
off the mouth of the Rio Del Oro, about 23° north of the
equator, nearly across the continent to the Egyptian Nile. In
the Portolano Laurenziano of 1351, the outline of Africa is given

[1] " Narrative and Critical History of America," Vol. I, p. 55, Justin Winsor.

Dutch Ships of the XVIIth Century.

an approach to reality that is highly remarkable, but it is clearly a happy stretch of guesswork.[1]

All of the region south of Cape Non was practically unknown to the adventurers of the fifteenth century.[2] Their ears were filled with doleful tales of the calms and storms, the

Dutch Ships of the XVIIth Century.

[1] "Life of Prince Henry of Portugal, surnamed the Navigator, and its Results," R. H. Major, London, 1868.

[2] Chief of the charts in the fifteenth century were those of Andrea Bianco, "Atlas," 1436, and "Carta Nautica." Justin Winsor, "Narrative and Critical History of America," Vol. I, p. 55.

mud-banks and the fogs, of the Sea of Darkness. If by any
stretch of daring they might cross the equatorial line, they were

Dutch Ship of the XVIIth Century.

burdened with the fear that they would begin to slide down an
inclined plane with a rush that would pitch them headlong into

Dutch Ships of the XVIIth Century.

some bottomless abyss.   The only assurance of a happier issue
was the bare tale of old Herodotus of some nameless Phœnician

sailors who had skirted the coast south from the Red Sea in the days of Pharaoh Necho (610–594 B.C.), and returned nearly three years later through the Pillars of Hercules and the Mediterranean. These sailors brought back, with their load of ivory, feathers, and gold, the report that during a considerable part of this voyage they had the sun on their right hand. It is this detail that now chiefly confirms the story, but this was beyond the credence of Herodotus,[1] and it would seem that this ancient mariner's tale was soon generally disbelieved, for the special

Dutch Ships of the XVIIth Century.

searches made in the Alexandrian library by Eratosthenes and Marinus of Tyre in the third and second centuries B.C. brought to light no other records or traces of the voyage. So it was not with reliance on this alleged circumnavigation that the adventurers of Portugal groped painfully for seventy years along the coast, until the daring Dias set his stone crosses at Angra Pequena and Algoa Bay and sighted the turning point of the path to the Indies in the frowning Cabo de Todos los Tormentos. King John was quick to see the promise in the land

[1] " Herodotus," Bk. 4, 42, Rawlinson.

of Dias and change the Cape of Storms to Cabo de Boa Espe-
ranza, but ten years passed before Vasco da Gama followed
down the trail and rounded the Cape in the immortal voyage
that reached the long-sought Indies six years after Columbus
had touched the island hem of the new world.[1]

Dutch Ships of the XVIIIth Century.

The completed circling of Africa by European adventurers
was a no less memorable achievement of Da Gama.   He touched
at Mozambique on the first of March, 1498, and there saw gold,
in the hands of Arabs, that had passed up the coast from Sofala.
Nearly twenty years before, a Portuguese courtier, Pedro de
Covilhao, had reached Sofala in an attempt to pass to India by
way of Egypt.[2]

For many years and possibly for many centuries there had
been a trickle of gold from Sofala through Arab traders, and
Da Gama saw enough of it to move his king to lay his hands
upon it.   In the expedition of Cabral, which followed in the
wake of Da Gama in 1500, the great captain, Bartholemeu

[1] " Prince Henry the Navigator," C. Raymond Beazley.
[2] " The Portuguese in South Africa," George McCall Theal.   " South Africa
from Arab Domination to British Rule," R. W. Murray, editor, London, 1891.

Dias, was specially commissioned to seek the source of the gold stream. Dias was drowned in the storm which sunk four ships of this fleet, but Cabral took a vessel carrying gold from Sofala and sailed to Kilwa, where the Arab Ibrahim and his forefathers had been drawing gold from Sofala for a long term of years. Upon the report of Cabral, Da Gama turned out of his way to Mozambique in his second voyage, in 1502, to enter Sofala and take possession of Kilwa, and three years later Pedro da Nhaya sailed from Lisbon with six ships and built a fort and trading station at Sofala.

Behind this persistent push to Sofala there was more than the actual showing of gold. Here was one of the traditional gateways to King Solomon's mines, and the Portuguese were quick to embrace the tradition. They gave the glittering name of Ophir to their fort. South of the fort there runs a river, called by the Arabs Sabi, and this was pounced upon as a

Dutch Ships of the XVIIIth Century.

probable twist of the old Hebrew Sheba. From those days Fort Ophir was the starting point of Portuguese adventurers in search of the fountain head of Solomon's treasures.

The Portuguese then had uncommonly sturdy sea-legs and asked nobody to show them the way over the ocean foam, but

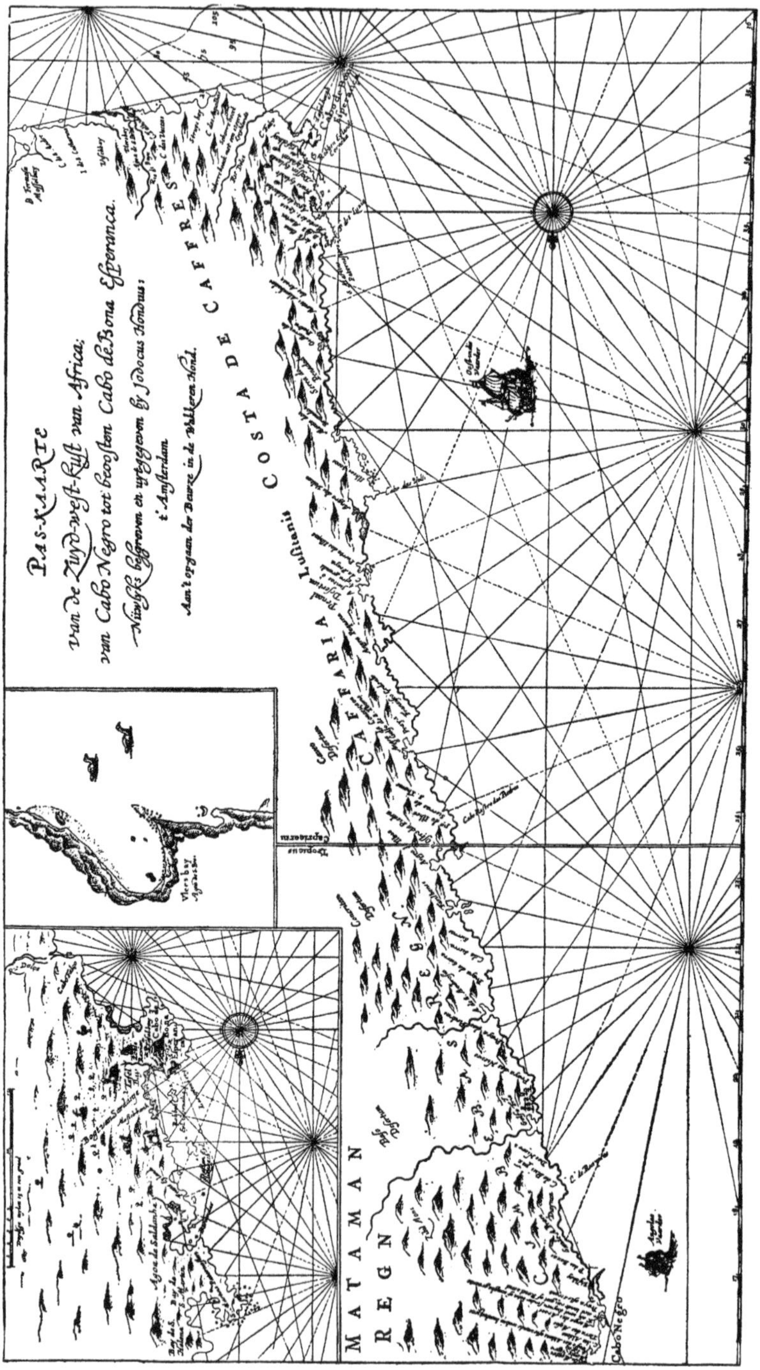

CHART SHOWING METHOD OF SURVEYING COAST LINES.

they were far less ready to weary their legs with trudging over
mountain ridges or scrambling through the dense thickets of
the rugged land west of Sofala.   The Arab traders were more
ready to venture inland, but there is no evidence to show that
any of them went farther than a few hundred miles, at most,
from the seacoast.   It was an exceedingly difficult country to
penetrate, and the savage natives were jealous of any approach,
if they did not stubbornly bar the way and murder intruders.

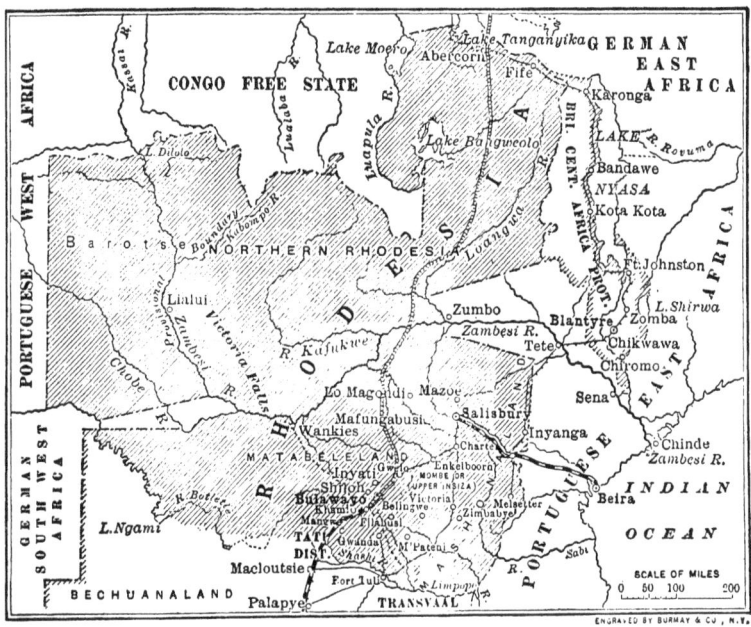

Map showing the Position of Ancient Ruins in Rhodesia.

The horrid death of the first Portuguese viceroy was a warning that struck deep into the hearts of the earlier adventurers.
Francisco d'Almeida, returning with his fleet from India in
1510, touched the African coast near the first landing of Diaz.
To resent some little clash with the nearest native tribe he led
a troop of soldiers inland to surprise their village, but was way-
laid in the bush and his troop was put to flight by a hail of
darts and stones.   D'Almeida put his ensign in the hand of a
trusty follower, but in the next moment he was stabbed in the
throat by an assagai and his head was crushed by the swing

of a knob kerrie.   Sixty-five of his picked swordsmen fell with
him and the rest only saved their lives by abject flight, chased
to the shore by a little band of naked negro dwarfs.

This was the greeting of a weak and puny coast tribe.   What
then might be feared from the rallying of the fierce and stalwart
blacks of the Bantu tribes, under some ruthless chief, in the
fastnesses of the mountain land encircling the gold of Ophir?

Insiza Ruins.

Still there was an enticing trickle of gold dust and nuggets from
inland mines to Sofala, and the flow of resplendent stories was
vastly bigger than the golden stream in sight.   So in 1569 it
was resolved to make an extraordinary effort to penetrate to the
source of the gold.   The East African coast was placed under
command of a governor independent of the viceroy of India.
Francisco Barreto was made the first governor, with instructions
to raise a force of a thousand men and lead them on to the
capture of Ophir.   The young cavaliers of Lisbon flocked
eagerly to Barreto's standard.   He led the way up the Zambesi
with a high-spirited troop, but the gay soldiers were soon

scorched by the sun, torn by thorns, and cast down by fevers. The Kalangu tribe was then the strongest of any living between the Sabi and Zambesi, and Barreto sought to win the good will of its head chief by offering to beat his rival. This offer made him

Insiza Ruins.

welcome, and he kept his promise, but he was soon after obliged to appoint Vasco Fernandez Homem to the command of his troop and to return to the coast. Homem soon followed him with the dispirited remnants of the adventurers. Barreto did not live to see the return of his broken expedition, and Homem

E

succeeded him as governor. Then the new governor tried another way of approach to the gold field, and finally pushed a party
through from Sofala to the foot of the mountain which the
Kalangu tribe called Fara and the Arabs Aufur, transmuted
forms, it was thought, of the Hebrew Ophir. Near the base
of this mountain were placers yielding nuggets worth from two

Insiza Ruins.

to three thousand dollars, but the ordinary toil of placer washing was so disgusting to the Portuguese visionaries that they
gloomily turned their backs on the mines of Abasia and the
rock mark of Ophir and wearily made their way back to Sofala.[1]
This disappointment dulled the glitter of some old stories, but
there were plenty of new ones to dazzle men's minds.

It is likely that the most accurate, as it certainly is the fullest extant, account of the mining in Ophir land is given in the
story of the old Spanish author, Joano de Barros, whose life
spans the first three quarters of the sixteenth century.[2] It is
too much to expect that his " Da Asia " should be free from
the coloring of the ardent fancy and the myths of the age, but
underlying his narrative there is, at most points, a credible
basis of personal observation and the current reports of many
witnesses.   He held several high offices in the Indian and

[1] " The Portuguese in South Africa," Theal.   " Conferencias Celebradas na
Academia Real das Sciencias de Lisboa, Acerca dos Descobrimentos e Colonisa-
coes dos Portuguezes na Africa."   [At Lisbon, 1892.]

[2] " Da Asia," Joano de Barros (1496–1570).

African establishments of Portugal, and had exceptional opportunities for preparing his remarkable memorial.

In his description the "mines of Manica" are placed "some fifty leagues west of Sofala." The Portuguese league was 3.84 English miles, and De Barros was as loose as contemporary writers in the measure of distances. "All gold found there is in dust," he writes, "and the workers have to carry the earth which they dig to some place where water can be had. Nobody digs more than six to seven spans deep (four to six feet), and if they go to twenty, they come to hard rock."

Beyond the Manica placers, in positions not defined, were the mines of Boro and Quiticui. There nuggets were found

Khami Ruins.

"embedded in reefs — some already cleared by the winter torrents; hence, in some of the pools, such as remain in summer, the miners dig down and find much gold in the mud brought up. In other localities, where are some lagoons, two hundred men set at work to drain off about half the water, and in the mud which they sift they also find gold, and so rich is the ground, that if the people were industrious, great quantities could be had; but they are so indolent that stress of hunger alone will keep them at work. Hence Moors (Arabs) who

visit those districts have recourse to a ruse to make them diligent. They cover the negro men and women with clothes, beads, and trinkets in which they delight, and when all are pleased trust everything to them, telling them to go to work

Khami Ruins.

the mines, and on their return, they can pay for those advances; so that in this way, by giving them credit, they oblige them to work, and so truthful are the negroes that they keep their word.

"Other mines lie in the district called Toroa, ruled by a vassal of Benomotapa. These are the oldest known in that region. They are in a plain, in the middle of which stands a square fortress, all of dressed stones, within and without, well wrought and of marvellous size, without any lime showing the joinings. The walls of this fortress are over twenty-five spans high (18 to 19

Gold Ornaments found in Ancient Ruins.

feet) but the height is not so great compared with the thickness. And above the gateway of that stronghold there is an inscription which some learned Moorish traders who were there could not read nor say what writing it was. And around this building are others on some heights, like it in the stonework, in

which is a tower twelve bracas (72 feet) high. All these structures the people of the country call Symbaoe, which with them
means a royal residence. They stand west of Sofala, under
latitude 20° and 21° south, one hundred and seventy leagues
more or less in a straight line. . . . In the opinion of the
Moors who saw them, they seemed to be very ancient and were

Khami Ruins.

built there to hold possession of those mines, which are very old,
from which for years no gold has been taken owing to the wars."

The latitude and position of the Symbaoe of De Barros correspond closely with the site of the ruins of Zimbabwe, described
three hundred years later by the explorer Karl Mauch. Both
Zimbabwe and its antique form, Symbaoe, are plainly versions
of the local Bantu nzimba-mbuie, a house of the chief. It is
true that the Zimbabwe of Mauch is only two hundred and forty
miles west of Sofala, but the leagues of the old chroniclers were
not laid off with the tape line.

Who was this Benomotapa whose vassal was housed in such a castle? — the mighty black sovereign of whom Camoens sings —

" Vê do Benomotapa o grande imperio,
De Salvatica gente, negra e nua " ' ?

In dull fact Benomotapa was simply the corrupted plural form of Monomotapa, signifying Lord of the Mountain, or by a possible stretch of derivation, Master of the Mines.[1] This was one of the hereditary titles of the head chief of the Kalangu

Khami Ruins.

tribe, the largest and strongest of any then living between the Sabi and Zambesi. His dwelling was at the foot of Mount

---

[1] " The Portuguese in South Africa," George McCall Theal.

Bent says the name Monomotapa should be written Muene-matapa, or " lord of Matapa," simply " a dynastic name, just as every petty chief in Mashonaland to-day has his dynastic name, which he takes on succeeding to the chiefdom." " The Ruined Cities of Mashonaland," p. 285. Both titles have in fact the same meaning : the first components bena and mono being the still current Bantu words bwana, bana, muene, mwana, that is ' lord,' ' master,' ' chief,' ' ruler.' The second part, motapa, common to both, probably means a mine, from the Bantu word *tapa* = ' to dig,' ' excavate.' " Africa," Vol. II, p. 372. (Stanford's Compendium.) A. H. Keane.

Aufur, which was held in such traditional reverence that the chief would not permit the Portuguese to ascend it. There was

Zimbabwe Ruins.

nothing of imposing splendor in the huts of the chief who received the embassy of Francisco Barreto, but no lack of evidence could prevent romance from creating an African empire under the sway of Monomotapa. Some corner-stones for this structure were found in the remains of the works of a people of far higher civilization than any of the existing native tribes, and these relics were prizes to a fancy that clutched greedily at every drifting straw of report, tradition, and myth supplied by Arabs and negroes.

Zimbabwe Ruins.

Every one in the succession of romancers, in the sober cloak of historians, of South Africa would outdo his forerunners in inflating the balloon of the traditional empire. The old Dutch writer, Kleveer, finally puffed it up to the bursting limit by bounding it " on the east, south, and west by the Atlantic, and north by the kingdoms of Congo, Abyssinia, and the Zanzibar country. Even

Dapper,[1] whose really great work is by far the most important, comprehensive, and creditable presentation of the Africa of the seventeenth century, jots down gravely most fantastic details of the empire ruled by the royal line of Monomotapa.   He paints a mammoth palace with four grand gateways leading to a succession of halls and chambers, rivalling the handiwork of the slaves of the lamp of Aladdin.   All the ceilings of the rooms were gilt or covered with golden plates.   For the furnishing of sumptuous

Zimbabwe Ruins.

couches and chairs there was gilding and painting in rainbow hues and artful inlaying with enamel.   Ivory chandeliers, hanging on silver chains, filled these resplendent halls with light.   When his majesty deigned to rise from his imperial bed, he was clothed by his valets in garments of native silk.   All his servants approached him on bended knees and served him like dumb slaves.   His table service of the finest porcelain was decorated with wreaths of gold, cunningly wrought in the fantastic forms of natural coral.

[1] "Naukeurige Beschrijringe der Afrikaensche Gewesten," etc., Dr. O. Dapper, Amsterdam, 1668.

Zimbabwe Ruins.

Two pounds of gold was daily spent in perfume for the royal nose, and torches of incense flamed day and night around him. When he took an airing, he was borne in a gorgeous palanquin on the shoulders of four of his trembling nobles, and his head was shielded from the profaning sun by a canopy studded with precious stones. If he was impatient of this slow promenade, he

Zimbabwe Ruins.

might mount on an elephant's back, but on nothing meaner, for nobody in that wonderful country would ride on any other animal.

It is small wonder that the court of monarchs of this splendor, and their golden cities of Davaque and Vigiti Magna, were ardently hunted for by adventurers, thirsty for every romance gilding the dismal stretches of sand and thickets and rocks which encircled them with the threads of a trail to the glittering realm of Monomotapa. But the expeditions of Barreto and

Homem were so painful, costly, and discouraging that for many years no more explorations were undertaken by the Portuguese crown. The spirit of chivalric adventure drooped low after the gallant young king Sebastian fell in battle with the Moors in 1578, and even the spirit that had so greatly spread the commerce of Portugal was losing its vigor. There was a momentary arousal in the beginning of the seventeenth century, when some rich silver ore was sent to Lisbon by the governor of Mozambique. It was believed that this ore came from veins in a region called the

Zimbabwe Ruins.

Kingdom of Chicova, stretching north from the bank of the Zambesi; but there was no definite report of the location. Still there was such an impulse in the sight of this silver that the order was sent to despatch five hundred soldiers to Chicova. No such force could be mustered, but Nuño Alvares Pereira set out from Mozambique with a hundred men. Soon Pereira was the victim of jealous maligning, and was superseded in his command by Diogo Sinoes Madeira. This commander succeeded in placing a few

trading stations along the Zambesi, and made a pretence of
opening mines by shipping some little silver to Portugal; but

Zimbabwe Ruins.

after a dozen years of costly maintenance, it was shown by the
search of Pereira that the pretended discovery of silver was a

Zimbabwe Ruins.

fraud, and disgusted Portugal abandoned the enterprise in 1622.[1]

[1] "The Portuguese in South Africa," Theal.

From that year nothing of note was attempted from the stretch of seaboard loosely held by a few feeble garrisons. Beyond the vague traditions and romances there were no guide-books to the rich realm of any African monarch, and there was no point on the South African coast outside of the Portuguese strip where the least enticement was shown to any visiting ship. Nowhere was there any evidence of an approach to civilization, and there was not even the gilding of barbarism. The

Zimbabwe Ruins.

shore tribes were filthy, famine-hunted negroes, who had, at most, a little ivory or a handful of feathers to barter for trinkets. There was an intermixture of blood and a medley of tribes and tribal names that confounds any tracing of distinction beyond a few blurred divisional lines.

When the Dutch and English began to tread upon the heels of the Portuguese in Africa, in the opening years of the seventeenth century, the tribes of the extreme south and along the southwesterly Atlantic coast might be roughly grouped under the name of Hottentots, or, as they called themselves with monstrous conceit, Kwa-Kwa, men of men. In this assertion there is plainly to be seen the origin of the Arabic Vakvak, the name sketched in by Edrisi on his map beyond Sofala. The southeast African coast was held by tribes of the wide-spreading Bantu family, lumped together by the Arabs as Kafirs. Filtered in between the Bantus and Hottentots were the pigmy Sana, rudely bunched as Bushmen.[1] There was endless wrangling and fighting among the tribes, regardless of any common flow of blood, and the Bantus and Hottentots were continually clashing like wildcats. Their only union was in their hate of the Bushmen, who were hunted from cover to cover, to hide in crevices in the rocks or in holes in the desert sand, from which they might sally, wasp-like, with the deadly sting of their poison-tipped arrows.

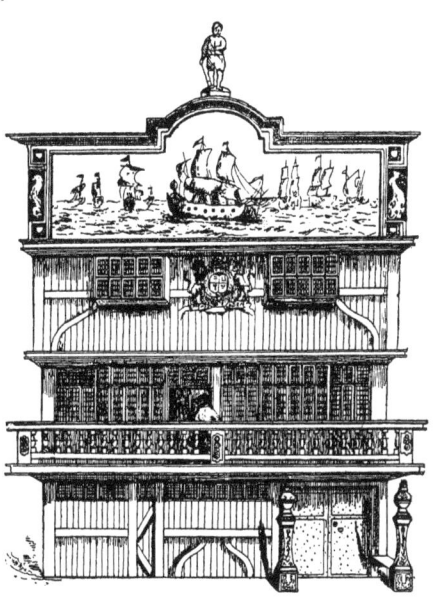

In view of the repulsive face of the South African coast lands it is not surprising that Francis Drake and many other bold voyagers circled the Cape of Good Hope without landing to seek for traditional treasures. But with the opening of the seventeenth century, Table Bay became a regular stopping place and refitting station for the ships of the English East India Company.

The Old East India House, Leadenhall Street, London.

For twenty years this slight hold on the continent was maintained, but it was so lightly prized that it was dropped in 1620 by a shift of the station to St. Helena. Thirty-two years later the Dutch East India Company took formal possession of the Cape and its adjoining bay without any challenging protest, and built their fort Good Hope as the first stronghold of the Dutch dominion in southern Africa. With this foundation the search for the golden realm of Monomotapa was vigorously and persistently revived.

Jan van Riebeeck, the leader of the Dutch colonizing expedition and the first commandant of the fort and settlement at

[1] "South Africa," George McCall Theal, London, 1888–1893. "South African Tribes," Sutherland.

Cape Town, was a man of ardent spirit and uncommon energy. He had entered the company's service as a surgeon's assistant, but his ambition and ability had soon pushed him to the front and marked him as a man to fix and strengthen the grip of the great trading company on the turning-point of the way to the Indies. In his portrait dark, sanguine eyes are set under a high, full forehead, crowned with thick waving hair of a cavalier cut, in keeping with his trim mustache. His well-moulded features and resolute chin have the stamp of refinement as well as action. He quickly put his hand to every practical device to make the new settlement productive and self-supporting. Nine months after his landing the first crop of wheat was reaped at the Cape. In the following year he set out vines from the Rhine. In his own vineyard the muscatel grape grew luxuri-

The Landing of van Riebeeck.

antly, and a few years later he made the first Cape wine, a high-flavored Constantia. In the same year, 1658, maize was brought to the colony from the coast of Guinea and successfully planted. To the introduction of the olive, particularly urged by the directors of his company, he gave unremitting pains, and succeeded in rearing a fine grove of fruitful trees on his own plantation at Wynberg. In his stretch of experiment he even tamed young ostriches and stocked the neighboring islands with rabbits.[1]

Such a man was not likely to be heedless of the chances for

[1] "South Africa," Theal.   "On Veld and Farm," Frances MacNab, London, 1897.

the possible enrichment of his company by penetrating to the
seat of the traditional empire and possibly to King Solomon's
mines. He reckoned that, in any event, his exploring parties
would be likely to succeed in uncovering ore beds of some useful metal, if not of gold and silver. But he seems to have had
great confidence in the traditions of Monomotapa, and it is
known that he had before him the highly colored work of
the Dutch traveller and author, Linschoten, as well as current

Portrait of Johan Antonyse van Riebeeck.
First Commandant of the Cape of Good
Hope.   Born 1618, died January 18,
1672.

Portrait of Maria de la Querellerie of Que-
rellerius, Wife of Johan Antonyse van
Riebeeck. Born October 28, 1629, died
November 2, 1664.

Portuguese books infused with the romance of Africa. His
calculation plotted the location of Davaque, the chief seat of
the splendors of Monomotapa, at a point 828 miles N.E. of
the Cape of Good Hope, and 322 miles W. from the Indian
Ocean, curiously near the present Witwatersrand. Davaque
was built by tradition on the banks of the river Spirito Sanctu,
flowing into the Indian Ocean at Delagoa Bay. Nearer still
to the Cape was another El Dorado, the city of Vigiti Magna,
which was confidently located on or near the meridian of 30° S.,
and not much more than three hundred miles from the Cape.

The first push into the unknown land north of Fort Good Hope was made in 1657 by a little party headed by Abraham Gabbema, Fiscal, and Secretary of the Council of the colony. Gabbema led the way to the first big beacon in sight, a peak with a grotesque flat top which the colonists had already christened Klapnuits, or night cap mountain. Skirting the base of this peak he pushed to the next conspicuous landmark, bearing toward the west, a mountain with bare rugged pinnacles of rock, which the explorers dully called Great Berg, and gave the same name to the river flowing below.

Constantia.

It was in the middle of October when the party set out, but this was the prime of the springtime in South Africa. On the lower slopes of the Great Berg herds were grazing that had never seen the face of a white man nor felt the sting of a bullet. Zebras capered over the hillsides, the unwieldy rhinoceros wallowed in the high grass, and hippopotami plunged and snorted in the turbid rivers. Every step of the way was a new wonderment to the explorers, and when the rising sun struck the mountain tops with its flame, two transfigured peaks gleamed like prodigious gems in their eyes, and were forthwith distinguished

Farmhouse on the Farm Constantia, near Cape Town. The Residence of
Jan van Riebeeck, and now owned by the Cape Government.

as Paarl and Diamant. These sunlit crests were the only things in sight, however, that had any glitter of the realm of Monomotapa, and after a little further advance into the unknown field, Gabbema's party turned back.

The next excursion was more daring. By promising rich rewards van Riebeeck formed a party of thirty volunteers headed by Jan Danckert. They took along a small stock of bread on three pack oxen, relying for their main supply of food on the game which they might kill on their way. These hardy volunteers plodded north, inclining to the west along the foot of the coast range. They saw whirlwinds of dust and a few roving Bushmen, but nowhere any trace of a monarchy except what they called " A Kingdom of Moles," where the burrowed ground sank under their feet and they could hardly flounder along. In December they reached a river flowing toward the Atlantic, on whose farther shore they saw a herd of more than two hundred elephants feeding. So they called the stream Olifants River, a name which it has borne since that day, and trudged back wearily to tell their story to the commandant at the Cape. Within ten days after their return, January 20, 1661, van Riebeeck, the untiring, mustered another party, of thirteen adventurers and two Hottentot attendants, and sent them away on the track of the discoverers of Olifants River.

Corporal Pieter Cruythof led off this party, which succeeded in crossing the river of the elephants and reaching the land of the Namaquas, a Hottentot tribe of the highest class. Here the explorers found natives who had rude copper ornaments twisted in tufts of their hair, and wore rings of copper and ivory on their arms. They entertained the white visitors with cheering hospitality and gave a grand dance in honor of the embassy. This was the nearest approach to the civilization of the traditional empire that had hitherto been reached by Dutch exploration, and the return of the adventurers on March 11, 1661, after forty days' wandering, was warmly welcomed by van Riebeeck.

Before two weeks had passed he had another excursion under

F

way led by Corporal Meerhoff, which penetrated into Namaqualand farther than any white man had ever gone, but brought
back bitterly discouraging reports.   It was learned that the
Namaquas had uncovered some veins of copper and iron ore
and had some crude process of smelting and working both
metals, but it did not appear to be practicable to undertake to
open mines at points so far from the Cape in a region that for
many months in the year was a torrid desert.   There was no
trace of gold or rumor even of any distant land of gold.   Over
every day's march was the hanging terror of death by thirst or
hunger or savage attack.

Still the unflagging commandant would not give up the
search, and in the following November Corporal Meerhoff
went back with another party of volunteers to Namaqualand, as
second in command under Sergeant Pieter Everaert.   This
expedition was better equipped for exploration than any previous one that had set out from the Cape, and it was three months
before it returned to Fort Hope.   Yet it had nothing new to
tell — only to repeat the same dreary story of painful tramps
over sun-scorched sands and jagged ridges of rock, of blinding
whirls of dust and the blare and clash and drench of terrific
thunder-storms, of sleep broken by nightly alarms, of lurking
Bushmen and prowling lions.   One of the party had been gored
and trampled to pulp by an elephant, and his comrades counted
themselves lucky in reaching the Cape fort empty-handed,
gaunt, and footsore.

Even after this sickening rebuff, the next year saw a renewal
of the attempt to reach the elusive empire of Monomotapa.
Then Sergeant Jonas de la Guerre set out with a little troop of
adventurers not yet disheartened.   But they were not able to
push their search into Namaqualand as far as former explorers
had gone, for they could not find a mouthful of water in the
desert sands, and were in imminent peril of dying from thirst.
This repulse was a crushing blow to the stubborn spirit that had
borne so many buffets.   The enterprising van Riebeeck had
been transferred to the government of Java in the previous year,

and his successor was a man of much fainter heart and energy. So for nearly a score of years the search for the traditional empire lagged, although there was a considerable show of less venturesome prospecting. One notable undertaking was the despatch of a party of expert assayers and miners from the Netherlands to Cape Town in 1669 by the Dutch East India Company, with instructions to search for any promising outcrops of ore in the region of the Cape. This party prospected for several years, but found nothing to inspire any investment in mining.[1]

A revival of the dazzling old visions came in 1681, with the appearance at the Cape of a party of Namaquas bearing pieces of rich copper ore. This exhibit spurred the East India Company to direct another exploration of Namaqualand. Then the commandant at the Cape was a man of the stamp of van Riebeeck, commander Simon van der Stel. He was quick to despatch a company of thirty soldiers, a draughtsman, and a reporter to make the venture so often tried in vain. Again, after months of struggle, the desert drove them back. Van der Stel then resolved to make an effort far surpassing any put forth before by adventurers from the Cape. He formed a party of forty-two white men, soldiers, miners, and draughtsmen, with ten Hottentot servants and guides. The expedition was provisioned for four months, and equipped with two boats, a train of wagons, several horses, and a herd of pack oxen. Ensign Olaf Bergh was put in command and led his company on to Namaqualand. But it was the same old story. No strength of men or oxen availed against the desert. No rain had fallen in the wilderness north of the Olifants River for twelve months, and the whole region was an arid waste without a trickle of moisture. So Bergh and his companions faced about in despair, and marched back to report their failure. Sergeant Izaak Schuyver and another forlorn-hope party tried their luck in the following year, and pushed over the desert a little farther than Bergh, but brought nothing back except a sack of copper ore on a pack ox.

[1] "South Africa," Theal, Vols. 1 and 2.

As a last resort the unflinching commander van der Stel resolved to head an exploring party himself. He obtained special permission from the directors of the East India Company, and his expedition was ordered in keeping with his distinction as the head of the Dutch power at the Cape, and with the labors and perils of the venture. He left the Castle of Good Hope, August 25, 1685, with fifty-six white followers and a troop of Hottentot attendants. Twenty-three wagons and carts were packed with supplies. Besides the draught teams, there were two hundred spare oxen, thirteen horses, and eight mules. For the dignity and comfort of the commander there was a coach, but this touch of parade was chiefly introduced to impress the native tribes and possibly a negro emperor with the grandeur of the sovereignty despatching such an embassy.

The time of year chosen for the start was precisely the same as that picked for the expedition of Bergh two years before, but the difference in the face of the country would amaze any one who had never seen the magic of rain-falls on South African deserts. Fresh, juicy grass and vernal flowers were sprouting from a soil of seemingly lifeless sand. Birds were building their nests in the leafy thickets, insects were creeping or buzzing in swarms, and a myriad of butterflies were fluttering their gay wings over the green sward and blossoms. After years of drought there had come a season of heavy rains. The arid sands were soaked, torrents foamed through the windings of the dry water-courses, and the region north of Olifants River, which had been an impassable barrier to so many explorers, was quite easily penetrated by the cumbrous procession of van der Stel. Van der Stel's farm and residence were near the present town of Somerset West and not far from Stellenbosch, which was named after him. His fine old house, "Vergelegen," is still one of the remarkable landmarks of these sturdy old Dutch settlers. They planted avenues of oaks, camphor trees, and pines, which to-day tend to make Cape Town and its environs one of the most charming spots on the face of the earth. The old picture of van der Stel's house, "Vergelegen," shows it partly hidden by a huge camphor tree, which measures nine feet in diameter.

As the expedition advanced, it found various promising showings of copper ore, and the croppings were particularly rich in a range lying a little below the meridian of 30° S., where one peak was singled out as " copper mountain." Van der Stel had succeeded in reaching the line of the supposed location of the golden city of Vigiti Magna, and he pushed his search along

Vergelegen.

this line to the Atlantic, but he could nowhere pick up a trace of the traditional city or any other vestige of the realm of Monomotapa. He did not even meet with any strange monsters or romantic adventures, except perhaps the charge of a huge rhinoceros, which upset his coach and forced him to fly for his life. After six months of travel his notable exploring party came back to the Cape, without any tidings of good cheer to the founders of the colony. The only relic of the tradition of empire left in the lands it had traversed was the attaching of the name of Vigiti Magna to the great river first shown on any map in the chart of this exploration. It had found rich copper ore in Namaqualand, but the deposits were too far from the base of transportation and supply to warrant the undertaking of mining.[1]

Van der Stel was fitly rewarded, four years later, by an appointment as the first governor of the Cape Colony, in recognition of his exploring enterprise and other displays of energy; but his pricking of the painted bubble of Vigiti Magna was a bitter disappointment to the Dutch East India Company, and a grievous thing to all adventurers filled with the conceit of a century of tradition.   It was true that Davaque or some other glittering city might lie farther to the east and north than any point yet reached by Dutch explorers, but with the growing familiarity with the land and natives of southern Africa there was a swelling discredit of the fine tales of the Dutch and Portuguese romancers.   The myth of the realm of Monomotapa was practically starved to death at the close of the seventeenth century, and unfortunately the greatly persistent daring of the Dutch explorers grew cold with its impulse.   When adventurers began to disbelieve in the marvellous empire and even doubt the location of the mines of Solomon and the throne of Sheba, there was no very potent lure in the dusty karoos and rocky ravines of South Africa   No discovery of ore, except possibly of the precious metals, was likely to be of any reward to a prospector, and it was even questionable whether rich veins of gold or silver could be successfully opened and worked at any considerable distance beyond the narrow range of the Dutch settlement at the Cape.

So the credulous search for Ophir and the mythical realms in Africa came to an end, and for more than one hundred and fifty years there was little life in the tradition of King Solomon's mines, until its embers were rekindled by the daring advances and glowing fancies of the intrepid explorer, Karl Mauch.   In 1858 Mauch marked the Lydenburg district as a probable gold-

---

[1] "South Africa," George McCall Theal, Vol. I, pp. 370–380.

These copper mines came into possession of an English company known as the Cape Copper Company in 1853, since which time copper to the value of £11,000,000 has been produced.

field, and in 1871 he won the honor of reaching and first clearly describing the extraordinary ruins of Zimbabwe and its adjacent gold-fields. Unfortunately for his credit as an archæologist he insisted on the fancy that the old building on the hill was a copy of King Solomon's temple on Mount Moriah and that the lower ruins reproduced the palace inhabited by the Queen of Sheba during her stay of several years in Jerusalem.[1] This does not impair, however, the probable accuracy of his main contention

Boschendal.

that he had revealed part of the ancient workings of the people who furnished the flow of gold to Arabia and Judæa in the days of King Solomon.[2]

[1] "The Ruined Cities of Mashonaland," J. Theodore Bent, London, 1896.

[2] "It was really (Adam) Renders who first discovered these ruins three years before Mauch saw them, though Mauch and Baines first published them to the world, and they only described what the old Portuguese writers talked of hundreds of years ago." E. A. Maund, "Geo. Proc.," February, 1891, p. 105.

The extent of these old workings has been proved beyond doubt by the reports of Hartley, Mauch, Baines, Nelson, and later explorers, and a precise and graphic study of Zimbabwe and

Entrance to Boschendal.

other ancient structures in Mashonaland was made in 1891–92 by J. Theodore Bent and his associates in the expedition chiefly promoted by the Royal Geographical Society, the British Chartered Company of South Africa, and the British Association for the Advancement of Science. Bent's expedition located Zimbabwe in latitude 20° 16′ 30″ south, longitude 31° 7′ 30″ east; slightly differing from the position given by Mauch.[1] Bent holds that Zimbabwe is of Abantu origin and may be freely translated "Here is the great kraal," meaning the kraal of the native head chief of the district. This name, however, marked only the native occupation of the buildings, and Bent sees in the ancient ruins and workings "evidence of a cult known to Arabia and Phœnicia alike, temples built on accurate mathematical principles, containing kindred objects of art, methods of producing gold known to

Boschendal.

have been employed in the ancient world, and evidence of a vast population devoted to the mining of gold."

[1] " List of Stations in Mashonaland astronomically observed, with Altitudes," by Robert M. W. Swan.

BOTANIC GARDENS.

Without entering into the varied researches supporting the views of Schlechter, Keane, and other leading authorities, it may be observed that the main conclusions pithily summarized by

Lekkerwijn.

Professor Keane are strongly backed. Ophir was not a source of gold, but its distributer, as the port on the south coast of Arabia through which the flow of gold came by sea. It is identified with the Moscha or Portus Nobilis of the Greek and Roman geographers.

Havilah was the land whence came the gold of Ophir, a great tract in southeastern Africa, lying north of the Limpopo and largely identified with the range of the modern Rhodesia. The ancient gold workings of this region were first opened by the South Arabian Himyarites, who were followed (but not before the time of Solomon) by the Phœnicians, and these very much later by the Moslem Arabs. Tharshish was the outlet for the precious metals and stones of Havi-

Lekkerwijn.   (Back view.)

lah, and stood probably on the present site of Sofala. The Queen of Sheba came by land and not over the seas to the court of Solomon. Her kingdom was Yemen, the Arabia Felix of the ancients.

Bien Donné, Groot Constantia.

In a word, the "Gold of Ophir" came from Havilah (Rhodesia), and was worked and brought thence first by the Himyarites (Sabæans and Minæans), later by the Phœ- nicians, the chief ports engaged in the traffic being Ezion-geber in the Red Sea, Tharshish in Havilah, and midway between the two, Ophir in South Arabia.[1]

For sixty years from the opening of the eighteenth cen- tury there was no considerable exploration, or even prospect- ing of any consequence, in the region north of the meridian passing through the Olifants River. Yet even in this ap-

[1] A. H. Keane.

Overmantel and Old Dutch Relics. (Lekkerwijn.)

Bien Donné, Groot Constantia.

parent cessation of enterprise there was a continuous progress, almost essential to the successful advance of later exploration. The Dutch settlement at the Cape was expanding. Year after year pioneer settlers pushed out farther from the Castle, moving

Bien Donné, Groot Constantia.

up the river valleys, and clinging at first to the base of hill ranges where the essential supply of water was most surely attainable. After the taking up of the choice locations, later comers passed on over the open veld, and it was seen that there were large tracts of land, unsuited to agriculture, which would serve well as ranges for cattle and sheep.

For many years, however, the raising of wheat was of prime importance in the eyes of the

Dutch farmers; for this product fetched the highest price rela-
tively, and any surplus was eagerly called for by ships that
touched at the Cape or by the
demand for the supply of East
Indian settlements. In 1685
the first export of grain was
shipped, and strenuous efforts
were made to extend the area
of land in cultivation. A botanic garden had been one of
the early undertakings of the
company, to serve as a nursery
for European, East Indian, and
native plants, and under the
direction of Commander van der
Stel this nursery was made the
pride of the Cape as an exhibit
as well as a very serviceable
source of supply of seeds and

Farm House, Klein Drakenstein.

plants for the garden and farm lands. The growth of the olive
had been particularly urged, and it seemed at first to be likely

Doorway, Palmeit Vallei.

to flourish, but the
success of the grove
of van Riebeeck was
not attained by plant-
ers generally. There
was a considerable
advance in vine plant-
ing and the produc-
tion of wine, and in
1672 the distillation
of brandy was begun.
It was hoped that the
Cape wine could be made an export of consequence, but the taste
of the Dutch planters preferred a sweet, strong fermentation to
clear, light wines, and they lacked the skill or the strong desire

to modify their product to compete with French vine growers.[1]
So the only considerable consumption of Cape wine, outside of

the colony, was
from the crews of
visiting vessels.

There was no
lagging on the
part of the East
India Company in
efforts to stimu-
late the industries
of their colony.
Upon the revoca-
tion of the edict
of Nantes (Oct.

A Wine Farm at Klein Drakenstein.

28, 1685) by Louis XIV., the steadfast Huguenots were forced
to seek new homes in foreign lands, and many were cordially
encouraged and aided to pass over sea to the young Cape Colony.

Muller's Farm, Achter Paarl.

Their expert knowledge of the growth of the vine and olive was
highly valued, and it was also desired to bring in tanners, har-

[1] "On Veld and Farm," Frances MacNab.

Dutch Farm House.

ness makers, wheelwrights, metal workers, and other artisans of essential service to the spreading settlements of farmers. In the allotments of land special care was taken to distribute the influx of foreign blood so that it must necessarily fuse with the main body of settlers. This design was so well carried out that in a few generations the only absolutely distinct survival of this Huguenot migration was the perpetuation of the old French family names. But the combination of these two strong strains of blood made a compound of remarkable character.

Besides this promoted

Muller's Farm, Achter Paarl.

Palmeit Vallei, Klein Drakenstein.

immigration of men there was an equally shrewd effort on the part of the company to advance the breeding of horses, cattle, and sheep. Stallions were imported from Persia to improve the stock, which had been falling off in size and quality though increasing in number. Spanish rams were used to lay the foundation of the South African breed of merinos, and the Angora goats bore transplacing excellently, and soon browsed greedily on the coarse grasses of the Cape.

By the advances of the voortrekkers or pioneer farmers the range of settlement was extended so far in 1761 that the start of

Muller's Farm, Achter Paarl.

Mooi Kelder, Lower Paarl.

the first large exploring party since the return of the van der Stel expedition was made in that year from a rendezvous near the mouth of Olifants River. This party was led by Captain Hendrik Hop of the burgher militia, and was made up of seven-teen whites and sixty-eight half-breed Hottentot servants. It started in August and advanced on the track of the former expedi-tion, passing the Copper Moun-tains of Little

Plaisis de Merle, Groote Drakenstein.

Namaqualand, and reaching the river Vigiti Magna on Septem-ber 29. This river was familiarly called by the colonists the Groote (Great) River, and held this name until both the tradi-

G

tional and common names were supplanted by a new christening in 1779, when Colonel Robert Jacob Gordon, commanding the

garrison at Cape Castle, led another expedition up the river, and named it Orange in honor of the stadtholder.

Hop's exploring party met a troup of giraffes soon after crossing the Groote River, and won the distinction of furnishing the first skin of a giraffe from South Africa to the Museum of the University of Leyden. But excepting this novel chase there was little to attract the explorers. The sun scorched them relentlessly in the open desert, and

Donkerhoek, Groot Drakenstein.

they could nowhere find water except in the deep sand-pits dug by the roving natives. Sometimes there was a shallow puddle at the bottom of one of these pits, and even when the sand was barely moist, further digging to the underlying stone would sometimes yield a trickle of water. Still they pushed on stubbornly to the farthest point yet reached from

A Wine Cellar.   Herd of Cape Goats.

the Cape, in latitude 26° 18′ S., before turning back to bring home their discouraging story.

It was thirty years before this advance was outstripped by

LA RHONE, GROOT DRAKENSTEIN.

OLD LE ROUX.

Willem van Reenen, of the farm Zeekoevlei on the Olifants River. This adventurous farmer set out in 1791 with four fellow colonists and a number of Hottentot servants, and reached on the 18th of November the end of the trek of Captain Hop's party. Prowling Bushmen and lions beset their camps continually, and in January, 1799, they had to beat off a fierce swoop of a party of Namaquas. Yet they pressed on until March 14, when they came to a little oasis which they named Modder Fontein, or muddy spring. Then they turned back after a few days' rest, and plodded home to the farm Zeekoe-vlei, which they reached on the 20th of June. They had killed sixty-five rhinoc-eros and six giraffes, without reckoning their

Tatr, 1757.

bag of smaller game, and brought back exultantly wagon loads of copper ore, which they supposed to be gold until their hopes were blighted by assayers at the Cape.[1]

The depressing reports from these expeditions were not the least of the straws that finally broke the back of the Dutch East India Company. For nearly a century and a half their colony in South Africa had been a continual drain and burden. All

[1] " South Africa," George McCall Theal.

the expedients and efforts of the energetic directors of the company in the seventeenth century, and such faithful servants as van Riebeeck and van der Stel, had failed to develop any mines or any product for export of any considerable importance. With

An Old Farm House, Lower Paarl.

the beginning of the eighteenth century there was an evident drooping in the enterprise of the company, and a drift toward hopeless discouragement, which culminated in 1794 with the declaration of bankruptcy. The company's debt was £10,000,000 sterling, and its credit was utterly exhausted. It could no longer undertake even to maintain a feeble garrison at the Castle for the defence of its colony. Issues of depreciated and irredeemable paper had driven out all gold and silver from circulation at the Cape. Debts could be paid in this paper, which was legal tender, but nobody would receive it in exchange for goods except at such a discount that there was a general resort to barter. Internal trade was paralyzed, and a little wheat, wine, and tallow was all that could be squeezed out of the colony for export to Java and India. The straggling settlers on the northern frontier were continually fighting with the Ishmaelite

Farm House, Achter Paarl.

Bushmen, and the Kafirs on the northeast were still more harassing and formidable. Every kraal was a rude fort and every family a garrison. Ammunition was growing scarce and costly, and there was no hope of succor from the Castle at the Cape.

In view of this patent collapse, the stretching out of the

strong arm of Great Britain to seize the Cape in 1795 should have been as welcome as rescue to a wreck. Then for the first time a power took hold of the way station of East Indian trade, and its straggling offshoots, that had the strength and the skill and the far-reaching conception to do more than repress savage on-slaughts and defend grazing grounds, — to open great mines, to convert arid karoos into irri-gated plantations, to extend the network of railways, and stretch in time the steel band of civiliza-tion across the darkest zone of Africa. This Britannia has done and is doing, either in her imperial way, or by the hands of the sons who have labored to make her greater.

Brand Solder (Fire Loft). For the prevention of fire.

But the coming of this saving and transforming power had the appearance, at the time, of a hostile attack. The Netherlands, in 1793, were wholly under the thumb of the new French republic, and war was declared against Great Britain through controlling French influence. There had been some revolting against the further collection of taxes by officers of the East India Com-pany, but the colonists as a body did not want any foreign interference. So the little garrison in the Castle at the Cape put on a defiant front, and rallied to its support

Cape Cart.

a number of burgher volunteers when a strong British fleet sailed into Table Bay in the first week of September, 1795. It was ap-parent, however, even to the boldest Dutch defender, that resistance was hopeless, and Cape Town, with its castle and garrison, surrendered to Admiral Sir George Elphinstone and General Sir Alured Clarke, on the sixteenth of September. So was ended one hundred and forty-three years of rule of the Dutch East India Company, and from this date British ascendancy in South Africa began. There was a brief intermission, it is true, some years later, when the treaty of Amiens (1802) transferred the Colony to the Batavian Republic. But the breaking out of war again in the following year ruptured the treaty, and ex-

The Gate of the Castle.

posed the Cape Colony again to the hazard of capture, which actually followed early in January, 1806, when Cape Town was retaken by Major General David Baird. From that time the Cape was held continuously by the strong arm until the convention at London, August 13, 1814, when all claims of the Netherlands to South Africa were extinguished by cession, and Great Britain became the heir of all the Dutch advances from the Cape of Good Hope.[1]

[1] "South Africa," George McCall Theal.   "Précis of the Archives of the Cape of Good Hope," H. C. V. Leibrandt.   "South Africa," Augustus Henry Keane.   "Heroes of South African Discovery," N. D'Anvers (Henry Bell).

A View from the Kloof Road leading from the Upper Part of Cape Town.

CHAPTER III

HEN Lord Charles Somerset came to the Cape as the first Governor of the Colony after the cession, how slight and infirm was the hold of any civilization on the indurated barbarism of the vast expanse of Africa south of the equator! In the three hundred years that had passed since Vasco Da Gama made known the bounds of the continent, the outer rim of the traditional Ophir land had barely been pierced. From the Atlantic side the Portuguese had not pushed beyond a fringe of trading posts on the Lower Guinea coast, and were clinging feebly to insignificant stations along the shores of the Mozambique channel. The Dutch grip was more obstinate, in spite of all disappointments, but the range of their advance was only a few hundred miles from the Cape, and outside of Cape Town the population was a mere sprinkling on the face of the land. When the British first wrested the Cape from the Dutch, Earl Macartney, who held the government in 1797, defined by proclamation the bounds of the Colony. It only ran east to the Great Fish River and on the north to the Zuurberg Mountains and the southern edge of Bushman's land, trending up to the Kamiesberg, and thence along the coast to Buffels River in Little Namaqualand. The total extent was roughly 120,000 square miles, merely the extreme tip of South Africa, and the entire population, both white and black, was reported to be less than 62,000, or about one person to every two square miles. This was a petty fringe on the skirt of the dark continent.

Not only was the Colony weak in numbers, but it was seemingly without any uplifting leaven of enterprise and ambition.

For generations the Dutch settler had been treading in the footsteps of his forefathers without any wish to stride ahead. What they had done, he would do if he could. No new way of working or living or thinking was as good to his mind as the old way. The pioneer farmer and grazier had often been constrained to pack all his goods on the backs of oxen or in a wagon with his wife and children. A little hut of " wattle and daub " sheltered the family. Rude frames of wood overlaid with raw hide strips were their bedsteads, and sheepskins, their bedclothes. They cooked their food on the coals of wood-fires or boiled it in an iron pot. They cut their meat with clasp knives and drank from tin cups. A big chest served them for a table. Their house floor was the bare earth, unless a strip was covered with a wild beast's skin. Their children were brought up from their birth in this habit of life and the lack of comforts was not to them a privation. Their standard of living was scarcely higher than that of the imported Guinea slaves who worked for them, or of the native tribes that surrounded them. Their isolation from civilized society and their life in the wilderness in familiar contact with slaves and savages was inevitably degrading. When the English took the Colony, there was not a bookstore or a single good school in it, and outside of Cape Town almost the only tutors were soldiers who were allowed to live with the farmers.[1]

Still there was one sustaining and universal spirit which kept even the rudest grazier from sinking to the barbaric level. They clung to the God of Israel and to the Bible as God's revelation. They never wearied of searching the Scriptures, and they prayed with the fervor and faith of the old Covenanters. Their creed was the strait and narrow way of Calvinism and the synod of Dordrecht, and they turned to the Old Testament as confidingly as to the New for guidance. They recognized the holding of slaves as a practice permitted to Israel, and they made bond servants of the Hottentots in their apprenticeship contracts. In their eyes the Bushmen were Ishmaelites and the Kafirs Philis-

[1] " South Africa," George McCall Theal. " Handbook to South Africa," S. W. Silver & Co.

tines, who were cumbering the ground that might be occupied by God's favored people.[1]   But the settlers were phlegmatic and peaceful by nature, content with their bare living, and with no ardor for extending their bounds by conquest.   An extraordinary impulse was needed to convert them into adventurers and wanderers in the desert.

This impulse was given by the capture of the Cape, the influx of jostling immigrants from Great Britain, new and vexing legislation, and disasters to crops which exalted the comparative value of pasturage lands.[2]   At the opening of the administration of Lord Charles Somerset there was a marked effort on the part of the Home Government to promote the growth of the Colony.   A regular mail packet service was established between England and the Cape, and £50,000 were voted by Parliament in 1819 to be disbursed in aid of emigration to South Africa.   This contribution was a powerful stimulus, and it is estimated that nearly 5,000 new settlers of British birth were added to the population of Cape Colony from March, 1820, to May, 1821.

Unfortunately the South African climate in 1820 and the years immediately following was peculiarly aggravating.   In 1819 there had been a heavy wheat crop and the consequent temptation to farmers to extend their wheat growing.   So they did, but the crop of 1820 throughout South Africa was fatally blighted. The next year's crop fared no better, and thousands of farmers were ruined and brought even to the verge of starvation.   Rations were distributed by the Colonial Government in the fall of 1821 to those who had no means to buy food, but the unrelieved suffering was widespread.   Following hard on this scourge of blight came the prodigious floods of October, 1823, when it seemed to the colonists in the eastern districts as if the heavens were open for another deluge.   Rain fell in torrents for days without ceasing, and overflowing rivers ran foaming to the sea, carrying millions of tons of earth in their turbid floods as well as the shattered houses of settlers who had barely time to fly for

[1] "Impressions of South Africa," James Bryce.   "South Africa," Theal.
[2] "Annals of Natal," John Bird, p. 505.

their lives. These staggering rebuffs in the face of the new emigrants were greatly demoralizing. Some fled from the Cape in
despair, and many more wrote home to their friends that the Colony was hung between flood and famine, and that the greater part
of South Africa was a dismal Karooland. Still there was a notably
plucky rally and an immediate turning to other resources when
wheat cultivation was shown to be an uncertain reliance. Cattle
and sheep breeding was largely extended at once, and in 1828
hides and skins ranked only second to wine in the list of
exports.[1]

The failures in wheat growing and the resort to pasture land
were strongly moving influences urging on the advance of pioneer settlers from the southern river valleys north and east over
the veld into unclaimed territory. This natural flow of migration was greatly swelled and impelled by the clashing of the old
settlers with the newcomers from Great Britain, and by their
resentment of British control and administration measures. By
the census of 1819 the white population of the colony was
42,217, and outside of Cape Town this people was almost
wholly of Dutch descent or of the fused Dutch and Huguenot
strains. It was inevitable that a stock of such breeding and tradition should be impatient of any ordinances or ways except its
own. It was peculiarly irksome to bow to a nation which had
captured the Cape by the strong arm, and was only represented
by a small minority of the settlers. The inevitable heart-burn-
ing was aggravated by the contact and rivalries of the new and
old settlers. Neither faction had the knowledge or temper to
recognize the best traits in the other and show tolerance for dissimilar habits and prejudices. The Dutch boer has an old
Anglo-Saxon root and is simply correspondent to the German
bauer, a farmer or countryman; but in the English mouth all
the Dutch colonists were lumped as Boers, and in the English
eye Boer was too often confounded with the clownish boor.
The Boers faced this contempt with a glowing resentment that
burned like a slow-match.

[1] "South Africa," Theal.

In the new measures of government there was a succession of vexations also to colonists attached to the old customs and ordinances. The expense of the new colonial establishment was a grievance. The adjustment of the currency aroused bitter complaint. The substitution of English for Dutch in official papers, and the abolition of the old Dutch courts, were heavy humiliations. But the keenest resentment was excited by the measures designed for the protection of Hottentot bond servants and free natives, and the emancipation act of 1833. There had been a rapid increase in the importation of slaves from Guinea after the first conquest of the Colony by the British, but in 1807 the last cargo of slaves was landed at Cape Town, and the slave trade was formally brought to an end by law in the following year. Still the colonists continued to hold and breed slaves as their fathers had done, and there were 35,745 slaves in the Colony when the emancipation act went into effect on the first of December, 1834. These slaves were valued at £3,000,000, but only £1,200,000 were appropriated as compensation to their owners. The loss fell heavily on many owners already sinking under the weight of mortgages, and there were rumblings and outpourings of bitter indignation. The deficiency in compensation was called Imperial confiscation, and the Boers resented it sorely, not merely on the score of the loss measured in money, but as a crowning instance of their political subjection.[1] Alien Imperial rule was the deep-seated grievance which was the underlying and impelling cause of the extraordinary exodus from Cape Colony called the Great Trek.[2]

In 1835 Louis Triechard led out the first pioneer company of this migration, and his advance into the wilderness beyond the bounds of the Colony was followed by a succession of slow-moving caravans pushing northeast to the head waters of the Orange River and the terraces of Natal, and moving on, in course of years, across the Vaal to the Limpopo water-shed. This outpush of pioneers in large parties, overcoming all barriers of

[1] "Annals of Natal." "South Africa," Theal. "The Great Trek," Henry Cloete, her Majesty's High Commissioner for the Colony of Natal. [2] *Ibid.*

mountains and deserts, and fearlessly venturing into the strong-holds of the fiercest native tribes, undoubtedly hastened and secured the acquirement of the marvellous diamond and gold fields of South Africa. The march of the caravans and the winning of the land was a drama full of barbaric color and movement.

At the time when the Cape first fell into the hands of Great Britain, there was an insignificant tribe, the Amazulu, living in

Zulu Chief Cetawayo and Part of his Family.

kraals on the banks of the river Unvolosi, which flows into the Indian Ocean at St. Lucia Bay. In their name there was an arrogance of high descent, for its meaning is "the people of the sky"; but the Amazulu had then nothing else to brag of, and while their head chief, Senzanzakona, lived, there was no terror in the Zulu name. But there was a son born to Senzanzakona in or near the year 1783 [1] who made the Amazulus masters of a region far exceeding any bounds of the Kalangu Monomotapa, and stamped his name across it in indelible blood.[2]

The boy was called Tshaka or Chaka, which, in the Sechuana tongue, is "battle axe." There is another tracing of his name to Cheka, a wasting disease afflicting his mother. In either translation the name was ominous. But this chief's son had no deformity that an eye could see. When he came to manhood, a sculptor would have picked him as a model of his tall, athletic

[1] "South Africa," Theal. "Annals of Natal."   [2] Ibid.

race. He was more than six feet in height, and every inch was
pulsing with vigor. No rival could leap as high or hurl an

assagai as far. In later life his
shapely features were swollen with
ugly passions and debauch, and
his lithe body was overlaid with
fat, but he never lost the beauty of
his deep-set, brilliant black eyes,
fringed with their long, curved
eyelashes.

For some cause Chaka, while
only a lad, was forced to fly for
refuge to Dingiswayo, chief of the
Abatetwa, the master tribe of
the district. Under protection
of this chief he was made a soldier, and took by craft the headship of his own Zulu tribe when
his father died. Then he was

Zulu Prince, Dinizulu.

able to betray and put to death his protector Dingiswayo, and

spread his mastery by force or
terror over the surrounding
tribes. As he grew in power he
showed an unfolding genius for
war and command. He pressed
every young and strong man
within reach into his army. He
marshalled his men in impis or
regiments. He discarded the old
bunch of assagais and armed each
man with a single, short-handled,
long-bladed unkonto or spear,
and protected him with a shield
of oxhide. He aimed with his

Zulu Family.

weapon to make every fight hand to hand, where every man
must kill or be killed. If a soldier lost his spear he was

doomed to die, unless he could show another in place of it, torn from an enemy.

No barbaric figure was ever more terrific and martial than the Zulu soldier in war-dress. Chaka's hair was cut close, except on the top of his head where the thick, crisp locks were matted or moulded into a ring made of a tree gum and polished to the likeness of ebony. Thick folds of otter pelt were wound round his head and great earrings of carved sugar-cane hung from the cut lobes of his ears, which were covered with pads of

A Zulu and his Ten Wives.

jackal's skin. From this turban projected two feet or more a jet-black crane feather, waving with every toss of his head. A circlet of twisted monkey and genet skins hung over his breast and back, and from his waist a thick flexible kilt of twisted skins hung to his knees. Bands of short-cut white oxtails circled his legs and arms, and the ruffles round his ankles made his bounding feet oddly like the winged Mercury. In his right hand he grasped his spear and swung at his left side his oval shield of white oxhide. Now pin with thorns a dozen bunches of the red feathers of the louri in the crisp tufts of his crown and scat-

ZULU IN WAR ATTIRE.

Zulu Kraal and Huts.

ter some other brilliant feathers on a circlet above his breast, and see Chaka dressed for parade.[1]

Then fancy the marshalling of an army of men like him, for the chieftain in arms was one of ten thousand. When the leading division marched on in review, every man was more or less closely the image of Chaka. These picked men were his Unbalabale or Invincibles, scarred veterans who had never been beaten. They bore white shields marked, like their chief's, with a black spot, and behind them followed in grade of honor divisions with red-spotted shields, gray shields, and black shields. Only the Invincibles had kilts of skins, the others wearing instead a trapping of oxtails. As these fierce troops marched on before Chaka's keen eye, the men of chief mark would bound from the ranks and show a marvel of vaulting, darting to and fro, whirling of spears and mimicry of fight, in which few athletes could compare with the supple Zulu.

In formation for battle Chaka curved the van of his impis

[1] "Annals of Natal," pp. 90–100.

like a crescent.   He called the end his horns and the centre his breast.   This was the old array of the warring Bantu tribes, but Chaka greatly strengthened it by a formation behind in an oblong block of men held in reserve to repel any break in the crescent or reënforce it when wavering.   His force of disciplined soldiers ranged up to fifty thousand strong.

Zulu Hut in course of Construction.

With this prodigious engine of war shaped to his hand, he overran all the country from Delagoa Bay to the Unzimvulu River and far into the interior, scourging its face mercilessly. Some of the terrified tribes in his way were blotted out completely.   "There was a white mark from the Tugela to Thaba N'chu, and that was our bones," said an old Hlubi to Theal, the historian of South Africa.   Sometimes stragglers escaped to lurk in mountain recesses.   These wretched survivors of the scourge were covered by one new and pitiful name, Amafengu, because their first cry to strangers was Fenguza, "we want."   Only one tribe held Chaka in check, the warlike Amaswazi, which stubbornly guarded their mountain paths and cliffs.   Even the fierce Amangwane were forced to fly before Chaka's resistless impis; but they kept massed together, and in their retreat drove off or massacred most of the tribes between the Orange and the Vaal rivers.   Then the Amangwane, still hot pressed by the Zulus, began to rub against the frontiersmen of Cape Colony. This inroad was bravely met by a muster of a thousand soldiers and Boers under Lieutenant Colonel Somerset, who finally put the Amangwane to utter route in a sharp battle, August 27, 1828, near the banks of the Bashil River.[1]

Chaka was a warrior capable of measuring the efficiency of the white man's organization and firearms. When the Amangwane were thrown back, the Zulu chief withdrew his own impis without risking a collision with the whites. A few weeks later he was murdered by two of his half brothers and his best-trusted attendant. Dingaan, his half brother, and one of his assassins, grasped the headship of the Zulus, but his succession was dis-

Zulu Woman grinding Corn.

puted by the commander of one of the chief divisions of Chaka's army, the unruly Matabele. This revolting chief, Umsilikazi, was the model of a Zulu warrior, tall, sinewy, shapely, and, except in war dress, naked save for a cord around his waist from which leopards' tails dangled. A string of little blue beads was drawn about his sturdy neck, and three green feathers of a paroquet were stuck in his crisp hair. His followers were like him, and the wild charge of the legion of such men armed

[1] "South Africa," Theal. "Annals of Natal."

with their keen-bladed spears was a sight that would try the nerve of any white soldier.   How the rudely armed and undisciplined Boers would face it was soon to be tested.

Umsilikazi, revolting from Dingaan, led his Matabele division across the desert to fall upon the country north of the Orange River and west of the Drakensberg, the Dragon Mountains.   Much of this country had been ravaged before by the Amangwane, and the Matabele spared nothing that had escaped slaughter and pillage.   Dingaan sent an army of Zulus in 1834 to dislodge his rival, but the warriors of Umsilikazi

Zulu Women.

beat back the attack.   By the Zulu raids and massacres and wars, the whole country from the seaboard of Natal nearly to the junction of the Orange and Vaal was desolated, and the native tribes of the region almost destroyed.   Thus great tracts of land were opened to the advance of the migrating Boers, but the push of the trekking pioneers soon brought them in conflict with Umsilikazi and Dingaan.

Then the remarkable traits of this peculiar people stood out in high relief.   To English immigrants, jostling the old settlers, the ordinary Boer appeared a Dutch clodhopper, sullen and jealous, unkempt in person and dress, immovably set in his traditional ways, pig-headed in his obstinate prejudices, a block to every suggestion of progress, Pharasaical in his prayers, absurd in his

customs, and often clutching to the last penny.[1] There were
some true lines in this partial portraiture, with a natural warping

Zulus smoking Indian Hemp.

of prejudice and lack of insight.   In face of the foreign intrusion the Boer had something of the instinct of the turtle and

Old Zulu Women taking Kafir Beer to a Wedding.

[1] "The Great Thirst Land," Parker Gillmore. "South Africa," George
McCall Theal.   " South Africa ; a Sketch Book of Men, Manners, and Facts,"
James Stanley Little.

porcupine. But in the heart of the wilderness, in his venturesome trek over the pathless veld, and in the traverse of mountains and deserts, he showed what scornful eyes had not seen, —
the self-reliance, the fortitude, and the pluck of the true pioneer.

He packed his wife and children and all his needful supplies
in a huge, low-bodied wagon under an arched frame covered with

Zulu Girls.

waterproof canvas. To this stout
wagon sixteen strong oxen were
yoked to the chain or rawhide rope
forming a trektouw. Every ox was
a helpmate. Every one knew his
name and place and resented a
change in yoking. The Boer and
his Hottentot helpers spoke to them
all familiarly, and could cut at will
a fly from the ear of any one with
a flick of their long-lashed whip.
When these prairie-schooners lum-
bered off, creaking and swaying,
with a chorus of Dutch and native calls, the Boers and their
sons rode beside them on ungainly flea-bitten horses, trained to
herding and hunting, and often possessing uncommon bottom
and speed.

The Boer was by nature prudent and wary. For comfort
and safeguard the advance of the Great Trek was in companies,
camping at night on plain and hillside, with wagons ranged to
form a rough palisade and kraal. No morning or nightfall ever
passed without prayers and the reading or recital of Scripture.
For every step of his way he looked to his God for guidance,
and he felt that the old promises to the chosen people were
renewed to him. His faith in the literal inspiration of the Bible
was unwavering. He did not doubt that the sun stood still at
the call of Joshua, or wonder at the slaughter of Philistines with
the jawbone of an ass. In face of every privation and the direst
peril he was sustained by his certain reliance on the help of One
who could make a spring gush from the desert rock, or deliver

any heathen host into the hands of a few faithful servants. But with all this reliant devotion he never forgot "to keep his powder dry," and used every opportunity to perfect his skill as a marksman.

Back of his faith and prudence was an unflinching spirit. In the uncouth Boer smouldered the fire of an ancestry that charged at Ivry and starved at Leyden. Even the women and children were dauntless at the pinch of need. With her white grease-cloth wrapped about her face, the Boer's vrouw was an uncouth object, but with her eye on the sight of a rifle many a fat old woman was a guard to be feared.

Native Laborers in War Dress.

No impediments nor dangers stayed the advance of these pioneers. When a heavy wheel dropped into a deep gully or earth-crack or ant-bear hole, it was pried out with un-tiring patience. When thunder-storms changed the red soil to beds of mire and the wheels were clogged masses of mud from nave to felloe, the mud was laboriously scraped away and the wagons tugged to firmer ground. When the violent wrenches and strains snapped trektouws and wagon-poles and king-bolts like pack-thread, the same inflexible temper relinked the broken touws with riems of rawhide, chopped out new wagon-poles, and forged new fastenings with rude blacksmith's art. No karoo was so forbidding and no stream so swollen as to bar the onward march.

The tired Boer snored serenely at night behind the bulwark of his wagons, regardless of the wild beasts prowling and sniffing outside. The giggling calls of the gray and brown jackals, the doleful howl of the slinking hyena, even the deep breathing sough of the lurking lion, did not open his eyes, and it must be a fiercely menacing roar indeed that would lift his head. His only haunting dread was the crippling of his march by the deadly tsetse fly or the wasting diseases that made his horses and oxen the prey of the vulture.

Trekbok (Springbok) Hunting.

In the passage of these pioneers the destruction of wild animals of all kinds was enormous, partly for the sake of needful food, and partly for the skins, but much wantonly and wastefully, for the Boer would rarely let pass a living mark for his rifle. Of lesser game there was no attempt to keep tally, but by a common report thousands of lions were shot in the march to the Transvaal. Any such reckoning must be largely guesswork, though there is no doubt that few beasts within range escaped without the sting of a bullet. But a foe more formidable than any multitude of lions sought to bar the progress of the Great Trek.

The revolting Umsilikazi was the first of the great Zulu chiefs to try the temper and the arms of these pioneers. One

of the larger divisions of the Great Trek, led by Hendrik Potgieter and Gert Maritz, left the Cape Colony in August,
1836, and pushed north of the Caledon River.[1] Some of the
pioneers in this advance were cut off suddenly and killed by
Umsilikazi. Flushed with this bloodshed, he made a swoop
with six thousand men upon a part of Potgieter's trek — a company of a few score men, women, and children. But the startled
Boers were now on their
guard. They ranged their
big, white-tented wagons in a
square, lashing the wheels together with rawhide riems,
and filling in the chinks in
their barricade with thorny
mimosa bushes. In the centre of this laager a few wagons
were placed as a cover for the
women and children.

Zulu in War Dress.

Upon sight of the ad-
vancing Matabele, all knelt
and prayed. Then some of
the men rode out boldly to
meet the attack with their
heavy rifles. Their fire was
deadly, killing, at times, two
or three at a shot, when their
guns were loaded with slugs,
but the impis pressed on,
driving the Boers back to their laager in a sullen retreat, turning
to fire as fast as they could reload. Within the laager all was
made ready for a defence to the death. Back of every wagon a
little heap of powder and bullets was put on the ground, and the
women stood by to hand spare guns and reload. It was sternly
ordered that there should be no shrieking or crying by women
or children. In silence the rush of the Matabele was awaited.

[1] The Caledon River divides Basutoland from the Orange River Colony.

On came the impis in raging masses that dashed on every side of the laager like surf on a reef, wrenching at the wheels, clambering over the canvas, plunging through the thorns. The heavy wagons were shaken and swayed, but the lashed barricade held fast. The grim Boers met the shock with withering volleys, piling up the blacks in bloody heaps around the laager. Crouching behind the firing line, the women moulded bullets and helped to reload.

The firing was so deadly and the laager so impenetrable that the surges massed against it recoiled. But, after a moment of rallying, on came the billows of men, flinging their assagais, and howling like madmen as they crashed against the barrier which shielded the Boers. They stabbed and slashed at the canvas covers in frenzied efforts to cut their way over the wagons, and wriggled through the crevices packed with thorn bushes, until some, torn, bloody, and gasping, squirmed into the square, where the Boer women killed them with knives and hatchets. The Boers fired as fast as they could lift their rifles, not stopping to use their ramrods, but grabbing handfuls of powder to charge their guns, and dropping in slugs with scarcely any wadding.

So intense was the strain of that hour that even these men of iron nerve were entranced. "Of that fight," wrote one, "nothing remains in my memory except shouting and tumult and lamentation, and a dense smoke that rose straight as a plumb line upwards from the ground."[1]

Four times the black impis charged and four times their onset was beaten back before Umsilikazi drew off his men. The field around the laager was a fearful sight, and the white tops of the barricade were slashed into strips and dripping with blood. Seventy-two stabs were counted in the cover of one wagon, and eleven hundred and seventy-two assagais were flung through into the camp. But none of the stout defenders were killed, and all joined devoutly in a psalm of thanksgiving.

In retaliation for this attack Hendrik Potgieter and Pieter Uys led a troop of one hundred and thirty-seven in a swift

[1] "Annals of Natal," p. 375.

march and onslaught upon the main division of Umsilikazi.
The attack was so well timed and aimed that the array of fierce
impis was shattered and their chief was driven in flight to the
wilderness beyond the Limpopo. There, in the present Matabeleland, Umsilikazi brought together the remnants of his
people, and ruled in awe of the pioneers until his death in 1870.

Hard upon the defeat of
Umsilikazi came the greater
clash with Dingaan, when the
trekking Boers crossed the Drakensberg or Dragon Mountains
to the terraces of Natal. This
cunning and tricky chief made
smooth professions of friendship
to the Boers at first. He welcomed as allies the company
headed by Pieter Retief and received the commander at his
kraal. The chief's house was a
spherical hut about twenty feet
in diameter. Its floor was pol-

Zulu—Jim Cameel.

ished till it shone like a mirror, and its roof was supported by
twenty-two pillars of wood completely covered with beads.
Around this house were seventeen hundred ruder huts which
Dingaan used as barracks for his impis, and each hut would
cover twenty men.

After some parleying Dingaan signed a cession of the greater
part of the present territory of Natal to the Boers. To cele-
brate the compact he invited Retief to visit him again with his
companions. It was agreed as an exhibit of good faith that no
arms should be taken into the chief's kraal. So Retief and
some sixty other Boers, with forty Hottentot attendants, piled
their arms outside the kraal, and came in before Dingaan, who
was sitting in an arm-chair in front of his hut. Two of his
impis were formed in a circle about him. The Boers took their
seats on the ground within the circle, and cups of utywala or

native beer were offered them to drink. But when they put their lips to the cup, Dingaan cried out, "bulala amatagati," "kill the wizards." At this cry his Zulus fell on their helpless guests in overwhelming mass. A few Boers had clasp-knives, and the others met the rush with naked hands, but all were overpowered in a moment and dragged over the ground to a hill near by, called Hloma Mabuto, or the mustering of the soldiers. Here their heads were crushed with knob kerries, and their bodies were flung into heaps. Retief was forced to see the horrid murder of all of his companions. Then his heart and liver were cut out and taken to Dingaan, and the mutilated corpse was cast on the heap of dead.[1]

None of the Boers in the trap escaped, and after the massacre the Zulus poured out to raid the scattered camps of the pioneers. They were finally beaten back at Bushman's River, after they had killed many trekkers and carried off their cattle, and the mounted Boers followed their retreat for days. But the Zulus were quick to turn and strike again like fierce hawks, and within two months they swooped down upon the English settlers and native blacks of Natal and cut them off almost to a man.

The trekking Boers were hard pressed. Pieter Uys was killed in ambuscade, with his son, a boy of fourteen, and a number of his men. When Uys was fatally wounded, he urged his son to escape by spurring his horse, and the boy rode on to a place of safety, but turned and rode back deliberately to die with his father.[2] Potgieter drove back the Zulus after the fall of Uys, but he did not venture to hold his ground, and withdrew across the Drakensberg. Only a determined rally and crushing blow could free Natal from the hanging menace of the impis that Chaka had trained for the hand of Dingaan.

In December, 1838, a force of six hundred mounted Boers was mustered to strike this blow under the command of Andries Pretorius. It seemed an absurdly weak force for such an attack, but the count in numbers did not measure its strength. Every

---

[1] "Annals of Natal," pp. 214–218.   [2] *Ibid.* p. 374.

man was a master marksman with the heavy rifle that had so often broken the bound of the lion and stopped the charging rhinoceros when to miss was death. In every one's heart was a flame of hate for the ruthless Zulu. "Remember Retief" was a mutter that ran from man to man as the troop rode on. They longed for revenge as thirsty men crave water. They advanced, too, with the spirit of the Israelites of old and of Cromwell's Ironsides. They marched only between matins and evensong. They prayed in their saddles and lifted their voices in psalms. Surely the God of their covenant had the power to confound any might of the heathen and deliver their enemy into their hands.

When they drew near to the Zulus, Pretorius halted, and with all his men offered a vow to the God of their fathers, should He grant them the victory, "to raise a house in memory of His great name wherever it should please Him, and note the day in a book to make it known to latest posterity." [1]

With this simple confidence in Divine protection there was the shrewdest practical judgment in selecting the best possible post to offset their comparative weakness in numbers and intrench their little force. Their laager was pitched at the junction of a broad river reach, called a sea-cow hole, with a deep, dry water-course, covering both flanks. Here, on Sunday, the 16th of December, 1838, at five o'clock in the morning, they were attacked by a force of many thousand Zulus and fought for more than five hours. Impi after impi, reckless of life, charged up to the rifle front belching smoke, flame, and bullets, only to reel back before the deadly hail. When even this raging horde wavered, Pretorius with one hundred and fifty picked horsemen circled about and struck their rear with a charge so fiery that the Zulus were utterly routed. The Boers drove the blacks to the river, shooting and trampling them under the feet of their horses. "The Kafirs lay on the ground," said one horseman, "like pumpkins in a rich soil that has borne a large crop." The sea-cow hole was packed so full that "the water looked like a

[1] "Annals of Natal," pp. 246–249, 448.

pool of blood," and the stream thenceforward was known as Blood River.[1]   Three thousand six hundred Zulus were left dead on the field, and this decisive victory was gained without the loss of a single life to the Boers.   A few were slightly wounded, but they thought nothing of their hurts in the common thanksgiving.

This signal triumph and salvation were humbly taken as the answer of God to their prayers, and the vow before the battle was faithfully ful-filled, as the old Dutch Reformed Church of Pieter Maritzburg, the mother church of Southeast Africa, bears witness.   The flying Zulus were pursued and the kraal of Dingaan captured, February 3d, 1839, where the bodies of Retief and his companions were found and

A Zulu Laborer in War Attire.

mournfully buried in one grave.   The Boers called the place Weenan, the weeping, and so it is known to this day.

Dingaan fled north and hid himself in a concealed kraal which he built.   A Boer writer tells a story of his capture and death with grim delight.   Many of the tribes which had been pressed in with the Zulus made peace with the Boers.   One of the Swazi chiefs, Sapusa, who had bowed to the tyranny of Dingaan, found his late master's hiding-place.   "On the first day old Sapusa pricked his captive with sharp assagais, not

[1] "Annals of Natal," pp. 246–249, 448.

more than skin deep, from the sole of his foot to the top of his head. On the second day he caused him to be bitten by dogs. On the third day Sapusa said to Dingaan, ' Are you still the rain-maker, greatest of men ? The sun is rising, you shall not see it set.' Then he took assagais and bored Dingaan's eyes out, and when the sun set, Dingaan died, for he had had no food or water for three days. Such was the end of Dingaan." [1]

So the Boers finally stayed the sweep of the Zulu scourge which had laid waste a great stretch of land north of the Cape settlements. Upon the defeat and flight of Umsilikazi, the vic-torious commandant, Hendrik Potgieter, proclaimed that all the territory overrun by this chief was forfeited to the pioneer Boers. This claim covered the greater part of the late South African Republic, and half, at least, of what is now the Orange River Colony. In this assertion there was no recognition of any sovereignty of Great Britain or attachment to the Cape Colony   It was the view of the Boers that the land which they took was theirs by right of capture and forfeit, and that they were independent adventurers with no ties of allegiance. A simple form of republican government was established for the Boers, north of the Orange River, by a general assembly of the pioneers at Winburg in June, 1837, and a few years later, on the land won from Dingaan, on the other side of the Drakens-berg, the republic of Natalia was declared to extend from the Umzimbulu to the Tugela. Outside of these crudely organized political associations there were from sixteen to twenty pioneer companies, headed by field cornets, which were practically as independent as the native tribes north of the Drakensberg. Neither of the republican creations was recognized by Great Britain, and, in 1842, Port Natal and the seaboard of the republic were captured, though Andries Pretorius repulsed the first British attack at Congella with heavy loss. In the follow-ing year Natal was formally declared to be a British Colony, and several thousand British immigrants were brought in to take the

[1] Of the basic fact of the assassination of Dingaan by a Swazi there is no question.

place of the retiring Boers who recrossed the Drakensberg. In 1848, by proclamation of Sir Harry Smith, her Majesty's High Commissioner and Governor of Cape Colony, all the territory between the Vaal and Orange rivers and the Quathlamba division of the Drakensberg was formally declared to be part of the British dominions under the name of the Orange River Sovereignty. The Boers had been spreading out towards the Vaal in many trekking parties north of the Drakensberg, and the British supremacy was not recognized until it was forcibly asserted by arms in the battle of Boomplatz, July 22, 1848. Then part of the Boers sullenly submitted, but many, headed by Andries Pretorius, preferred to pass beyond the farthest assertion of English dominion by crossing the Vaal and entering the wilderness stretching to the Limpopo.

There was then not even a glimmer of anticipation that the great stretch of veld and karoo between the Orange and the Vaal contained by far the richest diamond fields in the world. The controlling ministry in Great Britain at the time did not even consider it worth the cost of keeping and defending, and on October 21, 1851, Earl Grey wrote to Sir Harry Smith that " its ultimate abandonment should be a settled point in imperial policy." The territory beyond the Vaal was rated still more cheaply, and on January 17, 1852, the local independence of the inhabitants of the Transvaal was formally recognized by the Sand River Convention, signed by two assistant commissioners for Sir Harry Smith, and by appointed delegates for the Transvaal pioneers. The state organization of these settlers was first christened Hollandsche Afrikaansche Republiek, but this name was changed to Zud Afrikaansche Republiek in September, 1853. In the preceding month of July, Andries Pretorius, the pioneer leader who broke the Zulu power, died, but his great service was honorably recognized in the choice of his eldest son, Marthinus Wessel Pretorius, as the first president of the new Republic, and in the establishment of its capital of Pretoria.

On March 31, 1852, Lieutenant General George Cathcart succeeded Sir Harry Smith as High Commissioner and Governor of Cape Colony. The Transvaal had been already disposed of by the Sand River Convention, but, immediately after his arrival, May 13, 1852, General Cathcart issued a formal proclamation confirming this convention. It appeared, too, that it might be desirable to shift the charge of maintenance and local defence of the Orange River Sovereignty to the shoulders of the pioneer settlers. This conviction was confirmed by the outbreak of a war with the Basutos, the most powerful native tribe in this territory, under a cunning chief, Moshesh. In November, 1852, General Cathcart led a little army of two thousand infantry and five hundred cavalry to the Caledon River, but in the following month his expedition was beset by an overwhelming force of Basutos at Berea Mountain, and the battle was in effect a repulse to the British. After leaving a garrison at Bloemfontein, General Cathcart withdrew under cover of a fragile proclamation of peace, but his report and the accompanying news were so discouraging that the Duke of Newcastle wrote to him that " her Majesty's Government had decided to withdraw from the Orange River Sovereignty." In pursuance of this conclusion a convention was signed February 23, 1854, at Bloemfontein, by Sir George Russell Clerk, special commissioner representing Great Britain, and by the delegates from districts in the sovereignty. By this convention the independence of the settlers in the sovereignty was guaranteed, and the administration was handed over to a provisional council, which took charge until the first sitting of the Volksraad, March 28, 1854, and the declaration of a republic in the following month under the name of the Orange Free State. This independent state covered the greater part of the territory comprised within the bounds of the Orange River Sovereignty, excepting the large division between the Caledon River and the Quathlamba Mountains, reserved to the Basutos, and smaller reservations on the Vaal held by the Griquas.

Within the limits of the whole district between the Orange and the Vaal rivers there were then not more than fifteen thousand whites scattered over a territory of many thousand square

miles.   Except in the Caledon River districts little of this great
expanse was capable of supporting any clustered population or
even available for agriculture.   The soil throughout was shallow,
and in the southern and western sections the rainfall was ordinarily light.   There were a number of widespreading karoos, and
in the dry months the greater part of the veld was little better
than the desert.   The so-called farms were chiefly cattle and sheep
pastures, where the yield of grass and herbage was so varying
that several thousand acres were needed for any fair assurance of

Nest of Social Grosbeak.

safety for a small herd.   The total number of farms secured by
grant was only twelve hundred and sixty-five, but they extended
over eleven million acres.   Of the farm owners only one hundred
and thirty-nine were Englishmen, and a number of these were non-
residents.[1]   In the abstract there was seemingly little attraction or
value to excite any flow of immigration or to make the province
a prize worth the cost of defending

Not only the prospects of the Orange Free State and of its
neighbor on the other side of the Vaal seemed dull and inconsiderable to most observers, but the condition of Natal and

1 " South Africa," Theal.

of Cape Colony itself was little more promising. In Great Britain the whole dependency was so lightly esteemed that it was determined in 1849 to utilize it as a dumping ground for convicts, after Australia had resentfully thrown off this burden. The convict ship *Neptune* was actually sent out, but the indignation of the colonists was so demonstrative that no convicts were landed, and the ship with its load was held for five months in Simon's Bay, the present Naval Station, a little south of Cape Town, until the recalling order was received, February 13, 1850. The colony had not sunk so low as to submit to this mark of contempt, but it was undoubtedly drooping in hopes and enterprise, and the progress of its industrial development was painfully slow. There had been a pronounced diversion from agriculture to cattle and sheep raising for reasons before noted, and wool had become the chief and almost the only export of consequence. Still the peculiar condition and vagaries of the South African climate and seasons were hard to provide for or overcome, and there were prevalent diseases that attacked horses, cattle, and sheep, and greatly checked the rise of the pastoral industry. Communication from one part of the colony to another was very slowly improved. The roads were few and bad, and in 1867 the only stretch of railway in all South Africa was a bare forty miles from Cape Town to Wellington. The total annual export of the Colony was a trifle over £2,000,000 in value, and there was no diversification of industries and no manufactures of any considerable extent.[1] This was the situation when the gloom was suddenly dispelled and the whole face of South Africa changed by the discovery of the Diamond Fields.

[1] "South Africa," Theal.

CHAPTER IV

THE DISCOVERY

EARLY two hundred years had passed since the memorable expedition of van der Stel made known to geographers the Groote River, which, a hundred years later, was christened the Orange. Before Great Britain took the Cape, the daring van Reenen had penetrated to Modder Fontein, unconsciously skirting the rim of a marvellous diamond field. Since the beginning of the century scores of roving hunters had chased their game over a network of devious tracks, traversing every nook of the land between the Orange and the Vaal, and often camping for days upon their banks. Then the trekking pioneer graziers and farmers plodded on after the hunters, sprinkling their huts and kraals over the face of the Orange Free State, but naturally squatting first on the arable lands and grazing ground nearest the water-courses. So, in the course of years, in the passage of the Great Trek, thousands of men, women, and children had passed across the Orange and Vaal, and up and down their winding valleys, and hundreds, at least, had trodden the river shore sands of the region in which the most precious of gems were lying.

On the Orange River, some thirty miles above its junction with the Vaal, there was the hamlet of Hopetown, one of the most thriving of the little settlements, and a number of farms dotted the angle between the rivers. Along the line of the Vaal, for some distance above its entry into the Orange, there were some ill-defined reservations occupied by a few weak native tribes, — Koranas and Griquas, — for whose instruction there were mission stations at Pniel and Hebron.[1] For centuries unnumbered the aboriginal tribes had been ignorantly trampling under foot gems of countless price, and for years Dutch and English hunters, pioneers, farmers, shepherds, and missionaries trekked as heedlessly over the African diamond beds.

After the revelation of this fact, there arose, it is true, an imposing tale of an old mission map of the Orange River region, drawn as far back as the middle of the eighteenth century, across whose worn and soiled face was scrawled: "Here be diamonds." [2] Even if this report were true, there was no evidence determining the date of the scrawl, which might more credibly be a crude new record than a vague old one.   In any event, it does not appear that there was even a floating rumor of the probable existence of a South African diamond field at the time of the actual discovery of the first identified gem.

There is nothing surprising in this oversight.   When a spectator beholds a great semicircle of artfully cut gems sparkling on the heads, necks, and hands of fair women massed in superb array, and resplendent in the brilliant lights of an opera house, or when one views the moving throng glittering with jewels in grand court assemblies, it is hard for him to realize how inconspicuous a tiny isolated crystal may be in the richest of earth beds.   No spot in a diamond field has the faintest resemblance to a jeweller's show tray.   Here is no display of gems blazing like a Mogul's throne, or a Queen's tiara, or the studded cloak of a Russian noble.   Only in the marvellous valley of Sindbad are diamonds strewn on the ground in such profusion that they are likely to stick in the toes of a barefooted traveller, and can be gathered by flinging carcasses of sheep from surrounding precipices to tempt eagles to serve as diamond winners.

It needs no strain of faith to credit the old Persian tale of the discontented Ali Hafed, roaming far and wide from his

---

[1] "South Africa," George McCall Theal, London, 1888, 1891, 1893.

[2] "South African Diamond Fields and Journey to Mines," William Jacob Morton, New York, 1877.

charming home on the banks of the Indus in search of diamonds, and, finally, beggared and starving, casting himself into the river which flowed by his house, while the diamonds of Golconda were lying in his own garden sands.   It is probable that the diamonds of India were trodden under foot for thousands of years before the first precious stone of the Deccan was stuck in an idol's eye or a rajah's turban.   It is known that the Brazilian diamond fields were washed for many years by gold placer diggers without any revelation of diamonds to the world, although these precious stones were often picked up and so familiarly handled that they were used by the black slaves in the fields as counters in card games.

If this be true of the most famous and prolific of all diamond fields before the opening of the South African placers and mines, any delay in the revelation of the field in the heart of South Africa may be easily understood.   For it was not only necessary to have eyes bright and keen enough to mark one of the few tiny precious crystals which were lying on the face of vast stretches of pebbles, boulders, and sand, but the observer must prize such a crystal enough to stoop to pick it up if it lay plainly before his eyes.   To the naked native a rough diamond had no more attraction than any other pretty pebble.   There were millions of other white crystals and many colored pebbles on the river shores which were equally precious or worthless in his eyes.   The roving hunters were looking sharply for game bounding over the veld, and only glanced at a pebble-strewn bank to mark the possible track of their prey.   The stolid Boer pioneers would hardly bend their backs to pick up the prettiest stone that ever lay on the bank of an African river, even if it were as big as the great yellow diamond so jealously guarded by the Portuguese crown.[1]

It might be thought that some visitor to the fields would be more expert in judging its character than natives, hunters, and farmers; but there were few trained mineralogists in South

[1] "The Gold Regions of Southeastern Africa," Thomas Baines, F.R.G.S., London, 1877.

Africa, and it is doubtful if there was one who had ever examined a diamond field personally or compared one field with another. Even with this special experience an expert student of general mineral formations might survey this particular field closely without suspecting the existence of diamonds. This was demonstrated in the visit of the colonial geologist Wyley to the Orange Free State in 1856, when he investigated the alleged discovery of gold in thin veins of quartz lining the joints and crevices of the trappean rocks at Smithfield. In the course of his exploration he went to Fauresmith, where diamonds were afterward picked from the town commonage, and stood on the verge of the farm Jagersfontein, later the seat of a prolific diamond mine, yet it does not appear that he had even a surmise of the existence of diamonds in the field of his investigation.[1] It is but fair to him to observe, however, that the section which he visited had no such close resemblance to any known typical field as that which led Humboldt and Rose to the revelation of the diamonds of the Ural from the similarity of the ground formations to those of the Brazilian diamond districts.

As a matter of fact nobody who entered the Vaal river region conceived it to be a possible diamond field or thought of searching for any precious stones. Probably, too, there was not a person in the Orange Free State, and few in the Cape Colony, who was able to distinguish a rough diamond if he found one by chance, or would be likely to prize such a crystal. For the discovery of diamonds under such conditions it was practically necessary that a number of prospectors should enter it who would search the gravel beds often and eagerly for the prettiest pebbles. Were any such collectors at work in the field?

One of the trekking Boers, Daniel Jacobs, had made his home on the banks of the Orange River near the little settlement of Hopetown. He was one of the sprinkling of little farmers who was stolidly content with a bare and precarious liv-

[1] "Among the Diamonds," by the late John Noble, Clerk of the House of Assembly, Cape Town.

ing on the uncertain pasture lands of the veld.   Here his children grew up about him with little more care than the goats that browsed on the kopjes.

A poor farmer's home was a squalid hovel.   It was roughly partitioned to form a bedroom and kitchen, lighted by two small windows smudged with grime.   Dirty calico tacked on the rafters made its ceiling.   Its bare earthen floor was smeared weekly with a polishing paste of cowdung and water.   Father, mother, and children slept together on a rude frame overlaced with rawhide strips.   The only other furniture in this stifling bedroom was a chest of drawers and a small cracked mirror. There was no washbowl or water pitcher, but in the morning one after another of the family wiped their faces and swabbed their hands on the same moistened cloth.   Then they drew up chairs with rawhide seats to a rough wooden table and ate corn meal porridge, and sometimes a hunk of tough mutton boiled with rice, and soaked their coarse unbolted wheat flour bread in a gritty, black coffee syrup.[1]

When the sheep and goats were turned out of the kraal to graze on the patches of grass and the stunted thorns of the veld, the children ran away after them and roamed over the pasture land all day long like the flocks.   There was no daily round of work for them.   The black servants were the shepherds of the flocks, and did the slovenly housework, under the indolent eye of the Boer and his vrouw, for the poorest farmer would not work with his own hands except at a pinch.   His boys and girls had never seen a doll or a toy of any kind, but the instinct of childhood will find playthings on the face of the most barren karoo, and the Jacobs children were luckily close to the edge of a river which was strewn with uncommonly beautiful pebbles, mixed with coarser gravel.

Here were garnets with their rich carmine flush, the fainter rose of the carnelian, the bronze of jasper, the thick cream of chalcedony, heaps of agates of motley hues, and many shining

---

[1] "Life with the Boers in the Orange Free State," by a resident English physician's wife, New York, 1899.

rock crystals.[1] From this party-colored bed the children picked whatever caught their eye and fancy, and filled their pockets with their chosen pebbles. So a poor farmer's child found playthings scattered on a river bank which a little prince might covet, and the boy might have skimmed the face of the river with one little white stone that was worth more than his father's farm. Fortunately for the future of South Africa, he did not play ducks and drakes with this particular stone, which he found one day in the early spring of 1867, but carried it home in his pocket and dropped it with a handful of other pebbles on the farmhouse floor.[2]

A heap of these party-colored stones was so common a sight in the yard or on the floor of a farmhouse on the banks of the Orange and Vaal, that none of the plodding Boers gave it a second glance. But when the children tossed the stones about, the little white pebble was so sparkling in the sunlight that it caught the eye of the farmer's wife. She did not care enough for it to pick it up, but spoke of it as a curious stone to a neighbor, Schalk van Niekerk. Van Niekerk asked to see it, but it was not in the heap. One of the children had rolled it away in the yard. After some little search it was found in the dust, for nobody on the farm would stoop for such a trifle.

When van Niekerk wiped off the dust, the little stone glittered so prettily that he offered to buy it. The good vrouw laughed at the idea of selling a pebble. "You can keep the stone, if you want it," she said. So van Niekerk put it in his pocket and carried it home. He had only a vague notion that it might have some value, and put it in the hands of a travelling trader, John O'Reilly, who undertook to find out what kind of a stone the little crystal was, and whether it could be sold. He

[1] "The Diamond Diggings of South Africa," Charles Alfred Payton, London, 1872.   "South Africa Diamond Fields," Morton, New York, 1877. "Diamonds and Gold of South Africa," Henry Mitchell of Kimberley, London, 1889.

[2] "Among the Diamonds," 1870–1871.   "South Africa," Theal, London, 1888–1893.

showed the stone to several Jews in Hopetown and in Colesberg, a settlement farther up the Orange River Valley.   No
one of these would give a penny for it.   " It is a pretty stone
enough," they said, " probably a topaz, but nobody would pay
anything for it."

Perhaps O'Reilly would have thrown the pebble away, if it
had not come under the eye of the acting Civil Commissioner
at Colesberg, Mr. Lorenzo Boyes.   Mr. Boyes found on trial
that the stone would scratch glass.

" I believe it to be a diamond," he observed gravely.[1]

O'Reilly was greatly cheered up.   " You are the only man

John O'Reilly.

I have seen," he said, " who
says it is worth anything.
Whatever it is worth you
shall have a share in it."

" Nonsense," broke in
Dr. Kirsh, a private apothe-
cary of the town, who was
present, " I'll bet Boyes a
new hat it is only a topaz."

" I'll take the bet," re-
plied Mr. Boyes, and at his
suggestion the stone was
sent for determination to the
foremost mineralogist of the

colony, Dr. W. Guybon Atherstone, residing at Grahamstown.
It was so lightly valued that it was put in an unsealed envelope
and carried to Grahamstown in the regular post-cart.

When the post-boy handed the letter to Dr. Atherstone,
the little river stone fell out and rolled away.   The doctor
picked it up and read the letter of transmission.[2]   Then he
examined the pebble expertly and wrote to Mr. Boyes : " I
congratulate you on the stone you have sent to me.   It is a

---

[1] Lorenzo Boyes (statement furnished to author), 1899.

[2] W. Guybon Atherstone ; Lorenzo Boyes, 1899.   " Among the Diamonds,"
1870–1871.

veritable diamond, weighs twenty-one and a quarter carats, and is worth £500. It has spoiled all the jewellers' files in Grahamstown, and where that came from there must be lots more. Can I send it to Mr. Southey, Colonial Secretary?"

Mr. Lorenzo Boyes.

This report was a revelation which transformed the despised Karooland as the grimy Cinderella was transfigured by the wand of her fairy godmother. The determination was so positive and the expertness of the examiner so well conceded that Sir Philip Wodehouse, the Governor at the Cape, bought the rough diamond at once, at the value fixed by Dr. Atherstone and confirmed by the judgment of M. Henriette, the French consul in Cape Town.[1] The stone was sent immediately to the Paris Exhibition, where it was viewed with much interest, but its discovery, at first, did not cause any great sensation. The occasional finding of a diamond in a bed of pebbles had been reported before from various parts of the globe, and there was no assurance in this discovery of any considerable diamond deposits.

Dr. W. Guybon Atherstone.

Meanwhile Mr. Boyes hastened to Hopetown and to van Niekerk's farm, to search along the river shore where the first diamond was found. He prodded the phlegmatic farmers and their black servants, raked over many bushels of pebbles for two weeks, but no second diamond repaid his labor. Still the news of the finding of the first stone made the farmers near the river look more sharply at every heap of pebbles in the hope of finding one of the precious "blink klippe" (bright stones),

[1] "South Africa," Theal. Lorenzo Boyes, 1899. "Diamonds and Gold of South Africa," Theodore Reunert, 1893.

as the Boers named the diamond, and many bits of shining rock crystal were carefully pocketed, in the persuasion that the glittering stones were diamonds.   But it was ten months from the time of the discovery at Hopetown before a second diamond was found, and this was in a spot more than thirty miles away, on the river bank below the junction of the Vaal and Orange rivers.   Mr. Boyes again hastened to the place from which the diamond had been taken, but he failed again to find companion stones, though he reached the conclusion that the diamond had been washed down stream by the overflowing of the Vaal.[1]

From the Orange River the search passed up the Vaal, where the beds of pebbles were still more common and beautiful.   The eyes of the native blacks were much quicker and keener in such a quest than those of the stolid Boer, who scarcely troubled himself to stoop for the faint chance of a diamond.   But no steady or systematic search was undertaken by anybody, and it was not until the next year, 1868, that a few more diamonds were picked up on the banks of the Vaal by some sharp-sighted Koranas.[2]   The advance of discovery was so slow and disappointing that there seemed only a faint prospect of the realization of the cheering prediction of Dr. Atherstone, which was scouted by critics who were wholly incompetent to pass upon it.   Even the possibility of the existence of diamond deposits near the junction of the Orange and Vaal was flatly denied by a pretentious examiner who came from England to report on the Hopetown field.   It was gravely asserted that any diamonds in that field must have been carried in the gizzards of ostriches from some far-distant region, and any promotion of search in the field was a bubble scheme.

To this absurd and taunting report Dr. Atherstone replied with marked force and dignity, presenting the facts indicating the existence of diamond-bearing deposits, and adding : " Sufficient has been already discovered to justify a thorough and extensive geological research into this most interesting country, and I think for the interest of science and the benefit of the

[1] Lorenzo Boyes, 1899.   " South Africa," Theal, London, 1888–1893.

Colony a scientific examination of the country will be under-taken. So far from the geological character of the country mak-ing it impossible, I maintain that it renders it probable that very extensive and rich diamond deposits will be discovered on proper investigation. This I trust the Home Government will author-ize, as our Colonial exchequer is too poor to admit of it."[1]

There was no official response to this well-warranted sug-gestion, for it had hardly been penned when the announcement of a remarkable discovery aroused such an excitement and such a rush to the field that no government exploration was needed. In March, 1869, a superb white diamond, weighing 83.5 carats, was picked up by a Griqua shepherd boy on the farm Zendfon-tein, near the Orange River.[2] Schalk van Niekerk bought this stone for a monstrous price in the eyes of the poor shepherd, — 500 sheep, 10 oxen, and a horse, — but the lucky purchaser sold it easily for £11,200 to Lilienfeld Brothers of Hopetown, and it was subsequently purchased by Earl Dudley for £25,000.[3] This extraordinary gem, which soon became famous as " the Star of South Africa," drew all eyes to a field which could yield such products, and the existence and position of diamond beds was soon further assured and defined by the finding of many smaller stones in the alluvial gravel on the banks of the Vaal.

Alluvial deposits form the surface ground on both sides of this river, stretching inland for several miles. In some places the turns of the stream are frequent and abrupt, and there are many dry water-courses which were probably old river channels. The flooding and winding of the river partly accounts for the wide spreading of the deposits, but there has been a great abrasion of the surface of the land, for the water-worn gravel sometimes covers even the tops of the ridges and kopjes along the course of the river.

This gravel was a medley of worn and rolled chips of basalt, sandstone, quartz, and trap, intermingled with agates, garnets,

[1] W. Guybon Atherstone, 1868.   [2] " Among the Diamonds," 1870–1871.
[3] *Ibid.* (Accounts of this discovery differ somewhat.) Vide Theal's " South Africa," Reunert's " Diamonds and Gold," etc.

peridot, jasper, and other richly colored pebbles, lying in and on a bedding of sand and clay.   Below this alluvial soil was in some places a calcareous tufa, but usually a bed rock of melaphyre or a clayey shale varying in color.   Scattered thickly through the gravel and the clay along the banks were heavy boulders of basalt and trap which were greatly vexing in after days to the diamond diggers.[1]

For a stretch of a hundred miles above the Mission Station at Pniel the river flows through a series of rocky ridges, rolling back from either bank to a tract of grassy, undulating plains. Fancy can scarcely picture rock heaps more contorted and misshapen.   Only prodigious subterranean forces could have so rent the earth's crust and protruded jagged dykes of metamorphic, conglomerate, and amygdaloid rocks, irregularly traversed by veins of quartz, and heavily sprinkled with big bare boulders of basalt and trap.   Here the old lacustrine sedimentary formation of the South African high veld north of the Zwarte Bergen and Witte Bergen ranges has plainly been riven by volcanic upheaval.   The shale and sandstone of the upper and lower Karoo beds have been washed away down to an igneous rock lying between the shale and the sandstone.   It was along this stretch of the river that the first considerable deposit of diamonds in South Africa was uncovered.[2]

For more than a year since the discovery of the first diamond there had been some desultory scratching of the gravel along the Vaal by farmers and natives in looking for " blink klippe," and a few little rough diamonds had been found by the Hottentots, as before noted ; but the first systematic digging and sifting of the ground was begun by a party of prospectors from Natal at the Mission Station of Hebron.   This was the forerunner of the

[1] "Diamonds and Gold of South Africa," Reunert, Cape Town, 1893. "The Diamond Diggings of South Africa," Payton, 1872.   "Among the Diamonds," 1870–1871.

[2] "Diamonds and Gold of South Africa," Reunert, 1893.   "Among the Diamonds," 1870–1871.   "South Africa," Theal, 1888–1893.   "On Diamonds," Sir William Crookes, London, 1897.

second Great Trek to the Vaal from the Cape, a myriad of adventurers that spread down the stream like a locust swarm, amazing the natives, worrying the missionaries, and agitating the pioneer republics on the north and the east.[1]

The first organized party of prospectors at Hebron on the Vaal was formed at Maritzburg in Natal, at the instance of Major Francis, an officer in the English army service, then stationed at that town. Captain Rolleston was the recognized leader, and after a long plodding march over the Drakensberg and across the veld, the little company reached the valley of the Vaal in November, 1869. Up to the time of its arrival there had been no systematic washing of the gravel edging the river. Two experienced gold diggers from Australia, Glenie and King, and a trader, Parker, had been attracted to the field like the Natalians by the reported discoveries, and were prospecting on the line of the river when Captain Rolleston's party reached Hebron.[2] Their prospecting was merely looking over the surface gravel for a possible gem, but the wandering Koranas were more sharp-sighted and lucky in picking up the elusive little crystals that occasionally dotted the great stretches of alluvial soil.

It was determined by Captain Rolleston to explore the ground as thoroughly as practicable from the river's edge for a number of yards up the bank, and the washing began on a tract near the Mission Station. The Australian prospectors joined the party, and their experience in placer mining was of service in conducting the search for diamonds. The workers shovelled the gravel into cradles, like those used commonly in Australian and American placer washing, picked out the coarser stones by hand, washed away the sand and lighter pebbles, and saved the heavier mineral deposit, hoping to find some grains of gold as well as diamonds above the screens of their cradles. But the returns for their hard labor for many days were greatly disappointing. They washed out many crystals and brilliant pebbles,

[1] "South Africa," Theal, 1888–1893. "Among the Diamonds," 1870–1871. [2] "Among the Diamonds," 1870–1871.

but never a diamond nor an atom of gold dust. Then they pushed down the river more than twenty miles to another camp at Klip-drift, opposite the Mission Station at Pniel. Here too they washed the ground for days without finding even the tiniest gem, and were almost on the point of abandoning their disheartening drudgery, when finally, on the seventh of January, 1870, the first reward of systematic work in the field came in the appearance of a small diamond in one of the cradles.[1]

This little fillip of encouragement determined their continuance of the work, and a party from British Kaffraria joined them in washing the gravel in places that seemed most promising along the line of the river. It was agreed that the first discovery of rich diamond-bearing ground should be shared alike by both parties, but there was nothing to share for some weeks. Then some native Koranas were induced to point out to the Natalians a gravel-coated hummock or kopje near the Klip-drift camp, where they had picked up some small diamonds. When the prospectors began the washing of the gravel on this kopje, it was soon apparent that a diamond bed of extraordinary richness had been reached at last. Good faith was kept with the company from Kingwilliamstown, and the combined parties worked to the top of their strength in shovelling and washing the rich bed. The lucky men kept their mouths closed, as a rule, and did not intend to make known their good fortune; but such a discovery could not long be concealed from visiting traders and roaming prospectors, and before three months had passed some prying eye saw half a tumblerful of the white sparkling crystals in their camp, and the news spread fast that the miners had washed out from two hundred to three hundred stones, ranging in size from the smallest gems to diamonds of thirty carats or more.[2]

Then a motley throng of fortune-hunters began to pour into the valley of the Vaal. The first comers were those living nearest to the new diamond field, — farmers and tradesmen from the cattle ranges and little towns of the Orange Free State.

[1] " Among the Diamonds," 1870–1871.   [2] *Ibid.*

Some of these were stolid Boers, drawn to the fields as a novel and curious spectacle, but disdaining the drudgery of shovelling and washing from morning till night for the chance of a tiny bright stone. They stared for a while at the laboring diamond seekers, and then turned their backs on the field contemptuously, and rode home sneering at the mania which was dragging its victims for hundreds of miles, over sun-cracked and dusty karoos, to hunt for white pebbles in a river bed. Still there were many poor farmers who caught the infectious diamond fever at sight of the open field and a few sparkling stones, and they camped at Klip-drift or went on farther up or down the river, to join, as well as they knew how, in the search for diamonds.

Following this influx from the Free State came swarming in men of every class and condition from the southern English Colony, and from the ships lying in the coast ports. The larger number were of English descent, but many were Dutch, and hardly a nation in Europe was unrepresented. Black grandsons of Guinea coast slaves and natives of every dusky shade streaked the show of white faces. Butchers, bakers, sailors, tailors, lawyers, blacksmiths, masons, doctors, carpenters, clerks, gamblers, sextons, laborers, loafers, — men of every pursuit and profession, jumbled together in queerer association than the comrades in the march to Finchley, — fell into line in a straggling procession to the Diamond Fields. Army officers begged furloughs to join the motley troop, schoolboys ran away from school, and women even of good families could not be held back from joining their husbands and brothers in the long and wearisome journey to the banks of the Vaal.[1]

There was the oddest medley of dress and equipment: shirts of woollen, — blue, brown, gray, and red, — and of linen and cotton, — white, colored, checked, and striped; trim jackets, cord riding-breeches and laced leggings, and "hand me downs" from the cheapest ready-made clothing shops; the yellow oilskins and rubber boots of the sailor; the coarse, brown corduroy and

[1] "Among the Diamonds," 1870-1871.

canvas suits, and long-legged, stiff, leather boots of the miner; the ragged, greasy hats, tattered trousers or loin cloths of the native tribesmen; jaunty cloth caps, broad-brimmed felt, battered straw, garish handkerchiefs twisted close to the roots of stiff black crowns, or tufts of bright feathers stuck in a wiry mat of curls; such a higgledy-piggledy as could only be massed in a rush from African coast towns and native kraals to a field of unknown requirements, in a land whose climate swung daily between a scorch and a chill, where men in the same hour were smothered in dust and drenched in a torrent.

It is doubtful if a single one of this fever-stricken company had ever seen a diamond field or had the slightest experience in rough diamond winning, but no chilling doubt of themselves or their luck restrained them from rushing to their fancied Golconda. Their ideal field was much nearer a mirror of the valley of Sindbad than the actual African river bank, and it was certain that many would be as bitterly disappointed by the rugged stretch of gravel at Klip-drift as the gay Portuguese cavaliers were at the sight of the Manica gold placers.

Everything in the form of a carriage from a chaise to a buckwagon was pressed into service, but the best available transport was the big trekking ox-wagon of the Boer pioneer. This was a heavily framed, low-hung wagon, about twenty feet long and five and a half feet broad. In this conveyance more than a dozen men often packed themselves and their camping outfit and food. An exceptionally well-equipped party carried bacon, potatoes, onions, tea, coffee, sugar, condensed milk, flour, biscuits, dried peas, rice, raisins, pickles, and Cape brandy. The total weight of load allowed, including the living freight, was limited to seven thousand pounds.[1]

East London, the nearest port, was something more than four hundred miles from the diamond field, and Cape Town nearly seven hundred. Natal, Port Alfred, and Port Elizabeth were almost equally distant, as the crow flies, approximately four

[1] "The Diamond Diggings of South Africa," Charles Alfred Payton, London, 1872.

hundred and fifty miles; but the length of the journey to the Vaal could not be measured by any bare comparison of air-lined distances. The roads, at best, were rough trampled tracks, changing, after a rainfall, to beds of mire. Their tortuous courses rambled from settlement to settlement, or from one farmhouse to another over the veld, and were often wholly lost in the shifting sands of the karoo. It was a tedious and difficult journey by land even from one seacoast town to another, and fifty miles from the coast the traveller was fortunate if his way was marked by even a cattle path.[1]

When the rain fell in torrents with the lurid flashes and nerve-shaking crash of South African thunder-storms, the diamond seekers huddled together under the stifling cover of their wagons, while fierce gusts shook and strained every strip of canvas and water drops spurted through every crevice. In fair weather some were glad to spread their blankets on the ground near the wagon, and stretch their limbs, cramped by their packing like sardines in a box. On the plains they had no fuel for cooking except what they could gather of dry bullock's dung. Sometimes no headway could be made against the blinding dust-storms, that made even the tough African cattle turn tail to the blasts, and clogged the eyes and ears and every pore of exposed skin with irritating grit and powder. Sometimes the rain fell so fast that the river beds were filled in a few hours with muddy torrents, which blocked any passage by fording for days and even weeks at a time, and kept the impatient diamond seekers fuming in vain on their banks. Payton's party was forty-six days in its passage from Port Elizabeth to the Diamond Fields without meeting with any serious delays, and journeys lasting two months were not uncommon.[2]

Still, in spite of all obstacles, privations, and discomforts, the long journey to the fields was not wholly monotonous and unpleasant. As there was no beaten way, the prospectors chose

---

[1] "South Africa," George McCall Theal, 1888–1893.

[2] "The Diamond Diggings of South Africa," Payton. "South Africa Diamond Fields and Journey to Mines," William Jacob Morton, New York, 1877.

K

their own path, riding by day and camping at night as their fancy led them.   In ascending to the tableland of the interior from Natal, there were shifting and stirring visions of mountain peaks, terraces, gorges, and valleys.

On the higher terraces there was not the luxuriance of the coast, — the huge tree ferns with feathery fronds, the towering masses of palms, the drooping festoons of climbing vines, the exquisite flowers: spiked ansellias with their pale yellow blossoms, barred and spotted with red, pure white, sweet-scented clusters of mystacidium, and orchids of marvellous variety and hue, — but even the highest upland tree growth had beauties of its own.   On the slopes of the Drakensberg the wild chestnut, the Natal mahogany, the white pear and iron wood grow sturdily, and the common yellow wood, stink wood, bogabog, and sneeze wood flourish in spite of their rude names.[1]

Amid this varied scenery they could linger and wind about as they pleased, and every turn of their path revealed new charms of line and color.   As they descended the mountain flanks some marked how the lacustrine deposits of past ages had overspread the face of the land with their covering of sandstone and shale, even skirting the summits of the highest peaks at a height of more than six thousand feet, as was plainly shown on the Compassberg.[2]   On the plateau below they saw how the craggy hills, pointed spitz-kopjes, and columnar ridges of the trappean rocks projected above the sedimentary cover of the karoo.

Throughout the Orange Free State, but especially in the neighborhood of the valleys of the Orange and Vaal, these volcanic rock elevations are common, sometimes massed in irregular rows and often rising in the most jagged and fantastic shapes. " When we see them at the surface," wrote the geologist Wyley in 1856, "they look like walls running across the country, or more frequently form a narrow, stony ridge like a wall that has been thrown down.   The rock of which they are composed, greenstone or basalt, is known by the local name of iron stone,

[1] "The Colony of Natal," J. Forsyth Ingram.
[2] "Among the Diamonds," 1870–1871.

from its great hardness and toughness, and from its great weight. The origin of these dykes is well known. They have been produced by volcanic agency, which, acting from below upon horizontal beds of stratified rock, has cracked and fissured them at right angles to their planes of stratification, and these vertical cracks have been filled up with the melted rock or the lava from below. The perpendicular fissures through which it has found its way upwards are seldom seen, nor should we expect to see much of them, for it is precisely along the line of these that the rocks have been most broken up and shattered and the denudation has been greatest."

Even in the crossing of the karoos there were curious and awesome sights to attract and impress the mind of a traveller beholding for the first time these desert wastes so widely spread over the face of South Africa. They differ little in appearance except in size. The Great or Central Karoo, which lies beneath the foot-hills of the Zwarte Bergen range, has a sweep to the north of more than three hundred miles in a rolling plateau, ranging in elevation from two to three thousand feet. Day after day, as the diamond seekers from Cape Town plodded on with their creaking wagons, the same purpled brown face was outspread before them of the stunted flowering shrub which has given its name to the desert, spotted with patches of sun-cracked clay or hot red sand. To some of the Scotchmen this scrub had the cheery face of the heather of their own Highlands, and homesick Englishmen would ramble far through the furze to pick the bright yellow flowers of plants that recalled the gorse of their island homes.[1] These common bushes, rarely rising a foot in height, and the thick, stunted camelthorn, were almost the only vegetable coating of the desert.

Straggling over this plane ran the quaint ranges of flat-topped hummocks and pointed spitz-kopjes, streaked with ragged ravines torn by the floods, but utterly parched for most of the year. Shy meerkats, *Cynictis penicillata*, weasel-like crea-

[1] Special correspondence *London Chronicle* and other English journals, November, 1899.

tures with furry coats, peered cautiously from their burrows at the strange procession of fortune-hunters, and from myriads of the mammoth ant-hills that dot the face of the desert innumerable legions of ants swarmed on the sand along the track of the wagons.   Sometimes at nightfall the queer aard-vark lurked upon the ant-heap and licked up the crawling insects by thousands.   Far over the heads of the travellers soared the predatory eagles and swooping hawks, harrying the pigeons and dwarf doves that clustered at daybreak to drink at the edge of every stagnant pool.[1]

Even in the earliest years of the Dutch advance into South Africa, when wild beasts browsed in troops on every grassy plain and valley and the poorest marksman could kill game almost at will, the karoo was shunned by almost every living creature except in the fickle season of rainfall.   The lion skirted the desert edge warily, unwilling to venture far from a certain waterbrook or pool.   There was nothing on the bare karoo to tempt the rhinoceros from his bed in green-leaved thickets, and only the wide-roaming antelopes (trekbok) rambled for pasturage far over the sparsely coated and parched desert waste.   If this was true in the days when the tip of Africa was swarming with animal life, it is not surprising that the diamond seekers in 1869 and 1870 rarely saw any living mark for their rifles when they journeyed over the desert.   Rock-rabbits, akin to the scriptural coney, scampering to their holes, were often the largest game in sight for days at a time, and it was counted remarkable luck when any hunter put a bullet through a little brown antelope, a grysbok, or springbok.[2]   The springboks still haunted the Great Karoo, for they were particularly fond of its stunted bush growth, and in the rainy season many droves of these antelopes could be seen browsing warily or flying in panic from the spring of the cheetah, the African hunting leopard ; but most of the bigger game, blesbok, haartebeest, koodoo, and wildebeest, that used to feed

[1] " A Breath from the Veld," John Guille Millais, London, 1895.   " Among the Diamonds," 1870–1871.

[2] " The Diamond Diggings of South Africa," Payton, 1872.

greedily on the same pasture, had been killed or driven away by the keen hunting of the years that followed the taking of the Cape by the English.[1]

Sometimes the clear sky of the horizon was blurred by the advancing of monstrous swarms of locusts, the "black snow-storms" of the natives, sweeping over the face of the land like the scourge of devouring flames, chased by myriads of locust birds, and coating the ground for miles around at nightfall with a crawling, heaving coverlet. Then might be heard the hoarse trump of the cranes winging their way over the desert and drop-ping on the field strewn with locusts to gorge on their insect prey. Or the travellers saw the slate-white secretary bird stalk-ing about with his self-satisfied strut and scraping up mouthfuls with his long horny bill.

More marvellous than the locust clouds were the amazing mirages that deceived even the keen-eyed ostriches with their counterfeit lakes and wood-fringed streams, so temptingly near, but so provokingly receding, like the fruits hanging over the thirsting Tantalus. Sometimes hilltops were reared high above the horizon, distorted to mountainous size and melting suddenly in thin air or a flying blur. Now a solitary horseman was seen to swoop over the desert in the form of a mammoth bird, or a troop of antelopes were changed to charging cavalry. No trick of illusion and transformation was beyond the conjuring power of the flickering atmosphere charged with the radiating heat of the desert.[2]

When the prospectors crossed the karoo and entered the

---

[1] "A Breath from the Veld," John Guille Millais, London, 1895.

[2] Despatches of Julian Ralph and other special correspondents to London jour-nals, October–December, 1899. "Sketches and Studies in South Africa," W. J. K. Little, London, 1899. "Portraits of the Game and Wild Animals of Southern Africa," W. G. Harris, London, 1840. "The Large Game and Natural His-tory of South and Southeast Africa," W. H. Drummond, Edinburgh, 1875. "Travel and Adventure in Southeast Africa," F. C. Selous, London, 1893. "Kloof and Karoo," H. A. Bryden, London, 1889. "Days and Nights by the Desert," P. Gillmore, London, 1888. "Gun and Camera in South Africa," H. A. Bryden, London, 1893.

stretches of pasture land which the Dutch called veld, the scenes of their marches were much more lively and cheery. Little farmhouses dotted the plains and valleys, rude cottages of clay-plastered stones or rough timbers, but hospitable with fires blazing on open hearths, big iron pots hanging from cranes and simmering with stews, and broad-faced, beaming vrouws and clusters of chunky boys and girls greeted the arrival of an ox-wagon from the coast as a welcome splash in the stagnant stream of their daily life.[1]

At some of the halting places on the banks of streams, or where plentiful water was stored in natural pans or artificial ponds, the extraordinary fertility of the irrigated soil of South Africa was plainly to be seen in luxuriant gardens, with brilliant flower-beds and heavy-laden fruit trees and vines. Here figs, pomegranates, oranges, lemons, and grapes ripened side by side, and hung more tempting than apples of Eden in the sight of the thirsting, sunburnt, dust-choked men who had plodded so far over the parched karoos. They stretched their cramped legs and aching backs in the grateful shade of spreading branches, and watched with half-shut eyes the white flocks nibbling on the pasture land, and the black and red cattle scattered as far as the eye could see over the veld. Tame ostriches stalked fearlessly about them, often clustering like hens at the door of the farmhouse to pick up a mess of grain or meal, apparently heedless of any approach, but always alert and likely to resent any familiarity from a stranger with a kick as sharp and staggering as any ever dealt by a mule's hind leg.

The interior of the homes in these oases was not so inviting, for the rooms, at best, were small and bare to the eye of a townsman. But some were comparatively neatly kept, with smoothly cemented floors, cupboards of quaintly figured china and earthenware, hangings and rugs of leopard, fox, jackal, and antelope skins and brackets of curving horns loaded with hunting arms and garnished with ostrich feathers. For the guests

[1] "Among the Diamonds," 1870–1871. "The Diamond Diggings of South Africa," Payton, 1872. "South Africa Diamond Fields," Morton, 1876.

there was probably the offer of a freshly killed antelope or sheep ; but the farmer's family was often content with "biltong," the dried meat that hung in strips or was piled in stacks under his curing shed.

Near every house was the accompanying kraal or open-walled circle for the confinement of the flocks at night, built of stones, and usually so bedded and filthy with fresh dung that a heavy percentage of the farmers' sheep died yearly from foot-rot or scab.[1] Close to the kraal was the water reservoir for the flocks and the household use, unless the farm lay on the bank of an unfailing stream. These collections of water were commonly hill drainage, stored in long, narrow ponds by rough dams across ravines, or the drainage and rainfall filling shallow natural basins which the Boers call "pans." In the early morning the birds flew from all quarters to these ponds. Wild ducks, geese, plover, sandgrouse, and flocks of pigeons and doves hovered over the pools and splashed and dabbled in the water, while the blue-gray Kafir cranes stalked warily along the brink.

These basins are quite numerous in the country lying between the Orange and the Vaal, as well as throughout the Transvaal. The light earth washed down the hill slopes was largely calcareous, and incrusted the grasses and roots of the basin in a calc-tufa which is almost impervious to water. So the pans became excellent natural reservoirs, though there was, of course, a heavy loss from evaporation. No calamity is so dreaded by the graziers as the failure of their water-supply, for it has often caused the loss of a flock and the ruin of the poor owner. Therefore the pans are highly valued and strictly reserved, and the dams are daily inspected lest a burrowing land crab should open the way for a rush of water that would empty the reservoir.[2] When a settler was fortunate in getting a tract of land with a pan or a water-spring, he almost invariably gave the name to his farm, as Dutoitspan, Dorstfontein, Jagersfontein,

[1] "On Veld and Farm," Frances MacNab, London, 1897. "South Africa Diamond Fields," Morton, New York, 1877.
[2] "Among the Diamonds," 1870–1871.

Bultfontein, — names of inconsiderable little patches on the face of South Africa, which were destined to become memorable by approaching revelations.[1]

Attracted by the good pasturage and water and the sight of flowers, fruits, and birds, even the eager diamond seekers were not loath to linger for a day at one of these oases and rest themselves and their cattle before pushing on to the Vaal. As they drew near to their goal the face of the country began to change. After passing the Modder River, the grassy plains stretched out wider and longer and more gently undulating, and the mirage was more greatly magnifying and illusive. Herds of wild game, chiefly springbok, blesbok, hartebeest, wildebeest, and koodoo, were now frequently seen, and the ears of the travellers were tickled with the cheery karack-karack of flying korhaan and the pipes of red-legged plover. There are black headed or veld korhaan and bush korhaan. These birds, which are very plentiful along the Vaal River and about Kimberley, belong to the smaller bustard species. The cock bird of the veld korhaan has a black head with white spots on the sides. The top of the head or crest is of a reddish gray color. The back is also reddish gray, the markings of the feathers being in rings or stripes. The wings are black-and-white, and the legs yellow. The hen birds have reddish gray heads, but otherwise are similar in feather to the cock bird. The bird derives its name from the Dutch word knor, to scold, and haan, hen or bird, on account of the scolding noise made by the male bird as it rises from the ground. The original word, knorhaan, has been corrupted into korhaan. The bush korhaan has a gray head with a light blue patch on the crown, just back of which is a pink-brown crest an inch and a half long. The back is covered with brown-and-white feathers with diamond-pointed markings. The lower part of the leg is yellow and the upper part blue. The Dutch call one variety

[1] "Achtzehn Jahre in Sud Africa," E. J. Karrström, Leipzig, 1899. "Seven Years in South Africa," Emil Holub, London, 1881. "South Africa," A. H. Keane, London, 1895. "South Africa of To-day," Captain F. E. Younghusband, London, 1898. "Ten Years in South Africa," J. W. D. Moodie, London, 1835. "South Africa," George McCall Theal, 1888–1893.

of birds somewhat resembling the bush korhaan rudely "dik-
kops," thick heads, from their appearance when wounded ; but
they are none the less handsome birds, and they were eagerly
shot and eaten by the diamond seekers on the way to the fields
and in the camps on the river. There were great numbers, too,
of the paauw or cape bustard near the Modder River, and red-
winged partridges and Guinea fowl that gave a welcome variety
to the meals of the travellers.[1]

Over the rolling ground the prospectors pressed rapidly to
the Diamond Fields and soon reached the river border where the
plains ran into the barrier of ridges of volcanic rocks. Jolting
heavily over these rough heaps and sinking deeply in the red
sand wash of the valleys, the heavy ox-wagons were slowly tugged
to the top of the last ridge above Pniel, opposite the opened
diamond beds of Klip-drift, where the anticipated Golconda was
full in sight. Here the Vaal River winds with a gently flowing
stream, two hundred yards or more in width, through a steeply
shelving, oblong basin something over a mile and a half in length
and a mile across. A thin line of willows and cotton-woods
marked the edge of the stream on both banks. On the descending slope toward the river stood the clustering tents and wagons
of the pilgrims waiting to cross the stream.

In the dry season the Vaal was easily fordable by ox-wagons
at a point in this basin, and the ford, which the Boers call
"drift," gave the name to the shore and camp opposite Pniel,
— "Klip-drift," "Rocky-ford." When the river was swollen
by rains, the impatient fortune-hunters were forced to wait, fuming, in sight of the diamond diggings until the flood subsided ;
but, a few months after the rush began, a big, flat-bottomed ferry-
boat, called a punt, was constructed to carry over the wagons
and cattle, while the men crossed in rowboats, making regular
ferry trips between Pniel and Klip-drift.

How stirring were the sights and sounds from the ridge at
Pniel to every newcomer while the swarming diamond seekers
were crossing the river and spreading out over the northern

[1] "Among the Diamonds," 1870–1871.

bank! — the confused clustering at the ford — the rambling of stragglers along the shore — the gravel cracking and grinding under the hoofs of the horses and ponies racing along the bank and rearing, plunging, and bucking at the check of the bits and prick of the spurs — the outspanning and inspanning of hundreds of oxen — the swaying and creaking wagons — the writhing, darting lash of the cracking whips of the drivers — the sulking, balking oxen, driven into long, straining lines that dragged the ponderous, canvas-arched " prairie - schooners " through the turbid water and over the quaking sands — the whistling, shouting, yelling, snorting, neighing, braying, squeaking, grinding, splashing babel — the scrambling up the steep Klip-drift bank — the scattering of the newcomers — the perching of the white-topped wagons and the camp-tents like monstrous gulls on every tenable lodging place on bank, gully, and hillside — the scurrying about for wood and water — the crackling, smoking, flaming heaps of the camp fires — the steaming pots and kettles swinging on cranes — the great placer face, pockmarked with holes and heaps of reddish sand, clay, and gravel — the long stretches of the miners' rockers and troughs at the water's edge — and chief of all in interest, the busy workmen, sinking pits and throwing out shovelfuls of earth, filling buckets and hauling them up with ropes, loading and shaking the rockers, driving carts full of heavy gravel to the water troughs, returning for new loads, scraping and sorting the fine, heavy pebbles on tables or flat rocks or boards spread on the ground!

No labored, crawling recital can compass and picture in print any approach to the instant impress on the eye and ear of the moving drama on the banks of the Vaal. Observer after observer groped vainly for graphic comparison. " Klip-drift is a swarm of bees whose hive is upset," said one, "a bank lined with ant-hills," wrote another, prosily; "a wild rabbit warren, scurried by a fox," ventured a third; "an insane asylum turned loose on a beach," sneered a fourth. It was a mushroom growth of a seething placer-mining camp in the heart of the pasture lands of South Africa. To old Australian and American miners it had a patent likeness to familiar camps and diggings, but its local coloring was glaringly vivid and unique.[1]

[1] "Among the Diamonds," 1870–1871. "The Diamond Diggings of South Africa," Charles Alfred Payton, London, 1872. "South Africa Diamond Fields," Morton, New York, 1877. "To the Cape for Diamonds," Frederick Boyle, London, 1873. "Diamond Fields of South Africa, by One who has visited the Fields," New York, 1872.

Pniel Diggings.

CHAPTER V

EFORE calling to view the spreading of the diamond seekers along the line of the Vaal River, the rearing of successive camps, and the growing pursuit of gems in the gravel, it is essential to trace the progress of diamond mining from its original development on the water-shed of the Indus, and to account in great measure for the blundering, confusion, and failures in the new Diamond Fields by showing how crude and imperfect were any known methods of winning the precious stones at the time of the South African discoveries.

From earliest history there had been no change and no prospect of change in the diamond mining of India (described in Chapter 1). In the Deccan diamond fields, as in the other congested districts, there was such an influx of poor natives that no labor-saving contrivances were sought for, and the diamond-bearing gravels were lifted and washed by hand as they had been by the first generation of workers. There had been no competition with the Deccan field, and no considerable production outside of it, until the diamonds of the Brazilian fields were made known to the Portuguese in the year 1728. As soon as the Home Government learned of this discovery, the diamonds in Brazil were declared to be State property, and for a hundred years diamond mining was a Crown monopoly. This condition was a clog to any possible advance in the methods of mining. There was a constant drain on the industry without any effort to develop it systematically, thoroughly, or economically.

The chief deposits were found, at first, in river beds and ravines in a breccia of clay, quartz pebbles, and sand, charged with oxide of iron. Some of the richest beds were opened along the rivers Jequetinhonha and Pardo in the valley of Sejues, and on the line of the rivers Aboite, Andaja, da Serreno, da Prata, and San Francisco in the province of Minas Geraes.[1]

The diamond-bearing ground was worked under government agents or leased to contractors. Quick returns were the first object. So gangs of slaves were put on the grounds, regardless of loss, if only the cream of the fields was skimmed. In the dry season the beds of the smaller sierran streams were nearly or wholly dry. Underlying the surface wash of sand in the bed was the formacao or cascalho, heavy diamond-bearing gravel intermixed with boulders. The alluvial soil was generally from eight to twenty feet thick, a silicious sand chiefly, deep colored by ferruginous clay. The diamonds and other minerals of high specific gravity were held in the bottom layer of this alluvium, usually cemented in a coarse pudding-stone of quartz and itacolumite — the cascalho. The sand was rudely scraped away or carried off in pans, the boulders pried out, and the cascalho exposed. Then the gravel was collected laboriously in pans and piled in heaps to await the rainy season, when the streams filled the dry courses and there was water at hand for washing the gravel.

Bacus or shallow pits were sunk in the sand along the brink of the streams, and in these pits a few panfuls of gravel were thrown. The bottom of the bacu was made to slope so that the dashing of water on the gravel heap would readily wash away the clinging sand and the lighter and larger stones. The expert slaves washed the heaps in the bacus with splashes of water cast

[1] "The Diamond Fields of Brazil," Report of United States Minister Bryan, March 12, 1899, conveying report of American Secretary of Legation, Dawson. "A Treatise on Gems," Lewis Feuchtwanger, M.D., 1867. "An Account of Diamonds found in Brazil," James Castro de Sarmente, M.D. "Genuine Account of the Present State of the Diamond Trade in the Dominions of Portugal," a Lisbon merchant, London, 1785. "Travels in South America," J. J. von Tschudi.

from concave wooden plates with a peculiar whirl which hastened the separation of the heavier gravel. This concentrate,
containing most of the diamonds in the cascalho, was then
washed again in a batea, a wooden dish with a depression in the
centre.   By dexterous shaking and whirling motions of the batea
filled with water and a few handfuls of gravel, the lighter gravel

Delports' Hope, Vaal River Diggings.

was carried to the rim and washed or scraped away, and diamonds mixed with heavier pebbles were collected in the hollowed centre of the dish.   A gentle tilt of the batea drained off
the water, and the precious stones were picked from the other
pebbles by hand.

Sometimes the formacao was deposited in an inclined movable trough or cradle on whose face fifteen to eighteen pounds
were spread out at a time.   Then a carefully regulated stream
of water was allowed to run through this deposit into a lower
trough and gutter while the cradle was rocked continually.
When the water ran off clear from the lower trough, the working negro would pick out the stones in the cradle with his fingers, until only the finest pebbles remained, which he scraped
over and examined with the closest attention to detect the possible presence of diamond crystals.[1]

[1] " A Treatise on Gems," Feuchtwanger, 1867.   Report of United States
Minister to Brazil, March, 1899.

This was a slow and tedious process, at best. The percentage of precious stones won from the gravel necessarily depended on the care, expertness, and eyesight of the workers. Experience proved that fairly expert gold placer miners were not equally competent in handling diamond-bearing gravel, and slave labor was not diligent or trustworthy. The loss was increased by the greedy pressure for big and quick returns, and the premium set on the extraction of large stones.

When, in the course of mining, streams were diverted from their beds by dams and sluiceways, there was urgent need of hurrying, for the frail dams could not bear the rush of a flood in the rainy season, and it was necessary to remove the gravel from the stretches of river beds before the heavy rains fell.

Diggers' Camps on the Vaal River.

Often the formacao was buried under thirty feet or more of sand, and all this overlying mass had to be scooped up and carried off as well as the layer of gravel. As the slaves had nothing better than pans for this work, the beds were covered with swarms of negroes bearing pans on their heads and nibbling away at the ground like ants in the effort to reach the gravel before the floods came. In the reckless haste many tracts of diamond-bearing gravel were buried under ground too deep for profitable working, or covered by the waste of flooded rivers.

As the mines advanced up the hillsides, following the course of the mountain streams, it was seen that there were gupiaras or deposits of diamond-bearing gravel along the steep slopes of the ravines, and these were worked by carrying the gravel to the banks of streams, or by cutting sluiceways to the deposits. Finally, on the sierran ridges and plateaus the conglomerate beds were reached, from which the deposits in the river beds had been washed by the mountain streams. This conglomerate was chiefly itacolumite,[1] a micaceous sandstone, accompanied by mica-schist and penetrated irregularly by quartz veins. This was the prevailing composite in the Serro de San Antonio, in which the Jequetinhonha rises in the Serro de Matta de Corda, the fountain head of the Rio Francisco.[1]

Here the diamonds were not as thickly sprinkled as they were in the cascalho concentrate, but the quantity was sufficient to make extraction profitable, if the conglomerate could be disintegrated and washed. This was effected by collecting rain water in pools at points above the conglomerate and carrying down the water through ditches into gullies cut in the beds. By the flow of the water, the formacao was separated from the mass of rocks and sand. This device worked well, but owing to the scarcity of water, the washing could only be continued for a few weeks, at most, in the course of a year. In 1832 mining in these fields was opened to the public, but the most accessible and prolific beds had been worked, and there was little apparent encouragement for the investment of capital in any large undertaking which might have advanced the science of diamond winning. It is said that more than half of the diamonds produced in Brazil were stolen by the workmen and sold to contraband dealers, by whom they were secretly sent out of the country.

Outside of the Indian and Brazilian fields no considerable source of supply had been discovered anywhere. Some diamond-bearing ground had been found in Borneo, which yielded for many years a dribbling return, and in 1829 the first-known diamond of Russia was discovered on the west flank of the Ural Mountains by Humboldt and Rose, in a gold placer field near the iron mines of Bissersk. Here the prevailing rock formation, like that in the upper diamond fields of Brazil, was itacolumite, with an admixture of mica and iron pyrites.[1] The débris washed into a few valleys beneath this range yielded a meagre return to the searchers, but there was nothing to inspire any ardent working, and in Bohemia, Australia, Mexico, and the United States, the picking up of a few isolated specimens was noted as a curious occurrence rather than as the foundation of any hope of a productive diamond field.[2]

So, at the time of the discoveries of diamonds on the banks of the Vaal River, there was no known method for the extraction of diamonds beyond the shovel of the Indian, the batea of the Brazilian, or the cradle of the gold miner. There was no anticipation, on the part of the diamond seekers, of any formation in Africa except the diamond-bearing gravel of alluvial deposits, and the prospectors of the first rush did not seek for diamonds beyond the gravel along the banks of the Vaal.

The Early Mining at Klip-drift, now called Barkly West.

The first waves of the influx from the southern country and coast towns were warmly greeted by the small parties at work on the Vaal. The diggers were squatters, without any legal title to an inch of the river bank, as they very well knew. But they relied on actual possession without contest, for their rocky field was so apparently worthless that no farmer had cared to secure it. They did not trouble their heads with any questioning whether the South African Republic covered their shore line, or whether any native tribe laid claim to it, but they were

[1] " A Treatise on Gems," Feuchtwanger, 1867. " Notices sur les Diamants de l'Oural," Parrot. " Transactions of the Imperial Russian Mineralogical Society," at St. Petersburg, 1842. " De Novis quibusdam Fossilibus quae in montibus Uraliis inveniuntur," Gustav Rose, 1839.

[2] " Gems and Precious Stones of North America," Kunz, 1890.

L

so weak in numbers that they had some fear of possible attack from the neighboring Koranas and Griquas, or other natives who might covet their oxen and arms and supplies, as well as their hard-won gems.[1]   In view of the abject state of the few surviving Hottentots on the Vaal, any dread of their hostility seemed absurd, but the miners did not know how weak the natives were, and their new-found treasure unsteadied their nerves.   So they were glad to see a rally of prospectors on the fields large enough to scare off any menacing natives.

The early comers picked out irregular patches of ground here and there, to suit their fancy, and dug and strayed along the river banks as they pleased, prospecting on any unoccupied spot.   There was no precise limit to the size of any claim.   One party would pounce on a whole hillock, like the prolific " Natal kopje," and another would occupy a hundred feet or more of shore line.   There was no apparent need of jostling one another, when any square rod for miles along a river bank was as thickly sprinkled with diamonds as another, so far as any of the prospectors could judge.   Still, the known yield of the Natal kopje drew preference to locations around it, and the product of other neighboring placers was so enticing that the mass of diggers concentrated at Klip-drift.

This massing made it necessary to agree on some defined limits of ground which a man could reserve for his own working, or combine with the sections assigned to companions.   To fix and make this assignment a " Diggers' Committee " was chosen by an informal mass meeting of the prospectors, which made simple regulations controlling the working of the river diggings.   It was agreed that the size of a location should be thirty feet square, and that title should be conveyed by a certificate from the supervising committee.   The water's edge along the river was open to anybody wherever it was possible to set a trough or a miner's cradle without interfering with other ground-washing fixtures already in place, but locations might begin a few yards from the river.[2]   So there was soon a close-set fringe of

[1] " Among the Diamonds," 1870–1871, John Noble.   [2] *Ibid.*

cradles and water-troughs at the bottom of the Klip-drift bank, and the ridged and gullied slope for hundreds of yards inland was pitted with holes from ten to thirty feet square, and ranging in depth from four to twenty-five feet.   If the river shore in line with the parallel claim was too thick set with cradles to admit a new washing machine, or if the claim was high up on the bank, water for washing was sometimes carried up from the river in carts to the working ground.   Alluvial soil covered the face of the basin, more or less thickly, for a stretch of half a mile from the river, lying even on the tops of the kopjes, except

River Diggings at Gong Gong, 1880.

where rugged boulders and blocks of basalt and trap protruded stiffly above the coating of gravel.

The choice of location was largely determined by fancy, rather than any solid reason.   Some preferred light colored patches of gravel to dark, but would have been puzzled by any call to justify their choice.   Others sought for tops of kopjes, with a supposition that the rains had washed the light gravel downhill and left the heavier deposit with the diamonds on the crown of the hillocks and ridges.[1]   It was generally observed, however, that diamond crystals were most plentiful in spots

[1] "The Diamond Diggings of South Africa," Charles Alfred Payton, London, 1872.

where garnets and peridot were thickly deposited in the gravel, and this observation was in accord with current accounts of mining in other diamond fields. So the occurrence of these red and green pebbles was commonly hailed as an assurance of the presence of diamonds, and gravel so charged was washed and sorted with exceptional care. But there was no concentrated deposit in this field like the cascalho in the Brazilian river valleys, and the labor of washing the thick mass of loose gravel was necessarily great.

There were no appliances for handling and concentrating the gravel marking any noticeable advance above the slow and laborious methods of the Brazilian and Indian placer workers. The deposit was a mass of gravel and sand, thickly sprinkled throughout with heavy boulders of basalt and melaphyre which were laboriously pried and dragged out of the shallow pits sunk by the miners.[1] The mixed gravel and sand was shovelled into wheelbarrows or carts and taken to the river's edge, where it was dumped into heaps on the ground, or in troughs sunk in the bank. Then the gravel was washed in cradles, with two or three screens of perforated iron, or zinc, or wire mesh, set to form partitions with discharge holes so graduated that the larger stones were held above the upper and coarser screen, while the sand and lighter gravel flowed out through the upper and lower screen holes. Meanwhile the cradle was more or less expertly shaken to cause a deposit of the gravel of high specific gravity on the bottom between the screens. The worthless stones in the upper part of the cradle were then picked and scooped out by hand and thrown away, while the concentrate was taken out carefully and carried to the sorting table, an ordinary deal stand, or any level wooden or iron structure, or to a flat stone. Here the deposit was spread out thinly and sorted over inch by inch with a short scraper of hoop iron, or any other thin strip, while the appearance of a diamond was more or less keenly watched for.[2]

[1] "The Diamond Diggings of South Africa," Charles Alfred Payton, London, 1872.   [2] *Ibid.*

Vaal River Diggings.

This washing machine was practically the same as the Australian gold placer miner's cradle, or the American rocker, and it
had been used for years on the Brazilian diamond fields, though
the screening of the Vaal was probably more exact. But the
Brazilian negroes had become far more expert by long practice
and training than the green workers on the line of the Vaal, and
the handling of the concentrate in their bateas was extraordinarily
deft. It has been demonstrated over and over again in placer
fields that inexperienced washers cannot compete with trained
hands in concentrating gold dust, and even expert gold placer
workers often failed to handle diamond-bearing gravel efficiently.
So it is not surprising that many of the awkward adventurers
in the new fields lost heart completely at their failure to extract
any diamond from the masses of gravel which they dug and
washed so laboriously ; and it is practically certain that the percentage of gems saved, at first, was below the average winning
from the Brazilian sands.

The irregularity of the distribution of diamonds in the shore
bed was greatly perplexing and disappointing to the groping
locaters. The precious stones were strewed in the gravel in a
scattering way that defied any calculation. Here and there was
a rich patch of ground, while tracts all around it, precisely similar in a surface view, held only a few small diamonds or were

hopelessly barren.   Even in the best placers there were apparent freaks of deposit that sorely puzzled the diggers, and almost provoked the belief in the dropping of the gems by whimsical genii rather than by the play of natural agencies.   One man, working side by side with another for weeks in adjoining claims, would not find one precious stone, while his neighbor was adding daily to his little sparkling heap.   Even when claims were so split up that a digger could hardly turn about without' brushing against a comrade there was the like insolvable contrast of gem-studded gravel and worthless pebbles.   Often, too, when a claim had been abandoned by an unlucky miner, the next man who jumped into the deserted hole would unearth in a day a superb diamond, and, perhaps, wash out in a week a score more of precious stones.[1]

The miners were, as a body, so orderly, so tenacious of their own rights under the established regulations, and so prudent in restricting the possible extent of monopolized ground, that there was little "claim jumping" or bitter wrangling.   The provision against loafing or the holding of unworked claims on speculation was sufficiently sharp.   The neglect to work a claim for three days consecutively forfeited the holder's license, and the ground was then open for the issue of a new certificate to the first claimant.   For many months all unoccupied ground in the Klip-drift camp was greedily pounced upon by newcomers to the fields.   So this part of the river basin was continuously covered with a busy swarm of workers, digging, washing, sorting, driving carts, and stirring in all the daily occupations of camp life.   Where one man lost heart and went off prospecting up or down the river, or plodded wearily homewards, another was ready to take his place in a moment and continue the unflagging round of work.

It was soon perceived that such diamond placer digging was inevitably a gambling speculation, and few complained loudly of their hard luck, or bitterly grudged the success of their neighbors.   When an unusually large stone was found,

[1] "Among the Diamonds," 1870–1871, John Noble.

there was commonly a shout and a rallying of exultant friends around the lucky finder, and all through the fields a redoubled fervor of work from the spur of the signal success. Every one felt that the good fortune of a comrade might be his own the next moment, and, if this hope was cast down, the diggers toiled on with indomitable pluck and sanguine spirit, ever lifting the glittering image of better luck some day. So the rasping of shovels, the splashing of gravel, the rumbling of carts, the dumping of loads, and the rattle of cradles went on incessantly with a lively din from morning till night.

River Diggings. Waldek Plant.

For the sorting of the concentrated gravel shady spots were chosen beneath spreading tree-branches, where tables were set, or under the cover of canvas screens stretched over posts. Here the miners bent over the thin layer of gravel, scraping along the pebbles bit by bit, and gluing their eyes to the sliding stones in anxious search for the coveted tiny white crystals; or stretched out at full length on their stomachs, they scraped the gravel over the face of the boards or iron sheets laid flat on the ground. In this branch of diamond winning, where keen eyes were essential, the native blacks were largely employed, sometimes under close watch of a white overseer, and sometimes without any oversight. Part of the black sorters were strictly

Pniel Diggings, Vaal River.

faithful and honest, as was shown by test after test. One boy brought straight to his master a diamond of eighty carats, which his quick eye detected in the roots of an old stump that he had been told to dump into the river. Another returned the counterfeit stones that his employer had purposely dropped in the concentrate.[1] But all were not equally trustworthy, and many fine stones were filched from the tables by nimble-fingered sorters, even under the eye of a wary overseer. When the Boer farmers came to the fields, they often brought their families with them, and it was a common sight to see father and sons digging and washing, while the mother and daughters sat on the ground industriously picking over a layer of pebbles. Sometimes, too

Klip-drift, Early River Diggings.

[1] "Among the Diamonds," 1870–1871, John Noble. "The Diamond Diggings of South Africa," Payton, 1872.

the wives and sisters of English miners, even women who had rarely soiled their white hands before, might be seen sorting river gravel as ardently as any prospector on the line of the Vaal.

When newcomers roamed about sight-seeing over the fields, they were surprised to note how rarely their presence drew even a fleeting glance. Scarcely any one of all the groping swarm of diggers, washers, and sorters, white or black, men or women, diverted an eye for a moment from the intent absorption of the search for the tiny crystals embedded in the vast stretches of gravel. Eternal vigilance is the watchword of diamond winning as well as of liber-ty. It was keenly felt by the dia-mond seekers that a fortune might slip through their hands in the shift-ing and twinkling of an eye. So wan-dering strangers threaded their way among the burrows in the

Gong Gong.

pitted bank and the diamond sorting tables without attracting any more attention than stray pebbles rolling down the gravel heap.[1]

Whenever any one of this curious swarm found a big stone he had a prize in his hands, for the precious crystals of the Vaal river beds are exceptionally good and free from fractures. There were few stones ranging over thirty carats, but ten carat stones were not uncommon, and even the tiniest stones of one carat or less were usually well shaped. Some were lightly tinged with yellow, detracting somewhat from their market value, but there was a large percentage of stones perfectly white, or so nearly

---

[1] " Among the Diamonds," 1870–1871. " The Diamond Diggings of South Africa," Payton, 1872.

colorless as to defy any scrutiny except that of experts. Deep orange yellow stones were occasionally found, and shades of yellow grading to the finest straw color were represented as well as pale blue, brown, and pink, and other hues; but any color except white or yellow was rarely to be seen. The commonest crystalline form was the octahedron, but perfect dodecahedrons were not unusual, and twin stones or a conglomeration of crystals sometimes appeared. There was no adhering film or envelope such as commonly dulls the lustre of the Brazilian diamond crystals. The stones of the Vaal are clear and bright.[1]

Digging for diamonds never becomes dull drudgery, for there is always the glittering possibility in the mind's eye of upheaving a king's ransom with the turn of a shovel, and it is far more exciting to a novice than mining for gold or any other minerals. But the diggers on the Vaal River fields soon learned that the actual disclosure of a diamond on the face of the gravel which he was shovelling was a very rare occurrence, for only the largest stones were likely to be seen in a mass of earth and pebbles, and few even of these were actually detected in the sinking of the pits on the river banks. So the miners were rarely so absorbed in their search that they worked without stopping to eat, but they clung to the last gleams of the sun as the miners have done in the rich gold pocket placers of America and Australia. The diggers and washers went to work usually at the same hour, about sunrise, took an hour off for breakfast, and for dinner or lunch, and stopped work when the sun went down. In the hotter weeks of the African summer season (the summer — November, December, January, and February — is the hot as well as the wet season) they did little or no work in the midday, and when heavy rain and hail storms swept over the fields, all sought for cover.

---

[1] "South African Diamond Fields and Journey to Mines," William Jacob Morton, New York, 1877. "The Diamond Diggings of South Africa," Payton, 1872. "Diamonds and Gold of South Africa," Mitchell, 1888. "To the Cape for Diamonds," Frederic Boyle, London, 1873. "Diamond Fields of South Africa, by One who has visited the Fields," New York, 1872.

THE AUTHOR'S COLLECTION OF DIAMONDS.
Drawn from Nature.

Camping on the banks of the Vaal was rarely unpleasant to any one accustomed to a life in the open country, and even the townsmen found little to grumble about. As soon as they reached the Diamond Fields, the prospectors looked about for good spots on which to lodge their wagons and pitch their tents. Some took to the fields small circular or "bell" tents, but the greater part preferred a square or oblong "wall" tent, commonly ten feet long and eight wide. From a central ridgepole, propped at each end, the canvas roof was stretched to side posts four feet high, from which flaps hung to the ground. This shelter served as a home for two or three men, and a storehouse for their stinted outfit. It was not spacious, but even a little tent was a welcome change from the cramped bunking in mass under a wagon cover, and the airy, clean, canvas chamber was

Washing Diamond Gravel by Machinery at Gong Gong, 1880.

much pleasanter than the ordinary farmer's sleeping room, as many of the prospectors remarked from experience. Even when the campers were obliged, for lack of tents, to sleep in their wagons, the big arched wagon did not suffer by comparison with any Boer's hut on the veld. The tents were pitched, sometimes under the cover of the larger trees lining the river bank, and sometimes on sheltered slopes, but the mass at Klip-drift were bare to the sun, and exposed to the blast of every storm that tore through the valley.

Often these storms were terrific, opening with the rising of a yellow streak above the horizon, and the rapid spreading over the blue sky dome of rolling masses of heavy, lurid clouds.

Lightning at Kimberley.

Then from the coppery bosom of this pall there came such blazing streams of lightning in sheets and contorted shafts, such rending explosions of thunder peals, that the awful flare and crash would shake the nerves of hardened men. With this

Day View, Same Scene.

appalling discharge there poured from the clouds torrents of rain, or a volley of huge hailstones rattling on the canvas roofs

and driving man and beast to the nearest shelter.[1]   As a safeguard from these electric bolts the miners commonly put iron
lightning rods alongside of their tent poles and insulated them
with the necks of glass bottles, but the insecurity of this shield
was evident in the occasional shattering of a tent and the killing
or maiming of its occupants.[2]

Except for these storms the climate of the Vaal valley was
generally agreeable.   The winter days were particularly pleasant,
for the sun soon warmed the air even when the nights were so
cold that ice formed on the face of water-troughs.   In midsummer the days were often exceedingly hot, the mercury rising as
high as 100 Fah. in the shade; but the dry air was not nearly
as enervating as the humid atmosphere of summer days in
Europe or America, and the lightly clothed miners, avoiding the
midday glare, suffered little.   There was a notable exemption
from sickness throughout the year, except for diarrhœa and
dysentery, and fever contracted in summer chiefly from the reckless use of unboiled and unfiltered river water.[3]

Plain food of some kind was plentiful and cheap, especially
maize meal, commonly called mealie meal, and mutton and game
were brought into the camp from the neighboring Transvaal and
Free State farming and pasture lands.   There were many wild
fowls, too, that flocked to the valley of the Vaal, and several
kinds of food fish abounded in the river, especially one resembling the voracious English barbel, or the catfish of America, and
the one which the miners called "yellow fish."   The chief lack
in the food supply was cheap and wholesome vegetables — for
the dearth of these and the excess of meat caused a mild form of
scurvy to appear in the camp.   Fuel for cooking was readily cut
from the trees along the river bank or from the thickets in the
ravines.[4]

When the choice locations on the Klip-drift bank were taken,
the influx, continuously moving to the new Diamond Fields from
the coast, spread up and down the river, and little camps sprang

---

[1] The *Diamond News*, Klip-drift, Nov. 4, 1871.   [3] *Ibid.*
[2] "The Diamond Diggings of South Africa," Payton, 1872.   [4] *Ibid.*

up at Gong Gong, Union Kopje, Delport's Hope, Forlorn Hope, Niekerk's Hope, Blue Jacket, Waldek's Plant, Larkin's Flat, and other placer diggings, extending from Hebron twenty miles northeast of Klip-drift to Sefonell's, sixty miles west.[1]   It has been estimated that ten thousand diggers, white and black, were stretched along the river in this string of camps, and in roving parties of prospectors.[2]   Any possible reckoning of the extent of a rush of thousands, which nobody could measure exactly or tried to measure, was of course a rough guess, but it seems probable that this guess was not very far from the fact.   Such an influx of restless adventurers, pouring along a river line in a thinly peopled territory in the heart of South Africa, as heedless as a locust swarm of any questions of state sovereignty, or native tribal reservations, or mineral right titles, was certain to raise a rumpus, if any official authority in South Africa undertook to drive them away, or exact heavy license fees, or even to hold them down under strict laws rigorously enforced.   The Australian gold fields had furnished some highly significant object lessons enforcing this certainty, but the little Boer Republics were not disposed to learn any lesson from the experience of English Colonies.

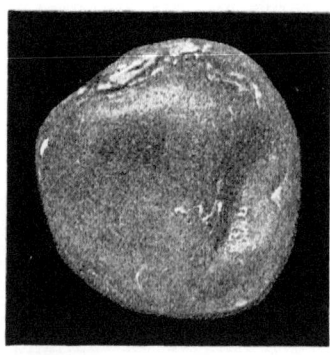

The Largest River Diamond ever found in South Africa. Weight, 330¼ Carats; Value, £3,500.

The South African Republic claimed the diamond placer border north and west of the Vaal as part of its territory, but it was content, at first, with the bare assumption that the diggers on the northern and western bank were within the confines of its domain, without caring to assert its right of control by any marked interference with the free proceedings of the diggers. It did not regard the upturning of gravel on its border line as any menace of serious intrusion within its territory, and the

[1] "Among the Diamonds," 1870–1871.
[2] "South African Diamond Fields," Morton, 1877.

neighboring Boer farmers were generally well pleased with the opening of ready markets for their produce. Representatives of the Republic were recognized as officers of the law at Hebron, but there was little attempt to impress any recognition of its authority on the camps farther down the Vaal.[1]

So the miners at Klip-drift went on digging and scraping the gravel, under their own simple regulations, month after month, until their busy camp burst suddenly into an uproar, when the news came in that President Pretorius and the Executive Council of the Transvaal Republic had granted to a firm of three privileged persons the exclusive right to search for diamonds in the territory of the Republic for a term of twenty years from June 22, 1870, subject to a royalty of six per cent upon the value of all diamonds discovered.[2] There were some old Australian placer miners on the Vaal River Diamond Fields, and they doubtless grinned at the thought of the reception that such a proclamation would have met with at Bendigo and Ballarat; but it was not necessary for an adventurer to have had a rearing on any gold placer field to fire his spirit to revolt against an edict of dispossession and monopoly. It is idle to debate the question of the technical legal right of the administration of the South African Republic to make this grant. This may be conceded without affecting the countering facts of its gross partiality, inexpediency, and practical futility. The whole regular army of the United States would have been too small to enforce any such disposition of its mineral lands after they had been occupied without protest for more than six months by squatting placer miners, and bare common sense would have sufficed to inform the administration of the little South African Republic that it could not give effect to its paper monopoly without a succession of fights that would add another " Blood River " to the face of South Africa.

The instant effect of the grant was a universal uprising and mass meeting of the Klip-drift camp, and the declaration of the foundation of another free and independent Republic on the Vaal,

[1] " Among the Diamonds," 1870–1871.
[2] " South Africa," George McCall Theal, 1888–1893.

of which Theodore Parker, one of the leading adventurers, was chosen president.[1] This was, on its face, a proceeding that smacked of opera bouffe, but, like Janus, it had another face. It was a flaunt of determination to cut off every shred of political connection with the South African Republic, and hold possession of a slice of rich mining land with a Colony which, at some future time, if not immediately, Great Britain might be disposed to welcome and incorporate with her imperial cluster on the coast. If this hope was not openly avowed at first, it undoubtedly existed in the minds of many of the diggers, and no time was lost in communicating the situation to her Majesty's High Commissioner at the Cape, Lieutenant General Hay.

It is, however, unlikely that there was any confident expectation of the endurance of the new Republic founded on a gravel bank whose precious contents were fast fleeting, but the organization was set up as a handy resort, on the spur of the moment, to make an imposing show of resistance to the authority of the South African Republic, and with the idea of shunning the penalty of forcibly contesting the execution of the monopoly grant within a recognized district of its domain. Whatever legal unsoundness there may have been in the construction of the Klip-drift Republic, and in the notions of its framers, the shaky ship of state served its main purpose. The administration of the Transvaal Republic realized their grave blunder too late, and being humane and peace-loving men, refrained from any attempt to maintain their grant or their contested authority by force of arms. But they complained earnestly to the British Colonial authorities of the intrusion and illegal occupation and insubordination of the squatting adventurers on the Vaal.

Meanwhile the diamond diggers did not concern themselves with the remote vexation of the Boer President and Council, but kept on ransacking the gravel. Early in the year there had been some straggling prospecting on the Pniel bank opposite Klip-drift, but the first continuous work on a south bank placer

[1] "South Africa," George McCall Theal, 1888–1893. "Among the Diamonds," 1870–1871.

was begun in June by a party from the Klip-drift camp.[1]   Their undertaking was an unwelcome intrusion on land claimed by the Pniel Mission, and the diggers were warned of their trespassing by the clergyman in charge.   The Mission Station was several miles from the diamond placer, and the diggers ignored the notice, as they were not interfering apparently with the mission work by washing river bank gravel.   The placer ground proved so rich that the diggers flocked to it rapidly, and the Berlin Society which maintained the missions at Pniel and Hebron was soon glad to obtain the license fee which it was generally able to secure from the diggers on the Pniel field.   The preferred locations on the Pniel bank were along a stretch in the middle of the rising ground, a few yards from the water's edge. In this tract diamonds were strewn so continuously as to suggest the existence of a flow or stream of them, in the red drift gravel between the boulders, to the eye of more than one observer. This strip was soon honeycombed with shallow pits reaching bedrock about twenty-five feet below the surface.[2]

The flow of prospectors continued to spread until the Pniel camp, in a few months, rivalled Klip-drift in size, and the two contained a population of four or five thousand people.   Small stone, brick, and iron buildings for stores and other business uses were quickly put up in rows along a main street in the heart of Klip-drift camp, which bore the name of Campbell Street, and a few others of the same durable materials rose from other spots in the fields, but most of the miners continued to live in their canvas tents, or in reed huts plastered with clay. The stone for building was readily obtained from neighboring hillsides, and was neatly cut and laid, so that Campbell Street soon compared favorably with any country town street in South Africa.   Butchers, bakers, and grocers opened shops; restaurants offered good, plainly cooked food at charges so moderate that it was reckoned that a man could be well fed at a cost of 2s. 6d. a day; a tavern and lodging-house, dignified by the name of hotel, accommodated travellers and regular boarders; diamond

[1] "Among the Diamonds," 1870–1871.   [2] Ibid.

M

brokers sat ready to judge and buy rough diamonds for export; a music hall had a rude vaudeville show every week-day night; members of the Masonic fraternity established a lodge; and a little brick church welcomed all comers to its Sunday services.[1]

Similar buildings were put up less regularly in the Pniel camp too, and both sides of the river showed the like medley of iron, brick, stone, light wood, and canvas stores and dwellings. The first mining-town newspaper in South Africa,

Views of Klip-drift.

the *Diamond News*, was started at Pniel, — a little four-page sheet that was chiefly filled with advertisements of local tradesmen on both sides of the river, and the local news and stir of the river diggings. Rowboats of an established ferry made regular trips across the river from one camp ground to the other, charging a passenger sixpence for crossing. So there was easy communication, and the two camps were one in their common appearance,

[1] "Among the Diamonds," 1870–1871. "The Diamond Diggings of South Africa," Payton, 1872.

work and sympathy, though the Pniel camp did not pretend to the dignity of an independent Republic, but submitted meekly to the payment of license fees to the Berlin Mission Society and to the assertion of the sovereignty of the Orange Free State, represented by a local magistrate, with the adjuncts of a canvas jail, whipping-posts, and stocks.[1]

Oddly enough, in view of the shallow gravel bed which was the sole support of these camps, the approach of collapse was not clearly foreseen. An observer of more than ordinary intelligence visited the camp at the close of the year 1870, and noted the exhaustion of the rich ridge gravel back of Campbell Street, where more than two thousand diggers were at work a few months before. Yet, while remarking the drift of prospectors to outlying placers, he wrote, " Notwithstanding this, Klip-drift flourishes, and together with Pniel will no doubt always continue to be a head centre of the diamond-digging community." For this sanguine view there was some justification in the general ignorance of the actual extent of the diamond beds in the alluvial deposit, and in the common declaration of a purpose to persist in searching for diamonds, even by those whose hard luck forced them to abandon the fields for a time. " Hope's blest dominion never ends " to the most unfortunate laborer. This visitor did not meet one of the many leaving the ground with empty pockets who did not protest his resolution to return to the diggings in the following March or April after the heat and storms of the summer season on the Vaal were past.[2] Fortunately for these luckless adventurers, there was a new and phenomenal development of other Diamond Fields, whose output soon dwarfed all the returns from the shallow River Diggings.

[1] " Among the Diamonds," 1870–1871. " South Africa," George McCall Theal, 1888–1893.

[2] " Among the Diamonds," 1870–1871.

CHAPTER VI

THE RUSH TO KIMBERLEY

HERE was a pretty green valley near the Free State settlement of Fauresmith, hardly a mile in width, but stretching for several miles to the northeast through ridges of volcanic rock kopjes. Fauresmith lay in the track of the stream flowing from the coast ports to the diamond-bearing valley of the Vaal, but there was no thought of a probable diamond field on the plateau so far from a river bed. So for months the adventurers passed on without pausing, except for a night's camp, on their way to the Vaal. A Boer settler, Cornelis Johannes Visser, had taken up a considerable part of the neighboring valley in his farm of Jagersfontein, where his house stood in the midst of a gay, blooming garden. He had died before the discovery of diamonds, but his farm was held by his widow, Jacoba Magdalena Cecilia Visser, and worked by an overseer in charge.

A little stream, flowing from the hills, ran through the valley in the rainy season, though for the greater part of the year its track was only marked by a spruit or dry water-course. De Klerk, the overseer, noticed that many small garnets mixed with pebbles of agate were sprinkled along the dry bed of this spruit, and learned that the diggers on the Vaal believed garnets to be an indication of the presence of diamonds. So he began prospecting one day in August, 1870, on the line of the spruit, awkwardly sifting the dry gravel and sand in a common wire sieve. At the depth of six feet he found a fine diamond of fifty carats, and the news of his discovery was soon widely spread throughout the Free State.[1]

[1] "Among the Diamonds," 1870–1871.

His neighbors flocked first to the farm, and the thrifty widow Visser was pleased to welcome them, and permitted them to dig in her spruit, on allotted patches of twenty feet square, for which each paid her a license fee of £2 a month. The phlegmatic Boers were not wildly excited by the prospect of fortune hid in the spruit, but diamond hunting was an agreeable diversion from dull farming, and they came with their wives and children in their big canvas-covered wagons, and spread out through the green valley like country folk at a picnic. The children delighted in their search for pretty pebbles and soon filled their pockets with garnets and agates; but the digging in the spruit was often so laborious that the farmers were content to squat on the ground and puff their long pipes while their black servants did the digging and rock heaving. When natives were not engaged as diggers, the farmers and their sons indolently shovelled out the gravel in heaps to be sorted by their wives and children.

Underneath the red surface soil filled with pebbles, there was a layer of calcareous clay, varying in thickness from a few feet to twelve or more, covering drifts and pockets of gravel thickly sprinkled with heavy boulders of greenstone and basalt. It was necessary to pry up and tug out these boulders in order to reach the underlying gravel, and this task was no child's play. Then the gravel was pitched out of the holes, rudely sorted by dry sifting in sieves, and picked over by hand in search of the precious stones. In some pockets there was quite a sprinkling of diamonds, garnets and peridot, mixed with coarse gravel, and the returns far exceeded the license charge; but the diamond deposit was scattered as irregularly as that of the Vaal River field, and many of the workers toiled for weeks on their claims without finding anything more precious than the jawbones and teeth of a hyena or jackal.[1]

Attention had hardly been called to the diggings at Jagersfontein when a still more remarkable discovery was made in the month of September, 1870, at Dutoitspan,[2] on the farm of Dorst-

---

[1] " Among the Diamonds," 1870–1871.

[2] The original and correct form of this name was "Du Toit's Pan," or the pan or pond of du Toit, the name of the man who first owned the farm. Both Du Toit's Pan and Dutoitspan are now used.

fontein, about twenty miles southeast of Pniel and Klip-drift on the Vaal.   Du Toit's pan was one of the curious natural land basins before described, receiving the wash of the surrounding ridges, and holding pools of water during the rainy season and sometimes during the year.   The title to the farm Dorstfontein was granted by the Free State Government to Abraham Pauls du Toit on the 4th of April, 1860.   Du Toit sold the farm to Adriaan J. van Wyk, who had built a little house near the side of the "pan," where he was living indifferent to the rush of prospectors to the Vaal River, until he was suddenly surprised by the finding of diamonds a short distance from his house.

When the news of this discovery spread, coming, as it did, so close upon the revelation at Jagersfontein, there was an instant rush of prospectors from the Vaal to the new field, swelled by the neighboring farmers and the influx still flowing from the coast towns.   Van Wyk demanded, at first, a royalty of one-fourth of the value of all diamonds found on his farm, from every prospector seeking to explore the new field ; but he soon concluded to issue licenses at a charge of 7s. 6d. monthly for every allotted claim of thirty feet square.   The Orange Free State government was aroused to assert its claim of sovereignty by the spread of the discoveries, and attempted to restrict the allotment of the claims on the farm land, for the benefit of its own citizens, by an ordinance prohibiting the issuance of licenses to any one except a Free State burgher or farmer ; but this requirement was easily evaded at Jagersfontein and Dutoitspan by the transfer of licenses granted to Free State citizens.   Furthermore, the spread of the news of the discovery and the resultant rush to the Diamond Fields was soon beyond any possible restriction imposed by this little Republic.[1]   Van Wyk was prevailed upon without much difficulty to sell his farm to the predecessors of the London and South African Exploration Company for £2600, a fortune far surpassing any glitter of pebbles in the ground, in the view of this simple farmer.

Side by side with the Dorstfontein farm lay the farm of Bult-

[1] "South Africa," Theal, 1888–1893.

fontein, divided by a public roadway.   The spread of prospecting soon passed naturally across the road to Bultfontein and to other neighboring farms.   Bultfontein was owned by a poor Boer, Cornelis Hendrik du Plooy, and before the discovery at Dutoitspan a thousand pounds would have been thought a grossly extravagant price to pay for the whole farm and its live stock.   But the luck of van Wyk put a new face on the scrubby farm lands near the Vaal, and an eager Free State speculator, Thomas Lynch, did not wait over Sunday to buy Bultfontein, but amazed the owner by driving out to his farm on the Lord's Day, November 14, 1869, and offering £2000 for his land.   Du Plooy accepted the offer on the spot, for such a sum in cash was vastly bigger in his eyes than any possible return from farming or picking up "blink klippe."   It is said that diamonds had been found on the farm previous to this sale, but Du Plooy was not aware of any actual discovery on his land, and preferred cash in hand to any gambling chances.   The story is told that Bultfontein mine was discovered by the finding of a diamond in the mortar used by du Plooy to plaster his house and the subsequent search for diamonds in the pit from which the sand had been taken.   It is true that diamonds were found as reported, but it was some time after the mine had been rushed.[1]

On the same day that du Plooy sold his farm to Lynch, he was beset by Leopold Lilienfeld and others, who advised him that the sale was illegal, being made on a Sunday, and eventually Lilienfeld gave du Plooy an indemnity against all damages if he would refuse to conclude the sale to Lynch. On November 16, 1869, the sale of the farm was concluded between du Plooy and Leopold Lilienfeld, Louis Hond and Henry Barlow Webb for the sum of £2000.   Hond sold his one-third interest to Webb, who, with Lilienfeld, Edgar Eager Hurley, and others, formed the "Hopetown Company."

Lynch brought action against du Plooy for £10,000 damages, and obtained a judgment for £500 and costs on August 19, 1872. In spite of his indemnity du Plooy was then obliged to sue

1 "Among the Diamonds," John Noble, 1870–1871.

Lilienfeld and his associates, and obtained judgment for £760 19*s*. 1*d*. and costs, February 12, 1893.   In 1876, when the Land Commission heard this case, the London and South African Exploration Company had been formed, and the title to the farm was granted to that company, as successors of the " Hopetown Company."

Bultfontein was linked to Dorstfontein by the acquisition of both farms by one holder, and transfer in a subsequent sale to investors associated as the London and South African Exploration Company.   The farm of Vooruitzigt, which lay bordering on Dorstfontein and Bultfontein to the north, was bought for £6000 shortly after by other speculative investors, — the firm of Messrs. Dunell, Ebden & Co., of Port Elizabeth.

The correct record of these farms is as follows : —

Bultfontein was originally granted by the British Government (then occupying the Free State under the name of " The Orange River Sovereignty ") to J. F Otto, December 16, 1848, under Warden certificate.

Dorstfontein was granted by the Free State Government to Abraham Pauls du Toit on the 4th of April, 1860.

Alexandersfontein was granted by the Free State Government to Johannes Cornelis Coezee on the 3d of December, 1862. That portion cut off by the Free State boundary from Griqualand West was granted to Philip Rudolph Nel and Willem Gabriel Nel on the 16th of January, 1880.

Vooruitzigt was originally a portion of Bultfontein, and was sold to D. A. and J. N. de Beer on the 18th of April, 1860.

At the time of these purchases the price paid for any ground outside of a short stretch on the Dorstfontein farm was wholly speculative.   There had been no considerable discovery of diamonds except along the top of a sloping ridge or long kopje lying north, at a distance of about a third of a mile, from du Toit's pan.   The total area of the three farms was about fifty-eight and a half (58½) square miles.[1]   The comparative ease of

---

[1] The total area of the farms, Dorstfontein (6579 acres), Bultfontein (14,457 acres), and Vooruitzigt (16,405 acres), is 37,441 acres, equal to 58½ square miles.

Kimberley, 1872.

working in the new fields was a pleasant surprise to the River Diggers, who had been obliged to sink pits in heavy gravel thick set with boulders. Now they found diamonds sprinkled through a light surface soil of decomposed yellow ground, and many stones were so thinly covered with earth that some little brilliant crystals were washed free from sand after every heavy rain, and lay shining on the ground, to be picked up by sharp-eyed diamond seekers.[1] The mines were not covered with basalt, but in many cases with a layer of rather hard limestone or calcareous tufa similar to that which covers a large part of the surface of the country in this neighborhood, which has been metamorphosed by the evaporation of water charged with carbonate of lime.

The first swarm of prospectors on the ground supposed that the diamonds of Dutoitspan were simply a sprinkling strewn through a sand wash like the river-shore deposit. When their shovels struck an underlying stratum of limestone with streaks of greenish shale, at a depth of two feet or less, they presumed that this corresponded to the known bedrock of the placers along the Vaal, and had no thought that it was a casing for any precious stones. So they simply dug through the soil and shovelled the ground into heaps to be sifted dry with common wire sieves of coarse and fine mesh. There were no boulders in this soil and few large stones, so that their claims could be rapidly worked.[2]

The ground contained a plentiful sprinkling of small yellowish diamonds and some larger stones, but the deposit was so shallow that it soon was exhausted. In the course of a week or two one digger with the help of a sorter shovelled and sifted all the ground of his claim, thirty feet square, and moved to another, or rambled off prospecting over the farm lands.[3] There seemed

[1] " The Diamond Diggings of South Africa," Payton, 1872.
[2] *Ibid.*   [3] *Ibid.*

no prospect to him that the Dutoitspan ridge still held anything to reward the labor of penetrating a rock bed. But after many prospectors had ransacked the soil of their claims and abandoned them, one of the workers on the ridge or elevated land had the fancy to see what might possibly lie under the stratum of limestone, and determined to cut a few feet, at least, through the rock. He found that the limestone soon grew so soft and rotten that it could be split easily by the stroke of a pick and the lumps crushed by the blow of a shovel. This rotten rock fused soon with a curious decomposed breccia of a yellowish color, and the sifting of this ground showed, to his amazed eyes, the presence of diamonds sparkling on his sieve or on the sorting table.[1]

With the spreading of this discovery there came another rush of diggers to the ridge that soon covered every patch of unoccupied ground on its slopes. Foot after foot the mining pits sunk through the soft cement, which was often so decomposed that the point of a pick pierced it like a mass of dried mud. Instead of decreasing in number, the quantity of gems in a claim often increased with the deepening of the pits, and the proportion of large rough diamonds was far greater below the depth of a fathom than in the surface soil or the crust of the limestone stratum. Payton says that fragments of volcanic rocks — green trap and basalt chiefly — were scattered through the limestone and yellow ground; but there were very few large boulders, and the work of mining was far less laborious than any pit-driving in the river bank at Klip-drift and Pniel.[2]

Some cut adits at varying angles in the slope of the ridge, and carried out their ground in buckets or wheelbarrows. This method of mining shunned the toil of lifting heavy buckets out of the pits, but it was dangerous from the frequent ground slides and rock falls, and caused many a wrangle when adit lines crossed or pits met the tunnels. Others opened their claims by cutting a series of descending stages, diminishing in size step by step, so that the pit bottom was reached by passing down a

[1] "The Diamond Diggings of South Africa," Payton, 1872.   [2] Ibid.

rude rock staircase. This was a rapid and convenient mode of opening ground at the start, but where claims were only thirty feet square, it was clear that no single claim-holder could go far down in this way without reaching a point where the bottom step of his staircase would cover the floor of his claim. For this reason many preferred to mine more slowly in small perpendicular shafts, in whose side little niches, familiarly known as toe holes, were cut, so that agile men could clamber up and down. Or the shaft bottom was reached by means of a knotted rope or riem of rawhide, dangling into the pit from a post set in the ground near the mouth of the shaft. When a bucket was filled with broken rock by a digger working on a pit floor, his mate hauled up the load by winding a rope stretching from the handle over a rude windlass, or by sheer lifting. When only one digger was holding a claim, he was obliged to clamber out of his pit and haul up his bucket whenever he filled it.

To extract the diamonds the broken rock was pulverized by beating with shovels and then screened in a common round sieve of coarse mesh, to separate the larger stones that were worthless. After this screening the ground passing through the coarse wire mesh was carefully sifted, a second time, in a rocking sieve of fine, strong wire. This sieve was set in an oblong frame, usually about three feet long and two broad, with handles at one end and deep notches at the other, gripping a narrow strip of rawhide stretched between two upright posts called sieve props. When this rocker was swung rapidly, all the sand and dust fell through the wire mesh, leaving a concentrate of fine chips and little pebbles of limestone, talc, basalt, and trap, carrying a sprinkling of garnets, peridot, and an occasional diamond crystal. This concentrate was then taken to a sorting table and scraped over in the same way as the river gravel.[1]

Diamond winning on the upland was easier, at first, than working the river placers; but there was one common annoyance which was much more irritating on the new fields than at

[1] "The Diamond Diggings of South Africa," Payton, 1872. "Among the Diamonds," 1870–1871.

the river diggings.   Hot winds blew the red dust from the surrounding veld in clouds over the workers, and these dust blasts
were mixed with the powdered white limestone and pulverized
cement of the ridge, shaken through the sieves and blown in
the faces of the miners, inflaming their eyes, clogging their
noses, and even coating their skin through their clothes.   So
fine was this powder and so sharply blown that it penetrated
even hunting-case watches, and few watches could be kept running after a month at the diggings of Dutoitspan.[1]

But this was comparatively a trivial concern to ardent diamond seekers, winning the precious stones so frequently.
Every day swelled the rush of adventurers to the pan, bargain-
ing for halves, quarters, and even eighths of a claim on the
ridge, and roaming over every foot of ground of Dorstfontein
and the neighboring farms of Bultfontein, Vooruitzigt, and Alexandersfontein in search of new diamond beds.   Oddly enough,
as the prospectors thought, no spot on the whole farm of Dorstfontein rewarded their search outside of the ridge near the pan,
and for months no better luck attended the hunting for diamonds over the neighboring farms.   But where one party of
the ardent seekers failed to find diamonds, another followed on
its track and scoured the face of the farms with shovels and
sieves, with a persistence that was certain to be rewarded, in
time, if any diamond surface beds existed outside of the ridge
at Dutoitspan.   In the frequent sinking of pits, also, in the
basins, for water, there was the further chance of piercing some
hidden bed of diamonds, for the search for springs was hardly
less keen than the quest for precious stones.

So, early in 1871,[2] diamonds were unearthed in the surface
soil close to the farmhouse of Bultfontein.   This discovery was
followed in the first days of May by the discovery of diamonds
on de Beer's farm, Vooruitzigt, about two miles from Dutoitspan.[3]   Two months later a second diamond bed was uncov-

---

[1] "The Diamond Diggings of South Africa," Payton, 1872.   [2] Ibid.

[3] Ibid.   [These dates differ somewhat from those given by Theal and others.
Payton was on the ground in July, 1871, and his account should be most accurate.]

The Homestead of the Farm Vooruitzigt on which are De Beers and Kimberley Mines.

ered on the same farm, lying on a gently sloping kopje, at a distance from the first location roughly reckoned at a mile. This kopje had been searched twice by prospectors, it is said, without success, and one report says that the deposit was finally discovered through the sinking of a well on the ground.[1] The diggers drove their well down seventy-six feet without finding water, but at this depth one was amazed to see a diamond of eighty-seven carats sparkling on the wall of his dry pit.

So many conflicting state- ments have been made as to the discovery of the first dia- mond at this location, called New Rush or Colesberg Kopje, and afterward famous as Kim- berley Mine, that I have been perplexed to decide to which story the most credence should be given. The difficulty in ob- taining trustworthy data arises from the fact that few of the original diggers are still alive, and that most of those who are still living are scattered to all parts of the world. More- over one cannot always rely upon the accuracy of the mem-

Mrs. Rawstorne.

ory of the old diggers now living upon the Fields as to dates and details after the lapse of more than thirty years. After diligent sifting of all reports and records, however, the following conclusion may be said to be well determined.

Through the courtesy of Mrs. Grimmer, the widow of Dr. Grimmer, a practising physician at Colesberg when the Diamond Fields were discovered, I was enabled to meet Mrs. Raw- storne, the mother of Fleetwood Rawstorne, then (1900) living at Cape Town. She is a fine-looking old lady, as her portrait

[1] "The Diamond Diggings of South Africa," Payton, 1872.

shows, well preserved after a long and eventful life in South Africa.   She was eighty-two years old at the time of our meeting. Her memory took her back to the days of the discovery, and she related the incidents of the Fields as clearly as if they had happened but yesterday.   The photograph, here reproduced, of the discoverer of Kimberley mine and his party was taken a few days after the discovery of diamonds on Colesberg Kopje. Fleetwood Rawstorne stands in the middle of the group (page 175), in the shade of a fine specimen of the camelthorn trees

which grew upon the mine. They had only begun to dig prospecting holes.   The cut on page 176 shows the primitive method of working the diamond-bearing ground.   I had the story of the discovery also from Mr. T. B. Kisch, who states that he is the only one now living of the first four locators.

Fleetwood Rawstorne, T. B. Kisch, and two other diggers were prospecting on this kopje during the month of July, 1871. Some of the party thought they saw "indications" of diamond deposits, and Rawstorne sent his Kafir servant to prospect thor-

Mr. T. B. Kisch.   (The only one now living of the first party who located claims on Kimberley Mine.)

oughly the spot in view.   The Kafir returned to his master with a diamond of about two carats weight.   This discovery was made known at once to the other members of the party, and all went immediately to the spot and marked and pegged off their claims ; Rawstorne pegging three, two as a reward for discovery and one as a digger.   After the claims had been pegged off Rawstorne went to the authorities and reported his discovery.   On the following day the government surveyor was sent to mark off the claims and allot them according to the existing law or custom.

Kimberley Mine just after the Discovery, July, 1871

The name of Colesberg Kopje was given to the hillock because the lucky diggers, headed by Rawstorne, came to the field from the town of Colesberg, near the Orange River. The instant flocking of people to the two Vooruitzigt farm diggings caused them to be roughly distinguished as " De Beers Rush " or " Old De Beers," and " De Beers New Rush," or the " Colesberg Kopje " — names which endured some months, until the " New Rush " was rechristened Kimberley in honor of the British secretary for the colonies.

This inroad of squatting prospectors was greatly vexing at first to the owners of the diamond-bearing farms. It disturbed the use of the ground for stock-raising purposes, and if there were any diamonds on the land, the purchasing speculators wanted to hold the beds for their own exclusive development and profit. But it was soon evident that this design was impracticable. The swarm that covered the ground could not be held in check by any force at command of the owners, and stiffly refused to recognize any assertion of legal claims that took the form of monopoly titles.   The first diggers on the Bultfontein farm were warned off by the owners for trespass.   There was a momentary hesitation till the rush was swelled by numbers so large that the forbidden ground was "jumped" in an hour, and diggers upturned the soil to the very door of the farmhouse.   Then the owners called on the Orange Free State police for help, and the miners were driven away for some days ; but the certainty of another irresistible rush was so ominous that, toward the end of May (1871), the proprietors opened the field to all comers on payment of a license of ten shillings a month for each claim of thirty feet square.[1]

In the grants of farms by the Dutch East India Company there had been no reservation of mineral rights, but from the time of the cession to Great Britain, MacNab says the grant of lands did not carry a title to " precious stones, gold, and silver," which were explicitly excluded, and in 1860 it was enacted in

Kimberley Mine just after the Discovery, July, 1871.

[1] " The Diamond Diggings of South Africa," Payton, 1872.   " South Africa," Theal, 1888-1893.

Parliament that no lands containing valuable minerals should be considered waste lands of the crown for purposes of sale. This did not apply to Griqualand West, as there was no mineral or precious stones act or ordinance in force in this territory until Ordinance No. 3 of 1871, of the Orange Free State Government.[1] Whether there were mineral reservations in the diamond-bearing-farm deeds was not questioned by the inrushing diggers. They would not suffer exclusion without a fight, but they were willing to pay small license charges to the farm owners for the privilege of working allotted claims. The size of these claims was fixed by agreement with representative "Diggers' Committees," chosen by the prospectors in mass meeting, and these committees determined also the simple mining regulations and camp rules. One committee had charge of the Dutoitspan and Bultfontein mining camps, and another directed the mining at De Beers and the Colesberg Kopje, pitching its official tent midway between these two diamond beds.[2]

Fleetwood Rawstorne.

The Orange Free State claimed the new diamond fields as part of its territory, but its right of control was not vigorously asserted in practice. There was a rising issue from the time of the discovery at Dutoitspan touching the ownership of the district containing the diamond-bearing farms and the diggings on the line of the Vaal. The South African Republic claimed

[1] "On Veld and Farm," Frances MacNab, London, 1897.
[2] "The Diamond Diggings of South Africa," Payton, 1872.

N

the ground north and west of the river, as before noted, but the miners at Klip-drift had continued to maintain their rude Republic or independent camp, drifting into a condition verging on anarchy, under the doubtful control of a factious " Executive Committee," until December 13, 1871, when the camp gladly submitted to the authority of a provisional magistrate, appointed by Lieutenant General Hay, her Majesty's High Commissioner.[1] This energetic official had his eyes widely open to the possible value and extent of the new diamond-bearing field, and was not only disposed to sustain the appeal of the river diggers against the monopoly grant of the Transvaal Republic, but wrote to President Brand, the head of the Orange Free State, in September, 1870, questioning the title of the Free State to the Dutoitspan fields and the river diggings at Pniel.[2]

At the time of the creation of the Orange Free State out of the domain included in the Orange River Sovereignty, there had been explicit recognition of reservations set apart for the Basutos, Koranas, and Griquas, — native tribes dwelling within the limits of the Sovereignty.   But there was an apparent lack of precision in the reservations or claims of the Koranas and Griquas especially, which was accounted of little consequence at the time, until the discovery of diamonds, on a tract otherwise not worth contesting, aroused rival claimants.   The Berlin Mission Society claimed the diggings at Pniel on the strength of a deed of sale of part of the Korana reserve.   Nicholas Waterboer and other Griqua chiefs, doubtless prompted by speculative agents, set up their claim to a considerable stretch of ground, covering Klip-drift and Pniel as well as the upper angle between the Orange and the Vaal, containing the diamond fields of Dutoitspan and the surrounding farms.   The Orange Free State did not dispute the right of the natives to hold such reservations as had been assigned to them by the British Government, but contended that the stretch of the native tribal claims was wholly unjustified, and that Pniel and Dutoitspan were clearly within the bounds of its domain.[3]

[1] " The Diamond Diggings of South Africa," Payton, 1872.
[2] " South Africa," Theal.   [3] Ibid.

Some of the Native Chiefs dealt with by Mr. Richard Southey, Lieutenant Governor of Griqualand West, during his Administration.

Sir Henry Barkly succeeded Lieutenant General Hay early in 1871 as her Majesty's High Commissioner and Governor of Cape Colony, and was expressly instructed by Earl Kimberley, the British secretary for the colonies (January 24, 1871), not to countenance any annexation of territory outside of the uncontested limits of Cape Colony, which the Colony would be unable to govern and defend with its own unaided resources. But the new High Commissioner — viewing the situation and the course of his predecessor, which he cordially approved — replied to his instructions bluntly that the British Government "had already gone too far to admit of its ceasing to support the cause of either Waterboer or the diggers." [1]   He concluded an arrangement, accordingly, for the transfer to Great Britain of the claims of the native chiefs, subject to the ratification of the Home Government, and his representations secured the consent of the Ministry, in the following May, to the transfer, and to the assertion of British sovereignty over the disputed territory, pending the final decision of the special court of arbitration which had been convened by the agreement of the contesting claimants.

The court had been opened, in the previous April (1871), in the village of Bloemfontein.   After considering the evidence

presented, the judges disagreed, and the disposition of the territory depended upon the award of the referee, Lieutenant Governor Keate, of Natal.   This was not rendered un-

The first Government House and Buildings of the Colony of Griqualand West.

til the 17th of October following, and it does not appear that the decision was hurried or improperly influenced.   But it was

[1] "South Africa," Theal.

PORTRAIT OF SIR RICHARD SOUTHEY,
Born April 25th, 1808; Died 1901.

warmly denounced as partial in sweeping aside the claims of the Orange Free State and the Transvaal Republic, and confirming the alleged title of Waterboer and other native chiefs to a tract covering 17,800 square miles, and including the Dutoitspan, Bultfontein, De Beers, and Kimberley diamond mines, as well as

Sir Richard Southey's Residence, Kimberley.

the diggings along the Vaal. Four days after this award had been made, Sir Henry Barkly proclaimed the grant to the native chiefs a part of the British dominions, as the Crown Colony of Griqualand West, which was placed under the administration of a Lieutenant Governor, Mr. Richard Southey.[1] Thus the control of the Diamond Fields was finally determined, and it is impossible to doubt that this settlement was greatly contributory to the extraordinary advance of diamond mining in these fields, as well as to the uplifting and development of the Colonies, and to the push of civilization into the heart of the dark continent.

It has been contended that the award was unjust to both of the Boer Republics, and this contention has been supported by the citation of a court decision rendered several years later, and the allowance of £90,000 to the Orange Free State by the London Convention of 1876, in compensation for losses sustained

[1] " South Africa," Theal.

through the creation of Griqualand West. But it has been fairly pointed out by the leading historian of South Africa, Theal, an earnest supporter of the rights of the Orange Free State and her sister Republic, that the claims of both contestants were weakly presented at the Bloemfontein court, and that Lieutenant Governor Keate cannot be reproached justly for any conscious unfairness in deciding the case upon the evidence before him, in a manner unsatisfactory to the Republics on the line of the Vaal.

There is, further, the practical view to present of the incorporation of the Diamond Fields in Griqualand West, — that this was the only feasible solution of the situation, at that time, which guaranteed to the irresistible rush of diamond seekers from the Cape and all parts of the world a government so strong that it could enforce its authority without recourse to arms and bloodshed. Klip-drift had already revolted at the first preposterous stretch of authority of the South African Republic, and maintained its independence until it submitted docilely to the British High Commissioner. The seething influx on the upland Diamond Fields was clearly on the verge of rebellion against any Free State regulations restricting their right of entry or supporting any monopoly title. Great Britain, with all her array of Imperial power, would not have ventured to assert such claims as had been set up by both of the Boer Republics, and could not have enforced them without an army on the spot. As a matter of fact, she prudently suffered the miners to occupy the land without any attempt to maintain crown reservations of mineral rights, even after her supremacy was undisputed through the formation of the Crown Colony. The Boer Republics, on the other hand, would have continued to blunder, almost certainly, as they had been doing, if control of the Fields had been turned over to them nominally by the decision of the referee.

It did not appear at that time, either, that there was any strong desire on the part of the authorities of these Republics to assume the cost and responsibility and prospect of collision which the supported assertion of control of the Diamond Fields would have involved. The founders of these states had sought only the plain homes of farmers and shepherds on the veld, under a government of their own choosing. Neither they nor their children were greatly stirred by the uncovering of diamonds, or the prospect of finding more on their lands. They disliked the spreading rush to the Diamond Fields, even when it was presumed that their own mines were developing. The plain, stolid farming folk, stiffly set in their old-fashioned ways, had little in common with the sanguine adventurers, delighting in stirs and surprises and novelties. Baines tells a story of the mobbing of the first surveyor who tried to use a theodolite in the streets of Potchefstrom, instead of stepping off the distance in the good old way of the "veld-valkt-meester." He avers, too, that he was himself made "vogel vrie," "free as a bird for anybody to shoot at," for the crime of concealing a sextant about his person.[1] This may be a fanciful stretch of fact, but there is no doubt of the ingrained conservatism of the Boers. How could such a people sympathize with the impetuous and ardent spirits that rushed to the Diamond Fields, and what prospect was there of the docile submission of the one to the other! It can scarcely be questioned, therefore, by a candid observer that the conclusion of Lieutenant Governor Keate was the best practical settlement, if not the most impartial and accurate.

It was not to be expected, however, that this significant departure from the halting policy of former ministries, this forward step of Greater Britain into the heart of a region hitherto indifferently resigned to the migrating Boers, should be viewed with resignation by the embittered Republics whose claims were disallowed. Resentment ran so high in the Transvaal that President Pretorius was forced to resign. His place was filled by a clergyman, Thomas François Burgers, and, after the short sharp war for independence in 1880–81, by Stephen J. Paul Krüger, a marcher with the Great Trek from the Cape to the Limpopo, a lion killer from boyhood as dauntless as David,

[1] "The Gold Regions of Southeastern Africa," Thomas Baines.

a crafty politician and a religious exhorter, a Covenanter of the Covenanters, a Boer of the Boers, uncouth, unschooled, conceited, bigoted, grasping, bristling with suspicion and prejudice, tickled with gross flattery, but a man of iron nerve, intensely loyal to his people and their push for independence, self-contained, self-reliant, bold, wary, cunning, ambitious, dominating, forehanded — masking his plans, biding his time, resolute in action, and far-seeing in shaping the future of his Republic. In the inclusion of the precious diamond-bearing province in Griqualand West, an inveterate antagonist of British Imperial extension was raised to power, whose keen forecast was almost able to overbalance the impulse of this great accession to the upbuilding of Greater Britain in South Africa.[1] On the coat of arms of the Transvaal Republic a lion lay crouching, ready to spring. From the day of Krüger's rise to head the Republic, the lion of the Transvaal has never shut his eyes nor feared to show his teeth.

While this protracted controversy for the control of the Diamond Fields was dragging on, the rush to the diggings had been spreading and moving from the ports of Australia, India, and China; from California, Canada, and the Eastern Atlantic states of the American Union — from Great Britain and Ireland and the countries of Western and Central Europe; from every region of the civilized world, at length, where men of restless and sanguine temper were living, who could command the price of the passage to diamond-bearing placers, unmeasured in number, extent, and richness. The virgin fields of California and Australia, once so glittering with gold and so potent in attraction, had lost their glamour with the scouring of their sands and the passing of their novelties. It had been demonstrated with plain, cold figures and dismal accuracy that the average farmer was getting far more from his wheat or potato patch than the average prospector from his scramble in a gold-field. But who could calculate, or even pretend to predict with any assurance, the pros-

---

[1] "South Africa," Theal. "Impressions of South Africa," James Bryce. "The Story of South Africa," William Basil Worsfold. "Cecil John Rhodes," Biography, "Imperialist."

STEPHEN J. PAUL KRÜGER.

(From a Photograph taken at Kimberley, 1884.)

pect of fortune in this African wonderland, so phenomenal in
character and so slightly explored! Here was a strange, luring
beacon in the heart of traditional Ophir, where river banks were
apparently lined with diamonds, where diamonds were strewn
on the face of farms, where children had diamonds to roll like
marbles, where wells were driven through diamond beds, and
huts were plastered with diamond-studded cement. Who would
not rush to a region so sparkling in promise, so embalmed in
traditions of resplendent empire, where another Koh-i-nûr might
be lying in wait in the dust for the first passer-by, and where a
lucky adventurer might stuff his pockets with gems far surpassing the hoard of any extortionate nabob, and return home with
a treasure that he could carry as lightly as a full purse!

The river placers had not drawn largely outside of the southern African colonies, but the discoveries at Dutoitspan, Bultfontein, De Beers, and Kimberley were so unexampled, and the
mines on the surface were soon shown to be so marvellous,
that their magnetic attraction was felt all over the globe. Who
can wonder, then, that the flying, inflated, distorted rumors
from this African hot-bed puffed up ardent fancy everywhere as
tongues of flames in tinder, and that men of all nations, call-
ings, and characters were swept along in the rush to the South
African Diamond Fields! Every sailing ship or steamer that was
bound for a South African port from any part of the world,
in 1871, bore some adventurers to the new fields. Some
had good outfits and supplies of money, while others had barely
been able to scrape together their passage costs. The seamen
on the ship caught the infectious diamond fever, and ran away
when the vessels were moored on the African coast, as their
mates had done, years before, in the ports of California and Australia. Nothing but actual bonds could hold back the diamond
seekers, and these would not serve if there was any chance to
cut cords and break irons.

The swarming of adventurers over mountain terraces, veld,
and karoo was more motley and ardent than the first rush to the
Vaal, and every one was consumed by the fear that others ahead

of him were dividing up the rich ground and a day's delay might cost him a fortune. So never before was there such a scurrying, reckless of lagging ox-teams and horses, blazing suns, and blinding dust. What a fuming there was, too, on the river banks when the sudden floods halted the rush with their impassable torrents, and the pilgrims on nettles watched the yellow water run surging, swirling, and whirling between them and their goal !

Most of the adventurers still plodded along with their bullock wagons, but some who could afford to pay roundly (£12) for transport were carried to the Diamond Fields by the wagons of the Inland Transport Company, an enterprising association

Coach leaving Kimberley for the Coast, 1875.

which undertook to run a regular coach-line to the Vaal from Wellington, the terminus of the short Cape railway in 1870. The carriage was a long, narrow wagon with five rows of seats for fourteen passengers and a driver. Only forty pounds of baggage could be carried by a passenger, but men who were anxious to reach the mines were ready to start without even a shift of shirts. Eight wiry horses dragged this rattling wagon over the rough track at a lively rate, changing teams at relay stations, from thirty to forty miles apart, and making the trip to the Vaal in eight or nine days when the way was not blocked by floods. By this stride of progress the journey from Cape Town was made in less than a quarter of the time required by the crawling ox-wagons from the other coast ports, although these towns were two hundred miles nearer the Diamond Fields. This was proudly noted as an advance of rapid transit, which promised greater developments, and was one of the many stirring impulses of the diamond discoveries. But as only one stage-coach started weekly from Wellington, the chief contribution of the new line to South Africa lay in its promise rather than its performance.[1] It was the first push of the enterprise which has followed its hoof tracks through the African desert with the tireless race of the iron horse.

While this swarm was gathering from India, Australia, Europe, and America, and pressing toward the diamond mines through the southern Colonial ports, another swarm was entering the fields from inland Africa. To the native tribesmen the opening of the diamond mines was a certain Golconda. For the shovelling of gravel under a burning sun, for the heaving of boulders, for the shaking of cradles in the midst of whirling dust, for the quarrying in pits and the scraping on sorting tables, — the wiry sinews, pliant muscles, nimble fingers, and sharp eyes of Africans, inured to the scorch of the sun, the pelt of the rain, and the blast of the sand, were greatly serviceable. So there was a cordial greeting of the influx of natives, ready to work for the barest pittance of pay while their masters lolled in the shade.

First came the neighboring Griquas, Koranas, and Batlapins, with Basutos from their southern reservation, followed by a stream of Zulus, Mahowas, Malakakas, and Hottentots, and Kafirs of one hundred tribes, ranging east to the Indian Ocean and far northwest into Namaqua and Bechuana lands and northeast into Matabeleland and the regions lying beyond the Limpopo and the Zambesi.[2] There was every shade of dusky color in this throng, from livid and tawny yellow to jet black. Some stalked proudly over the veld in the full plumage of the Zulu veteran, with flowing ox-tail girdles, armlets, and anklets, decked with

[1] "Among the Diamonds," 1870–1871.

[2] "The Diamond Diggings of South Africa," Payton, 1872. "South African Diamond Fields and Journey to Mines," William James Morton, M.D., New York, 1877.

NATIVES SEEKING WORK.

waving feathers and gleaming earrings and bracelets. Others vied with this show in greasy red shakos, faded blouses, and other cast-off equipments of soldiers and hunters. So the parade ran down to the barest loin cloth or utter nakedness, through leopard skin wraps, dirty karosses, ragged breeches, tattered shirts, and every other meagre covering of the native hunter or shepherd. Some of this drift to the mines tramped more than a thousand miles over mountain ridges and sun-scorched veld, swimming through rivers, scrambling down steep ravines, and plunging deep in mud and desert sand, to reach their goal, as many did, gaunt skeletons of men, with bleeding feet, and bodies scratched and sore and tottering with weariness and hunger.[1]

Diamonds were no temptation to them. They would not have walked a mile to pick up a Koh-i-nûr. But the white diamond seekers were willing to pay, for a few months' hunting for little white pebbles, enough to buy a cheap gun and a bag of powder and balls — most precious of all earthly things in the eyes of a roving African. Then the white camps were lively, humming social resorts, abounding with good food and tempting drink, where black men were welcome and well protected. So the natives swarmed in faster and faster as the mining progressed and the news spread to distant regions. Some of this swarm could be persuaded to remain at the mines for a year or more and work quite steadily ; but most drifted away, at the end of a few months, or as soon as they were able to get their coveted guns and powder pouches. Thus while many thousands flocked yearly to the Fields from their opening, the outflow kept the supply from swamping the demand. As this influx from the dark continent met and mingled with the rush from the outside world in the diamond-mine workings and camps, how greatly vivid, unique, and stirring were the kaleidoscopic shifts of this strange concourse! Europe, Asia, Africa, and America had boiled over into a hotchpotch, splashed on a diamond bed in the heart of South Africa.

[1] "The Diamond Diggings of South Africa," Payton, 1872. "South African Diamond Fields and Journey to Mines," William James Morton, M.D., New York, 1877.

CHAPTER VII

THE GREAT WHITE CAMPS

OW quickly and marvellously was the face of the little South African stock farms transformed by this influx! Open pasture land, where the eye saw one day only a few scattered cattle browsing on the thin grass and scratching their sides against a stunted camelthorn, was covered next day by swarms of roving prospectors, with shovels and sieves, upturning grass roots and shaking dry earth through their screens. White canvas camps, foaming with life, rose in a

Kimberley, before the Discovery of Diamonds.

night, with the seeming magic of Aladdin's palace, at the foot of kopjes where, before, a burrowing meerkat was the only tenant. Beyond the masses of tents ranged long straggling arches of wagon tops and tethered troops of bullocks, horses, and mules.

Only a few months from the day when the first diamond was picked up near du Toit's pan, the camp at Dorstfontein was proudly claiming the title of the "City of the Pan." A spacious market square was laid out on the ground between the pan

Dutoitspan. (From a very early Photograph.)

and the ridge covered with diamond diggers, and around this square were ranged the white walls of the aspiring camp. Streets radiating from the central square gave open access to the market-place, and the white tent blocks were soon dotted near the square with shops of brick and iron and wood, rivalling the pioneer diamond-digging town of Klip-drift on the Vaal.[1]

Klip-drift struggled on with the best face possible, making much of its position of vantage as the distributing market of all camp supplies from the South African Republic; but its day of ascendancy soon flitted away never to return. In September, 1871, its chief standard-bearer, the *Diamond News*, moved to the "City of the Pan," and there was no question from that time of the preëminence of the "dry diggings," although a rival paper, the *Diamond Field*, bore up for a time under the sinking fortunes of Klip-drift. Before the end of the year 1871, Dutoitspan boasted "many large hotels," "immense stores,"

[1] "The Diamond Diggings of South Africa," Charles Alfred Payton, London, 1872.

DUTOITSPAN.
(Dutoitspan Mine in the Distance.)

two churches, a hospital, and a theatre, and might have mentioned, besides, its less distinguished billiard room, "canteens,"
and dance halls.[1]   It was surely a wonderful birth of a smartly
growing infant city on the face of scrub-covered prairie in the
heart of South Africa.

The rise of the camps at De Beers and Kimberley was even
more rapid than the growth of the camp on Dorstfontein and
Bultfontein farms.   There was no regular working in the De
Beers diggings before May, 1871, but the diggers could buy

Kimberley, 1871.

Christmas presents that year in rows of brick and iron stores
on the main roadsides, intermingled with "hotels" and saloons,
and a great white canvas town was spread out in a picturesque
medley of tents and marquees, straggling far over the veld, and
seeking the shelter of some stubbornly rooted mimosa or camelthorn.[2]   Kimberley's growth was still more surprising.   Three
months after the rush began, the Colesberg Kopje was the centre
of an immense encampment in whose heart streets were irregularly
laid out, and neat stores built of iron and brick.   In December,
1871, there were, by actual count, on the lower street of Kim-

[1] "The Diamond Diggings of South Africa," Payton, 1872.   [2] *Ibid.*

o

berley, six stores, four hotels, and several butcher and shoe-
maker shops, besides a billiard room and saloon.   On the upper

Kimberley Mine, 1871.

or main street there were three hotels, several diamond merchants'
offices, a wholesale spirit and provision store, a bakery and con-
fectioner's shop, a drug dispensary, butchers' shops, eating houses,
bars, club and billiard rooms, and other miscellaneous shops and
resorts.   On the edge of these white-walled cities, and on the
slopes of all the neighboring hills, were scattered the huts of wood
or dirty canvas or mud-plastered stones, where the native blacks
huddled together.   When even this cover was lacking, some
slept in tents, or in burrows scraped in the hillsides.   How many
diamond seekers were massed in these camps at the height of
the rush can hardly be reckoned with any approach to exactness.
There may have been fifty thousand whites and blacks on the
Fields, for the flow to Dutoitspan is said to have mounted as high
as forty thousand shortly after the opening of the Vooruitzigt
farm mines.

When, after long weeks of plodding over rugged mountain ranges, parched karoos, and rolling prairie, a traveller suddenly saw rising before him these white camps, springing up like prodigious mushrooms in an African desert, even the dullest brain was strangely disturbed. It was hard to realize that these exotic plants were the work of men's hands, for they seemed rather the fantastic conceit of the trance of an opium eater. Here were such cities as the mirage shapes from clouds or as Solomon might have built with the help of his docile genii. When they lay outstretched and gleaming under the burning sun in the full splendor of noon, they were weird creations to amaze the beholder; but who can conceive their impress at night, under the towering sky dome sprinkled with stars, with their masses of

Around Kimberley Mine, 1871.

twinkling and sparkling lights on the black face of the veld, like the tail of a fallen comet.[1]

[1] "The Diamond Diggings of South Africa," Payton, 1872. "To the Cape for Diamonds," Frederick Boyle, 1873. "South African Diamond Fields," William Jacob Morton, 1877.

Outside of these three main camps tents were thickly sprinkled around the farmhouse of Bultfontein, in a field where a thousand diggers were at work in the first week of the rush, after the ground was opened in May, 1871. Immediately south of this diamond-bearing farm was the farm Alexandersfontein, where many prospectors were also turning and sifting the ground. By the determination of the limits of Griqualand West these diggings, as well as the chief camps, became part of the British Colonial domain; for the boundary line separating the new Colony from the Orange Free State ran just outside of this cluster of farms, Vooruitzigt, Dorstfontein, Bultfontein, and Alexandersfontein, — through the outlying farm of Benaauwdheidsfontein, where no diamond mine had, as yet, been discovered.[1] So all the known diamond fields of South Africa, except the Jagersfontein farm within the bounds of the Orange Free State and the shallow Vaal River placers, were bunched on a plateau four thousand feet above the sea level, within the angle formed by the junction of the Vaal with the Orange River, on a patch with a radius of 1.72 miles at the crossing of longitude 24° 46 east of Greenwich with latitude 28° 43 south of the equator.

The London and South African Exploration Company, by its purchase of Dorstfontein, Bultfontein, and Alexandersfontein, held a tight grip on the mineral rights comprehending the diamonds on all these farms, and leased the surface diggings under licenses of 10s. for every claim 30 feet square. Messrs. Dunell, Ebden & Co., of Port Elizabeth, held the farm of Vooruitzigt, and exacted the same license fee for working claims which were laid out in squares 30 by 30 Dutch feet, or 31 by 31 English feet.[2] Outside of the Colesberg Kopje or Kimberley mine all the diggings were at first a jumble of holes, pits, and burrows, with no attempt to secure any system or union in mining. But the objections to this helter-skelter opening of the ground were so apparent that a strict reservation of roadways to give access to all parts of the surface of the mine was insisted upon by the

[1] "Diamonds and Gold in South Africa," Theodore Reunert, 1893.
[2] "The Diamond Diggings of South Africa," Payton, 1872.

KIMBERLEY MINE, 1872,

Showing End View of Excavations.

Orange Free State Inspector of Mines, in the laying out of claims on the Colesberg Kopje. His authority was then so far recognized that his direction controlled the survey and opening, on July 21, 1871, of the diggings since famous as the Kimberley Diamond Mine.

Roadways, 15 feet in width, running approximately north and south, were carried across the longer axis of the diamond bed, at a distance of 47 feet from one to the other. Each road cut $7\frac{1}{2}$ feet of surface ground from the side of the bordering

Kimberley Mine, 1872.

claims, so that the working surface of each allotted claim was 31 by $23\frac{1}{2}$ feet. Fourteen of these roadways crossed the mine, whose ground surface permitted the laying out of about 430 claims of the allotted size, 31 feet square. A great many more claims had been granted to license-holders before the survey, for there had been no accurate measurement of the kopje, and there was a consequent overlapping and conflict of locations and spreading of claims beyond the limits of the diamond-bearing ground. In the settlement of contests the claims were split up by concessions, bargains, and sales, until there were not less than 1600 separate holdings of claims, and fractional parts running as small as $\frac{1}{16}$, or about 7 square yards. A lucky claim-holder would sell off parts of his claim or the whole at high prices ; for bidders were ready to pay large premiums beyond the license fee of 10s. exacted from every working owner, whether his claim was full size or a paring.   The competition for a share in the riches of the ground was only less keen at De Beers, and there was a like subdivision of claims there, and not infrequently at Dutoitspan and Bultfontein.[1]

It was obvious from the start, without any stretch of fore-sight, that these minute subdivisions of claims and individual working were only practicable in open cuttings whose depth must depend on the character of the ground and the coöperation of the miners.   But at the outset of the mining in these Fields no one could forecast the unknown continuance in depth of the diamond deposits, and few supposed that the new beds differed essentially from any before uncovered, and were vastly more im-portant than the shallow gravel wash along the banks of the Vaal.   It was commonly expected that some barren stratum would be reached not far from the surface, corresponding to the "bed rock " of the river diggings, and that this must terminate the hope of the diamond seekers.[2]   So the rush for the surface claims was the keener, in view of the belief that a few months' work at most would exhaust the precious deposit, and nobody paused to consider what he would do if he was unable to sink an open pit deeper.

Beneath the red surface soil at Dutoitspan a thin layer of calcareous tufa[3] had been exposed, below which lay the dia-mond-bearing breccia which the miners called "yellow ground" from its prevailing color.[4]   At De Beers and Kimberley there was comparatively little limestone beneath the red soil, for the

---

[1] "The Diamond Diggings of South Africa," Payton, 1872.   [2] *Ibid.*

[3] I look upon the calcareous tufa which covered the diamond mines as only the altered yellow ground which had been metamorphosed by the evaporation of water highly charged with carbonate of lime.   The calcareous tufa which covered the Premier mine was diamond bearing.   This is the only one of the mines whose surface ground has come under my personal observation.

[4] "The Diamond Diggings of South Africa," Payton, 1872.

rich "yellow ground" rose nearly to the surface under a thin coating of chalk. It appeared in exploring the yellow ground in most of the openings that the deposit was enclosed in an oval-shaped funnel of shale, or decomposed basalt resting on shale, which the miners called "reef." This reef contained no diamonds and marked the limits of any profitable prospecting. The surface area of the yellow ground within one of these funnels ranged from about ten acres at Kimberley to twenty-three acres at Dutoitspan, and on these patches all the diamond-bearing claims of the Fields were located.[1]

When the bottom of the "yellow ground" was reached at a depth of from fifty to sixty feet below the surface, it was supposed at first that diamond digging in the funnels had come to an end; but the hard underlying rock was cut by experimenters, and it was found, to the delight of the miners, that this also was diamond bearing. It was a breccia composite, essentially like the "yellow ground" above, but much more compact and hard, and of a prevailing bluish slate color, so that it was familiarly known as "blue ground."[2] Exposure to the air, sun, and rain decomposed it so rapidly that most of the rock could be readily pulverized after a few weeks, and its precious contents extracted by sifting. The whole mass of the ground in the funnels was diamond-bearing, in greater or less extent, except in occasional streaks and masses of barren shale, floating reef, floating shale, or non-diamond-bearing volcanic mud, and volcanic rocks. So the pit sinking was widened to the extreme limits of the claims, and the entire area of yellow and blue ground excavated in open quarries.

The work was pushed with feverish energy and remarkable rapidity in view of the bare hand labor and crude mining appliances, but there was no uniformity of method or extended coöperation. Every claim-holder cut down his patch with pick and shovel, and lifted the broken ground in a way that suited his individual notion. Some set stout windlasses in the surface ground near the edge of their claims, and hoisted buckets filled

---

[1] "The Diamond Diggings of South Africa," Payton, 1872.   [2] *Ibid.*

CENTRE BLOCK, KIMBERLEY MINE, 1874.

on the working levels. Others carried up buckets and tubs and rawhide sacks to the surface, climbing ladders resting on successive terraces, or mounting flights of steps cut in the rock, or trundling wheelbarrows up plank inclines. Around the edge of the mines there was a mustering of carts, and barrows, and carriers, to bear off the quarried ground to depositing places, where it was dried, pounded, and sifted.[1]

The open quarries, swarming with workers, buzzed like prodigious beehives. The upsetting of the tower of Babel would scarcely have poured out such a medley of tongues and sounds. From the vast amphitheatres scooped in the rock there rose in the air the clicking of picks, the rasp and clatter of shovels, the cracking of rock, the rattle of gravel, the thud of bucket-filling, the creaking of windlasses, the tramp over planks, the thump of wheelbarrows, the rolling of carts, the lowing of bullocks and braying of mules, mingled with calls and chatter and chants of whites and blacks in an indescribable din. Diggers in rough working dress, and natives almost stark naked, bent and heaved, and scrambled and climbed, side by side, reeking with sweat and grime, in an ever shifting, restless swarm that covered the face of the quarry like flies in some monstrous sugar bowl. The flocking in of the native African tribes — joined with the white diamond seekers in opening the strange funnels of crystal-sprinkled breccia — made a compound of color, feature, and character never before assembled in any mines on the face of the earth.[2] The sinewy negroes proved themselves such willing and sturdy workers in the dust and heat of the sun-scorched quarries, that the claim-holders were glad to hire them and confine their own work to the task of overseers, directing the digging and hauling, and the sifting and sorting. No blaze of the sun and no whirl of the dust could subdue their bubbling spirits, breaking out in wild whoops and chants, and yelling in pack when any big diamond was found, revelling in every chance diversion, — the fall of a bucket, the slip of a ladder, the

[1] "The Diamond Diggings of South Africa," Payton, 1872. "South African Diamond Fields," Morton, 1877. [2] *Ibid.*

THE ROADWAYS, KIMBERLEY MINE, 1871-1872.

tumble of a climber, and convulsed with laughter whenever they could set mules capering, or bullocks shying or balking by shrill whistles and screams, and mimicry of a driver's call: "Yek!" and "Trek!" "Ah now!" and "Whoa!" and so through the range of cries, Dutch, English, and African.[1]

Almost all the natives were barefooted, and most were bare-headed, barebacked, and barelegged, except in the coldest weather. Some had ragged trousers, and others ragged shirts, but few put both on together. A greasy, gaudy handkerchief twisted around

Kimberley Mine, 1872.

a black head, and party-colored bunches of rags, or moochies made of the tails or skins of wild animals, dangling from a waist-belt of rawhide, were a camp parade dress too precious to use in the quarries. Mingled with these wild Africans, the white miners worked soberly and arduously, bearing the pains of diamond digging stoically, in the hope of its rewards. Their working clothes were commonly plain suits of brown corduroy or other coarse cloth adapted to the season, and when the sun shone they wore generally broad-brimmed straw hats, or pith helmets, with light muslin "puggarees."[2]

It was long before there was any notable advance in the pro-

[1] "The Diamond Diggings of South Africa," Payton, 1872. [2] Ibid.

cess of separation of the diamonds from the ground, beyond the
cradle for dry sifting, constructed to take the place of the common hand-sieve at Dutoitspan.   Level spots were sought on
the veld near the mines, or patches of ground were levelled
sufficiently to serve as dumping places, where the broken diamond-bearing breccia was piled and spread out.   The "blue
ground" exposed to the air crumbled away by degrees, but the
miners were rarely patient enough to wait for this disintegration,
preferring quick returns by pulverizing the ground with their
shovels and mallets.   This was hard work and costly, from the
loss in imperfect pulverization.   But the diamond seekers were

Roads in Kimberley Mine, 1871–1872.

poor men who could scarcely afford to hold any stock of blue
ground for the sake of increased returns, even if they had been
able to guard their depositing floors from theft.   After pounding the broken rock it was sifted in the midst of dust clouds by
rockers swung on riems of rawhide, and the concentrate was
then scraped over and sorted.[1]   In July, 1871, a large cylindri-
cal revolving sieve, driven by a small steam engine, was put at
work by some American miners, and this sifting machine was
said to be an efficient and rapid separator.   The pulverized
ground was thrown into the upper end of the screen, which was

[1] "South African Diamond Fields," Morton, 1877.   "Diamonds and Gold
in South Africa," Reunert, 1893.

rapidly revolved, and the concentrate passed out through the lower end, falling upon a sorting table. The cylinder, covered with fine wire mesh, sifted out the dust thoroughly, and its operation was so rapid that thirty cartloads of diamond-bearing ground were screened daily. Its owners claimed to be able to sift all the ground in a claim thirty feet square to a uniform depth of thirty feet in three weeks. The machine attracted a curious crowd at first, when the steam whistle blew off and the cylinder began to throw off thick clouds of dust, but for some reason its

Kimberley Mine. (Showing workings in 1872. Subsidence cracks appear in the foreground.)

use was not long continued. Probably the fine mesh was too light to bear the strain and friction of the revolving rock fragments.[1]

The amount of ground which any one man could work, was, of course, very small, but there were so many workers on the Fields that the aggregate extent of ground sifted was enormous, and the breccia in spots was so thickly sprinkled with crystals that many miners won rich rewards. When Payton was leaving the field in November, 1871, it was estimated that from forty to fifty thousand pounds' worth of diamonds were taken

[1] "The Diamond Diggings of South Africa," Payton, 1872.

weekly from the Colesberg Kopje alone, and he states that the best claims had risen in value from £100 or less to £4000.[1]

It was clearly shown, too, that even the highest price paid for a claim might be cheap, for one poor Dutchman, "Smuts," who bought half a claim for £50, is said to have found diamonds in two months' working to the value of £15,000 or more. Another digger found, in a few months, no less than 730 stones in his claim, one of which weighed 156 carats.[2] Such great good fortune was rare in the other mines, and many miners won little or nothing from months of hard work in their claims, but in the Colesberg Kopje, or Kimberley mine, the prizes were so common and exciting that every foot of ground was covered by diamond seekers. When the rubbing of shoulders was too close for comfort, one or more of the partners in a claim would be pressed to sell out and start again prospecting. Sometimes a share in a claim, worth many hundreds of pounds, would be risked on the toss of a penny.[3]

In the heat of the search and extraction many fine diamonds were fractured, and many of the smaller stones ran through the sieves into the tailings, as was afterward demonstrated when the waste heaps were reworked with better appliances.[4] The Kimberley mine produced some stones of large size, running sometimes over one hundred carats, but the mass of crystals ran under five carats. A yellowish tinge was more marked in the diamonds of the uplands than in the river stones, and many otherwise superb crystals were so decidedly "off color" that their value was greatly impaired.

It was early noticed that the diamonds of one mine often differed materially from those of another, and even in the same mine diamonds of one section were unlike the yield of another. Thus, in the west end of the Kimberley mine the diamond crystals were exceptionally perfect octahedrons, or exceptionally white "glassy stones," as the miners called them; while elsewhere in the mine the crystals had, more commonly, rounded and bevelled

---

[1] "The Diamond Diggings of South Africa," Payton, 1872.   [2] *Ibid.*   [3] *Ibid.*
[4] "Diamonds and Gold in South Africa," Reunert, 1893.

ROADS IN KIMBERLEY MINE, 1872.

Kimberley Mine,

edges, and, more or less, a yellowish tinge, and there was a large proportion of split, flawed, and spotted stones, and boart.   The De Beers mine crystals resembled the Kimberley stones, but their quality ran a little below the Kimberley mine.   Dutoitspan produced comparatively few stones, but the average weight was notably large, and the crystals were of fine color.   Bultfontein stones differed greatly from those of the other mines.   Here the diamonds were chiefly small, rounded octahedrons, many of them so pocked and spotted that the crystals had a cloudy appearance.[1]   These crystals were greatly inferior to the " glassy stones " of Kimberley or the large diamonds of Dutoitspan; but the Bultfontein surface ground yield was so uniform at first, that many diggers held and worked claims for the sake of sure, if small, returns to defray their expenses, while they counted on their Dutoitspan claims for the occasional large stones that richly rewarded a lucky digger.

All the crystals in the blue ground were encased in a smooth bed of the same material which did not adhere to the diamonds, so that their lustre, when extracted, was quite bright or glassy.

[1] " South African Diamond Fields," Morton, 1877.

1872, looking South.

Amid the mass of white and light yellowish stones in all the mines were scattered some of varied color. Brown was the most common of these; next came the deeper yellow shades, and pale blue stones were sometimes uncovered, as well as the black diamond (boart) used for setting drill-crowns. Pink, mauve, and green diamonds were occasionally found, but were less common than in the river diggings.

As already mentioned, it has been estimated that the rush which built up these mining camps and covered the surrounding farms with prospectors brought fifty thousand men to the new Diamond Fields in the first year, though the shifting population of the Fields did not rise as high as that at any one time.[1] The influx of native Africans was not so large at first, but increased from year to year. Morton says that there was a flow of thirty thousand natives annually to the field for seven years after the discovery of the mines.[2] This is a credible estimate, at least, in view of the constant drifting away from the field of the native workers, after a few months' stay, when they had earned

[1] "The Diamond Diggings of South Africa," Payton, 1872.
[2] "South African Diamond Fields," Morton, 1877.

P

sufficient money to buy their coveted guns and ammunition, and wives, cattle, blankets, etc.

The bulk of the general merchandise was hauled slowly from the coast ports in ox-wagons. Algoa Bay (Port Elizabeth) was the chief port of supply at first, and the transport to the Diamond Fields was a trip ranging from thirty days, at least, to six months.[1] Certain kinds of food — beef, mutton, poultry, game, dried venison, commonly called " biltong," [2] and maize meal (mealie meal) — were furnished quite cheaply and plentifully from the neighboring Free State and the South African Republic, through the Klip-drift distributing market. Tobacco, butter, eggs, and honey were less freely supplied from the country, and commanded a ready sale. Ordinary beef and mutton sold for 4d. a pound in 1871, with an additional charge for choice steaks. A whole sheep could be bought at wholesale for 4s. Game, chiefly springbok, blesbok, and wildebeest, was as cheap as mutton. Chickens and ducks ranged from 2s. 6d. to 3s. 6d. apiece. The price of eggs ran high, ranging from 2s. 6d. to 4s. a dozen, and butter was sold at from 2s. 6d. to 5s. per pound. For " Boer meal," a coarse wheat flour, the charge was from 35s. to 50s. per muid, about 183 pounds; white flour brought 6d. a pound; rice 9d.; sugar and tobacco 9d. to 1s.; oranges and onions were sold at 10s. per hundred, and dried fruits at from 3d. to 9d. per pound.[3]

The most urgent calls were for fresh vegetables, and the supply was so meagre that the prices shot up to exorbitant figures. From 5s. to 7s. was freely given for a bucketful of potatoes, and the wholesale price for a bag of a hundredweight was from £2 to £2 10s. Half a crown (sixty cents) was often paid for a small cabbage or a handful of onions. Choice forage for the horses and mules was almost as costly as vegetables. A bundle of five pounds of unthreshed oat hay was sold for as high

---

[1] "The Diamond Diggings of South Africa," Payton, 1872.

[2] *Biltong* is made of meat of any of the antelope species, but that made from the springbok is considered the best.

[3] "The Diamond Diggings of South Africa," Payton, 1872.

as 2*s*. Dry cut fuel was as high priced as forage. Bundles of light sticks sold from 9*d*. to 1*s*. per bundle, and £3 was charged for a load of good firewood. There was a considerable forest growth on the hills near the Vaal River, and many thickets on the ridges nearer the camps, but the cost of cutting and hauling was so great that many diggers contrived to make their fires of dried bullocks' dung (buffalo chips as they were called by the emigrants crossing the American plains), as they had learned to do when crossing the karoo.[1]

Market auctions were the common and popular mode of selling food and ordinary miners' supplies. Criers swinging bells rang up the drowsy camps for the early morning market, where meat, eggs, butter, fruit, and vegetables were offered from wagons and stalls in the open market squares. These sales and gatherings of bidders and lookers-on formed one of the liveliest camp scenes, especially on Saturday, when thousands of whites and blacks flocked to the auctions, surrounding the stands with dense masses of jovial bargainers. How strange and curious to a newcomer's eye was the market show, — carcasses of big brown shaggy wildebeests hanging up in line with sides of beef, antelopes with slender legs stretched out stiffly among the slaughtered sheep and lambs, strips of biltong and freshly killed kids, little long-legged hares, party-colored bustards, red-wing partridges, red-legged plovers, guinea fowl, ducks, geese, and other wild fowl, mingled with the poultry from country farmyards Here were lines of huge tent-covered wagons filled with hides, and wool, and meal, and wood, driven to market by the stolid Boers or Hottentot servants grinning from ear to ear. Potatoes, and beets, and carrots, and onions, and cabbages were piled in heaps, tempting the last shilling of scurvy-haunted men. The gobbling of turkeys, the crowing of cocks, the quacking of ducks, swelled the chorus of chatter and laughing and singing and badinage, that smothered, at times, the brisk calls of the auctioneers and the offers of the diggers and the hotel and shop keepers.[2]

[1] "The Diamond Diggings of South Africa," Payton, 1872. [2] *Ibid*.

MARKET SQUARE, KIMBERLEY.

In the afternoons special sales of tents, miners' tools, guns, and general merchandise were frequently made by auction, and large stocks were sometimes sold off completely in this way. Often in the flurry of competition these goods brought absurdly high prices, when the market was overstocked with like articles in the stores. It was observed as a curious fact that scarcely a bid could be got for revolvers, which many adventurers had supposed to be an indispensable part of their outfit. There were very few outbreaks of ruffianism in the camps, where the great

Market Square, Kimberley.

body of miners was disposed to be orderly, and occasional sprees were the chief disturbances. The swaggering bullies, and cheating gamblers, and lurking garroters, who infested the seething camps of Nevada and Colorado, rarely drifted as far as these isolated Diamond Fields, and the few who came in were held in check.

The crying need of the camps was good water. The Dutoitspan basin did not always hold out through the dry season, and besides, this pan was filled by drainage and was not wholesome ; but two rude dams were built that served to store up drainage water longer than the natural reservoirs. To eke out the supply the " Diggers' Committee " at Dutoitspan and Bultfontein sunk several wells which furnished some additional water, and a digger was licensed to draw two bucketfuls daily upon the payment of one shilling a month for his water rights. This privilege was so keenly sought for that there was always a little crowd of men with buckets, waiting their turn, at the mouth of a well in the daytime. The water was muddy, but it was nevertheless eagerly drunk, and the stinted supply was too precious for washing. Following this push of the committee, prospecting water shafts were sunk by private enterprise, and when water was reached, the well was opened to a limited number of subscribers upon payment of a monthly fee of four shillings.[1]

At Kimberley, water, for months, was so dear that it was sold for threepence a bucket, and a daily washing of face and hands was a stretch of luxury. A stinted bath at Dutoitspan cost two shillings and sixpence, and bathing at the other camps was rarely attempted. When the coating of grime grew unbearable, the best resource was a ride or tramp to the Vaal and a plunge in the river. In the dry season, when the air was full of floating dust from the claims and cradles, and when hot winds from the veld blew in clouds of red sand, the dearth of water was bitterly felt, and no joker was safe who ventured to recall the " old oaken bucket " and other vain visions of cool, bubbling springs.

Often the dust-storms passed beyond the aggravation of thirst and discomfort, driving sand-whirls so furiously in the faces of the workers that the hardiest men were forced to drop their picks and shovels, and buckets and cradles, and run to cover. Then, for hours, storms would rack the tents, straining every cord and stitch of canvas to the snapping point, and often tearing rents in the walls, or pitching over tent-poles and all in utter wreck. Even when the stout posts, braced and guyed against a hurricane, bore the strain unyielding, the sheltered miners had to swelter in a mist of dust that was blown through the crevices into every fold of bedding and clothing, and coated every inch of their skins with irritating powder.

Next to this pest of dust was the plague of flies, little and large, black and green, that swarmed over the camps in countless

---

[1] " The Diamond Diggings of South Africa," Payton, 1872.

myriads in the summer season, tainting every morsel of food, and settling on every bare face or body with a dash so bold and persistent, and a grip so malignant, that it hurt like a sting. No possible device could clear the tents completely, or keep out these swarms; but the miners armed themselves with big whisks of wildebeest and ox tails, and got some relief by constantly flicking and slashing, or when they were forced to use both hands at work on the cradles or sorting tables, "fly flappers" stood by to brush back attacks.

Hot days in the dry diggings on the bare veld were more keenly felt than the same days on the tree-fringed Vaal, and some midsummer days were too scorching even for the endurance of the seasoned black diggers. But, except at midday, few working hours were lost when the sun was shining. The swooping thunder-storms were scarcely less terrific than the storms in the river valley, striking the camps with drenching pelts of rain and heavy hail, hurled from cloud banks blazing and bellowing with monstrous forks of lightning and stunning thunder peals.

The clear winter days were greatly invigorating. At break of day it was often so cold that jugs of water were skimmed with ice and a hoar frost covered the ground. But when the bright sun mounted the sky, the chill air was so warmed in a few hours, and so pure on the breezy veld, that the miners gained fresh spirit with every breath, and went through their monotonous round of work with unflagging life and good humor. The actual record of a week at the mines, in August, 1871, gives a clear idea of the winter shifts of temperature.[1]

DAY.		NIGHT.
Highest Temperature.		Lowest Temperature.
Aug. 21	83° Fah.	40° Fah.
" 22	85° "	35° "
" 23	83° "	30° "
" 24	92° "	33° "
" 25	93° "	28° "
" 26	56° "	28° "

[1] "The Diamond Diggings of South Africa," Payton, 1872.

The health of the camps was usually good, except in mid-summer, when " low " fever, diarrhœa, dysentery, and colic were prevalent.   The impure drinking water was the most persistent cause of sickness and the most difficult to combat.   Inflammation of the lungs from the fretting dust, and mild scurvy, were the other common ailments, occasioned by the conditions of life at the diggings.[1]

It was not all work and no play in these diamond diggings. Saturday afternoon was commonly taken as a half holiday in addition to the Sunday rest and recreation.   In the springtime, or the beginning of the rainy season, fresh flowers sprang into bloom on the face of the veld, and birds built their nests in the grass and thickets.   Little dusky black-and-white birds, recalling the English linnet, were sweetly trilling songsters, and were so fearless and sociable that they flew everywhere over the diggings, in the midst of the dust and stir, perching on heaps of broken rock, or even on the diggers' cradles, comically fluttering their tails, and chirping so musically that the wearied men were charmed to watch and listen.   There was good shooting, too, for wild fowl and small game on the open veld; and not far from Dutoitspan there was a large stretch of thickets and scrub where korhaans and paauws and partridges and plovers and hares abounded.   The stately Kafir cranes shook their bluish gray plumes on the brink of the vleis, or water holes, where they came to drink, and were shot by the hunters who lay in wait. Their flesh was not unpalatable as a change from biltong, but the hunters who pushed toward the Vaal brought back better eating than cranes in their bags, — wild ducks and geese and guinea fowl, and even a nimble springbok or queer-headed wildebeest or hartebeest.   The swift leopard, too, was occasionally shot near the river bank, and the rambling diggers found some fun in unearthing porcupines, or chasing a jackal with dogs, or lying in wait for the shy head of a meerkat to pop out of a hole.   A string of fish, that could readily be caught in the river,

---

[1] "The Diamond Diggings of South Africa," Payton, 1872.   " To the Cape for Diamonds," Frederick Boyle, 1873.

made a welcome meal; and a run to the Vaal, with a few days' camping under the willows and long swims in the swirling current, was a gay streak of diversion from dusty diamond digging.[1]

Within the camping grounds there were always, in the day-time and evening, stirring scenes to attract the eye, — antics of ponies and mules; the passage of straining bullock teams with carts piled high with diamond-bearing ground, or wagons loaded with country produce; the rambling of pedlers with packs and trays; the groups of native tribesmen in trappings of skins and feathers or comical old clothing, chattering or singing or whooping or dancing; the clustering of black women, washing on the edge of the water pools; the rallies of amateur minstrels and travelling shows; the merry-go-round, ridden by laughing children or solemn country clowns; the rattling of pins in the bowling alleys and clicking of balls on pool tables, the crowded "canteens," and all the other lively sights and sounds of the fermenting camps.

Fortune-tellers and wizards, who professed to be able to see through the earth, did not need to dig diamonds. Credulous prospectors filled their laps with silver and gold. Payton tells of one whose tent in Dutoitspan was thronged, day and night, with eager dupes, showering shillings upon her, and her income was reckoned at £30 a day. Many of the Boers had unshakable faith in her. When she told a poor Dutchman that there was nothing in his claim, he could not be persuaded to dig any longer. If she promised diamonds and none were found, she had an easy defence: "The niggers must have stolen them." Then the wrathful claim-holder would pounce on his Kafirs and haul them to the "tronk," the police station of the camp, where the blacks were searched to the tips of their toes. Sometimes the wizard guessed right, for diamond stealing was common, and precious stones would be brought to light with joy to the owner and credit to the fortune-teller.[2]

[1] "The Diamond Diggings of South Africa," Payton, 1872.
[2] *Ibid.*

The whipping-post was soon set up in the new camps and freely used to chastise theft and other offences. The whipping was not very severe, and it was remarked that the "cat" was not as heavy as the one at Klip-drift. Many of the natives would steal anything that they could carry off, and put on a brazen face that would impose on a police court judge. A story is told of two Hottentots who took the sheepskins off a man's bed and tried to sell them back to their owner as soon as his eyes were open in the morning. They took fifteen lashes on their bare

Natives resting, on their Way to the Mines.

backs without a whimper. Small fines were imposed for slighter breaches of the camp regulations, and roaring drunkards were occasionally clapped into the "tronk," a weak little jail, but the whipping-post was necessarily the main dependence for punishment of natives.

Strangest and most interesting of all features of the camps to a newcomer were the habits and antics of the marvellous collection of savages, streaming into the Fields from the heart of Africa. No mining camp on earth before ever held such a motley swarm of every dusky shade, in antelope skins and leopard skins and jackal skins and bare skins, — with girdles and armlets of white ox-tails, and black crane plumes and gorgeous bird feathers, and dirty loin cloths, and ragged breeches, and battered hats and tattered coats. With and without the fire of rum they might dash off at any moment into some wildly whirling reel or savage dance, gabbling in a hundred dialects, whooping with weird cries, and chanting plaintive, gay, and passionate strains, now dissonant, now sweet. Whenever a new party of "raw" natives came in from the wilderness, weary, grimy, hungry, shy, trailing along sometimes with bleeding feet and hanging heads, and bodies staggering with faintness, a howl of jeers was a common greeting, and a pelting with rotten fruits and stones was likely to follow the scared troop up the street of the camp, though the natives were not churlish at heart, and might, afterward, share their last crust with the strangers.

Their savage habits clung to them long in camp. Some delighted to smoke in the old native way, by making a little funnel in the wet ground with a slender stick and sucking the smoke through one end while the tobacco leaves burned in a hollow at the other. As a rule all the natives from Delagoa Bay and districts to the north of that part smoked cigars with the fire end in their mouths. When sheep or bullocks were killed at market, the natives hung about and returned exulting if the obliging butchers gave them the entrails to hang in festoons about their necks and carry off smeared with filth. They fed content day after day on a few handfuls of mealies or ground maize with an occasional chunk of refuse meat. They had little use for water except to drink, and they much preferred Cape brandy. After working all day, and roving about and singing at night, they could sleep as soon as their heads touched the ground, on the bare earth, without shelter, or in a squalid hut with a dirty sheepskin wrapped around them. These quaint Africans, mingling in a kaleidoscopic show with adventurers coming from the ends of the earth, made a unique, moving drama on the stage of the Diamond Fields that cannot be forgotten by any spectator.

CHAPTER VIII

A N ever present danger hung over the miners from the very outset of their pit digging in the diamond-bearing funnels. The yellow ground was a breccia so loose and friable that it was constantly caving in upon the heads of the diggers. Then the pits were sunk so close together that the walls gave way and slipped, crumbling into the claims below. A loaded cart, passing along the edge of a road, would often topple over and sometimes plunge with driver and mule into the pit below.

Prospecting on the Alexandersfontein farm was not long continued; but the diggings at Dutoitspan, Bultfontein, De Beers, and Kimberley were ardently opened by swarms of diamond seekers. The surface area covered by claims was very much larger than the diamond-yielding ground, whose total extent was, approximately, seventy acres. When the claims were consolidated by purchase, many years later, the Kimberley open mine surface was figured to be 33 acres; De Beers, 22 acres; Dutoitspan, 45 acres; and Bultfontein, 36 acres. These measurements more than cover the extent of the original locations, which were as follows: Kimberley, 470 claims, equal to 10.37 acres; De Beers, 622 claims, equal to 13.72 acres; Dutoitspan, 1441 claims, equal to 31.79 acres; and Bultfontein 1067 claims, equal to 23.54 acres. Only a few scattered diamonds were found outside of the rim of "reef" enclosing the diamond-bearing craters.

To present clearly the progress of mining in the several funnels, it is desirable to trace the advance of each separately through the period of the open mine working, to show the different methods employed, and how one mine profited by the costly experience of another. The superior richness of the diamond-bearing ground in Kimberley mine urged forward its opening more rapidly than the development of the others, and this may properly be outlined first. The plan of mining, with the reservation of roadways determined by the Free State inspectors, proved a poor makeshift at best, before the sinking of claims had progressed many feet below the sur-

The Breaking-up of the Roads, Kimberley Mine, 1872.

face. The bordering-claim owners undercut the roadways crossing the mine, in working to the bounds of their allotments, and these reserved roads soon began to cave away in places to an extent that made the passage of carts very risky. It was doubtless convenient to have ready access to every part of the surface of the mine, and it was a moving spectacle when fourteen parallel roadways were covered with files of plunging mules and rumbling carts, goaded by the cries and whips of many hundreds of half-naked Kafirs or white drivers; but it was a pitiful burlesque of mining when the roadways cracked and crumbled, and crevasses were bridged with sliding planks, and mule carts

Miners going to Work.

and men staggered and slipped over the roadsides into abysses.
Yet in spite of all risks and accidents, the roads were patched
up and maintained in some shape long after it was evident that
they were doomed.   At length no possible patching and bridg-

The Hand Drums used for Winding-up the Blue Ground.

ing could arrest their fall. One after another, before the end of the year 1872, had crumbled away and slipped into the great pit. The mine was then an open, oval quarry, about a thousand

De Beers Mine, 1873.

feet in length and six hundred feet in extreme width. The broken blue ground on the face of the rough jumble of terraces had been hoisted to the surface usually in buckets, by means of a rope passing around a windlass and through a pulley fixed in a pole set near the edge of the claim, but in 1872 a simple device of haulage over two grooved wheels was largely introduced. One wheel was set on the pit bottom, and the other on the surface, with a handle attached by means of which one or more stout natives could wind up a rope passing from wheel to wheel, carrying up a loaded bucket and lowering an empty

Kimberley Mine, 1873.

one. This crude device served the purpose as long as a wheel could be set near the edge of a claim on unbroken ground, or along the roadway ; but when all the claims were at the bottom of one huge open pit, it was obvious that only the outer tier of claims could be worked by this method.

Then a haulage system of really remarkable ingenuity was contrived. A massive timber staging was set completely around the mouth of the mine, carrying two, and in parts three, platforms, one above the other. The upper platform was connected by strongly anchored ropes with the claims in the middle of the mine, and the lower platform in the same way, with the claims

Kimberley Mine, 1874.

nearer the margin of the mines. Where there was a third inter-vening platform, ropes were stretched to claims lying between the outer and inner circle. Windlasses were set on the plat-

Another View of Kimberley Mine, 1874.

Natives carrying Ground out of Dutoitspan Mine in Buckets.

forms, together with guide wheels over which hauling ropes passed, dragging the buckets swiftly from the bottom of the mine on little overhead runners, rattling over the stationary roped inclines. When the loaded buckets reached the platform levels they were dumped into chutes carrying the contents into bags, which were readily carted away to level depositing grounds, or " floors," as they were technically termed, where the blue ground was sifted and sorted. The empty buckets were easily

Back View of the Staging with Grooved Wheels, at Kimberley.

returned, running back by force of gravity over the ropes to the claims. The buckets were of rawhide, for this material was found to be more lasting than iron, and the ropes were at first largely of hemp or twisted rawhide; but iron and steel wire gradually replaced all other material.

So thickly together were these lines set, that the whole face of the vast pit seemed to be covered by a monstrous cobweb, shining in the moonlight as if every filament was a silver strand. Never has any eye seen such a marvellous show of mining as

Kimberley Mine, 1875.

was given in this grand amphitheatre, when the huge pit was sunk far below the surface level; when the encircling wreath of the chasm rose sheer and black like the walls of a deep, gloomy canyon, or the swelling round of a demon's caldron; when a downward glance from the perch of a platform made weak heads reel; when thousands of half-naked men, dwarfed to pygmy size, were scratching the face of the pit with their puny picks like burrowing gnomes; when thousands more, all grimy and sweating and odorous, were swarming around the pit's mouth, dragging up loads of diamond-sprinkled ground and carrying off their precious sacks; when hide buckets were flying like

Snow in Kimberley Mine, June 21, 1876.

shuttles in a loom up and down the vast warp of wires, twanging like dissonant harp-strings, with a deafening din of rattling wheels and falling ground; and where every beholder was won-der-struck at the thought that this weird creation in the heart of

Method of Hauling, De Beers Mine, 1873.

South Africa had been evolved by men for the sake of a few
buckets of tiny white crystals to adorn the heads and hands of
fanciful women.[1]

The First Horse Whim, Kimberley Mine, 1874.

With the deepening of the mine, " horse whims," first introduced in 1874, were gradually substituted for hand tackle in
hoisting and lowering the buckets, which were enlarged tubs

Hauling Gear and Jumpers, Kimberley Mine, 1878.

[1] When Lord Randolph Churchill visited the diamond fields, while looking at a
huge parcel of diamonds he remarked, "All for the vanity of woman."   A lady,
who heard the remark, added, "and the depravity of man."

A Nook in Kimberley Mine, 1874.

holding five or six cubic feet of blue ground. These whims
were timber wheels from fourteen to eighteen feet in diameter,
set near the edge of the mine, to revolve horizontally about

eight feet above the surface level. To turn the whim an iron hoop, projecting from the wheel, was attached to the harness of a horse or mule. The hauling rope was wound above the hollowed rim of the wheel, and each end of the rope was fastened to a tub, one hauling up the load of blue ground, and the other lowering the empty tub.

In the following year, 1875, the first steam winding engine employed at the mines was transported to Kimberley to take the place of horse power in moving the whim, and the first

The Horse Whims, Kimberley Mine, 1875.

application of modern mining methods to the South African Diamond Fields was made. This seemingly tardy development was due less to a lack of enterprise than to the heavy charges of freight transportation from the coast, ranging for years over £30 per ton, and to the scarcity and cost of fuel, combined with the lack of any positive assurance of the continuance in depth of the diamond-bearing ground. Such a deposit of diamonds as had been uncovered in the South African farm lands had never been opened before, and the erection of costly machinery for its extraction was naturally deemed an unwarranted risk.

Hauling Gear, Dutoitspan Mine, 1876.

But as the cutting passed farther and farther down through the reef-circled funnels without disclosing any barren stratum or break in the body of breccia, the surmise rose gradually to the point of conviction that the funnels were craters of extinct volcanoes, filled by successive eruptions of steam or gas under great pressure with a diamantiferous breccia, carrying fragments of volcanic and sedimentary rocks and crystals of many kinds of minerals.   This conclusion, however, was hardly more than one of several varying assumptions in advance of the thorough researches and analyses of later years, when the prosecution of deep mining works determined positively the existence of craters, the character of the breccia, and the composition of its encasing reef.   So the progress of mining on the Diamond Fields was long a hesitating and tentative advance, groping step by step into the depths of the blue ground.

After the device of staging and hoisting ropes had solved, for a time, the problem of open excavation in the Kimberley mine crater, and the caving of the blue ground was no longer a terror to the diggers, the collection of water in the pits was a serious annoyance.   Most of this water was surface drainage,

Surface Loading Boxes.

Aerial Trams and Surface Chutes, De Beers Mine, 1885.

flooding the lower levels in the rainy season, but never sufficient in quantity to have been any considerable obstacle, if the mine had been equipped with the ordinary pumping machinery erected in other mining districts.   The lack of any such machinery, compelling for years the bailing and hoisting of the water in buckets or tubs by hand or horse power, was no slight draw-back to the progress of sinking.   Hard upon this impediment came the much graver trouble occasioned by the crumbling, cracking, sliding, and falling of the encasing reef of decom-posed basalt and shale.   The unstable walls of these soft rocks caved rapidly upon exposure to air and moisture into the open pit, and the fracturing and slipping were aggravated by the imprudent vertical cutting of the mine, removing the entire body of blue ground without cutting away the reef in compara-tively stable terraces or slopes.   Obviously no single claim-holder would undertake the cost of removing the dangerous reef for the common benefit, and it was difficult to secure the general coöperation and subscriptions so urgently required for this work.

What is everybody's business in theory has too often been nobody's business in practice.   The mean and short-sighted

Hauling Gear, Kimberley Mine, 1885.

hope to be protected without cost by the enterprise of the more liberal and prudent! The central claim holders counted on the distance of their claims from the reef to assure their safety, and the outer circles of claim-holders hung upon luck to shield their ground. But the frequent recurrence of reef falls and slides,

The French Company's Sling Gear, 1885.

together with the gathering of troublesome water pools, so emphasized the necessity of combination that a Mining Board was organized in 1874 by general concurrence of the claim-holders, with power to levy a comprehensive assessment to cover the expense of keeping the mine clear of water and fallen reef. This board took the place of the original " Diggers' Committee " which had hitherto been charged with the execution of the crude code of mining regulations.

The creation of this new administrative board was a move in the right direction, but unfortunately it did not go far enough. The opening of so large a number of small separate claims by individual holders barred the essential application of system to the sinking of the great pit. The Mining Board lacked the means, if it had the foresight, to undertake the checking of the reef slides by cutting back the vertical reef walls, and it attempted little practically besides the removal of the drainage and spring water and the clearing away of fallen reef from the face of the blue ground.   This was slipshod mining at best, for the bare extraction of the reef, which had slid and fallen over the claims, actually exposed the mine to further reef slides, and this disaster was aggravated by the utter

Loading Tubs at Bottom of Kimberley Mine, 1885.

lack of system in clearing off the fallen débris. Every claim-holder was permitted to clear off his own claim independently, and credited with an allowance of 4s. for every load of 16 cubic feet of broken reef removed.   The clearing of the face of one claim or a cluster of claims was no security against repeated reef slides, and barred the possibility of developing any section of a mine in an economical and well-planned way.

The practical impossibility of opening a little claim, whose surface area was only 961 square feet, beyond a limited depth forced the consolidation of claims in spite of the original pro-hibition of " claim blocking."   The poorer sections of ground were the first to feel the pressure for the enlargement of hold-ings, and, to secure the continuance of working, permission was granted in 1874 by the Kimberley Mining Board for the hold-ing of ten claims by a single owner.   This concession led to

further combination and consolidation of claims in the hands
of partners and stock companies, but the comprehensive union
essential to the proper development of the mine was far too long
delayed.   The mining operations of a number of individual
claim-owners, firms, and companies — whether in keen rivalry
or in varying degrees of energy and listlessness without any sus-

The Standard Company's Claim, Bottom of Kimberley Mine, 1885.

tained concert of views and means — could not be prosecuted
efficiently and prudently within the small area of a diamond-
mine crater.   Unluckily for the advance of diamond mining
and the fortune of many struggling claim-holders, this irresistible conclusion was not made clear to the mass of miners until
it was demonstrated after long years of costly fumbling in the
diamond-bearing funnels.

In view of the subdivision of ownership, the incoherence of
the mining operations, and the lack of essential funds, the
Mining Board can hardly be charged with a great part of the

Bottom of Kimberley Open Workings.

burden of responsibility for the failure to save the mine from disaster through reef falls. The open pit working was not its design, but the inexperienced undertaking of a mass of diggers who could not be prevented from extracting the diamond-bearing ground in their own rude way. They scooped out the crater to a depth that made reef falls inevitable, and pushed on their cuts through the body of blue ground, in spite of all warning falls and slides, long after it was apparent to any mining engineer that the open pit sinking could not be continued with safety to the workers or with profit to the owners.

But it is impossible to approve the relief measures of the Mining Board. It could only check the reef falls at best, temporarily and partially, but it failed to do even this. It set up expensive hoisting machinery on the surface level at opposite ends of the mine, and sunk a large vertical shaft (Kendric shaft) in the reef at a point two hundred yards from the northeast edge of the crater, with the apparent intention of removing reef rock through this opening or determining the continuance of the blue ground by a drift to the crater below the pit bottom. The shaft was driven down to the depth of 286 feet, when a stratum of volcanic rock was reached, so hard that the work was abandoned. No use whatever was made of this costly shaft, and no considerable attempt was made to cut back the dangerous reef wall. Even with the stinted means at the command of the Board, something might have been done to preserve the mine, and an energetic and well-directed push to this end would have commanded at least the confidence and support of the more intelligent claim-holders. So, when the caving of the reef cast enormous heaps of débris upon the claims in the pit, the lack of foresight of the Mining Board was discreditably apparent. The cost of removing the reef rock was then vastly increased, and the burden was the heavier because the reef falls prevented the extraction of the buried blue ground.

Two of the larger companies, the French and the Central, holding claims in the mine, were the first to undertake the removal of the solid reef on any extensive scale, by sinking shafts, in 1878–1879, at points several hundred feet distant from the

Pumping Engine in Kimberley Mine, 1875.

north and south sides of the mine. By this means considerable reef was removed, and a third shaft was sunk in 1882 through the northeast reef border to check the imminent peril at that edge of the mine. To supplement the service of these shafts inclined tramways were opened on the west and east sides of the mine to cut back the upper reef walls, while wire tramways were stretched from the bottom of the mine to the surface edge to carry off the fallen reef in large tipping tubs, holding from 16 to 32 cubic feet of broken rock. At the end of 1881 tramways, aggregating 19 miles in length, had been constructed by

Incline Tramway for Hauling Reef, 1878.

the claim-holders and the Mining Board. Steam pumping engines had been put in to pump out the influx of water, and this obstacle was, at last, easily overcome. To hasten and cheapen the extraction of blue ground, drilling and blasting were substituted for hand labor with picks, and the work of mining was pressed with incessant energy. But the sliding, falling reef mocked every effort to withstand it.

The work of removal was undertaken too late. The reef slipped faster than the tram cars and tubs could haul it out. In 1878 more than a quarter of the surface of the claims in the mine was covered by fallen reef. The cost of removal, at the original allowance rate of 4s. per load of 16 cubic feet,

FLOATING REEF—KIMBERLEY MINE.

mounted so high that the Mining Board was constrained to cut down the allowance to 2*s.* 6*d.*, but even with the rate reduced the expenditure for reef work and drainage in 1879 and 1880 ran over £150,000 a year, and in 1881 it rose to over £200,000. Still, the need of stimulating extraordinary exertion was then so apparent that the rate was put up to 3*s.* 9*d.* a load in October, 1881, and for the eighteen months following fifty-six million cubic feet of broken reef were hauled away by the claim-holders

Hauling Reef, Kimberley Mine, 1873.

alone, at a cost to the Board of over £650,000, without reckoning the amount extracted by the operation of its own tramways.

    This stupendous charge was obviously too heavy to be borne even by the richest diamond mine, and no assessment scheme could sustain it. The Board struggled for months under the load, issuing notes when it had no cash in hand; but in March, 1883, its issue of outstanding notes or " reef-bills " was so great that its book showed a debit balance of over £250,000, and the local banks would extend no further credit. The Board was bankrupt, reef extraction was stopped, perforce, and the

R

claim-holders were face to face with an appalling situation ; for
in spite of all efforts and the outflow of money like a water-
spout, the resistless reef was unchecked. The mine walls continued to fall in faster than they could be hauled out, and even
central claims in the mine were buried. The gloomiest forebod-
ings fell like a black cloud on the spirits of claim-holders. In

Reef Falls, Kimberley Mine, 1881.

the judgment of
many observers, the
great Kimberley dia-
mond mine was
doomed beyond hope
of resurrection.

The open pit had
been sunk to the
depth of something
over four hundred
feet, in the lowest
working, at the end
of the year 1882. In
order to haul out one
million loads of blue
ground during that
year, three million
loads of reef had
been raised. The
cost of hauling was
increasing with the
deepening of the
mine, and owing to the reef falls, the production of diamonds
was disastrously sinking. In 1883 the lack of funds only permitted the lifting of one and a half million loads of reef at a cost
of £250,000, and the output of blue ground sunk to 350,000
loads. In November of that year a long portended reef slide
cast 250,000 cubic yards of shale upon the face of the pit, piling
its mass on the claims half across the mine. This was seemingly
a crushing infliction. It was, at least, a conclusive proof that

STEAM PUMPING ENGINE, DE BEERS MINE, 1879.

open pit sinking was no longer feasible even for the richest claim-holders.   About four million cubic yards of reef had been hauled at a cost of nearly £2,000,000, yet there was no check to the reef falls and slides.   At the close of the year the Inspec-

tor of Mines reported that "only about fifty claims had been regularly worked during the past year." The field for the operation of individual claim-holders was decisively closed. The only hope for the mine was in the prosecution of deep and extensive underground works by the combination of claims in hands able to conduct such operations successfully.

In advance of such an undertaking the yield of the mine was fortunately sustained by an expert makeshift. Mr.

The Central Company's Shaft, Kimberley Mine, 1885.

Edward Jones, a trained mining engineer, had been one of the leading contractors for the removal of reef, and had given close study to the problem of the continuance of the extraction of blue ground.   Through his design and insistent confidence, in spite of all doubts and sneers, a shaft was sunk through the mass of fallen reef at the bottom of the deepest part of the mine by lowering a square timber frame and shovelling out the loose rock from the inside of the enclosure.   The frame was constructed in sections on the plan of a coffer dam, adding section to section from the top until a stout timber shaft passed entirely through the broken shale and entered the underlying blue ground.   The shaft was then read-

ily extended, and drifts from this opening were made through
the blue ground. The peculiar service of this device was its
saving of hundreds of feet of costly shaft cutting through the
solid reef to reach the blue ground — a very desirable contribution at a time when the richest claim-holders were sharply pinched
by the failing mine and the discouragement of capital. The cost
of all development work was defrayed by the blue ground
extracted in opening the drifts and cross-cuts, so that there was
no further delay in resuming operations in the mine. The first
shaft had been sunk on the ground owned by the Central Company, and it was soon copied by a number of similar shafts in
other parts of the mine. This brought about a most welcome

The Bottom of Kimberley Mine, 1885.

revival of mining, and was so far highly beneficial to the laborers, claim-owners, and townspeople of Kimberley, though it was
not designed for permanent service.

While the blue ground was being removed through shafts
sunk in the bottom of the open mine, it was apparent to all that

the life of these shafts must be very short.   Preparation was
therefore made for future work by sinking shafts outside the
margin of the open mine, and at sufficient distance from it to
insure them against any probable caving of the surface ground
in their vicinity.   Vertical shafts were sunk by the Central and
French companies, and tunnels driven from them.   The plan
of Kimberley mine, 1883, shows these tunnels.

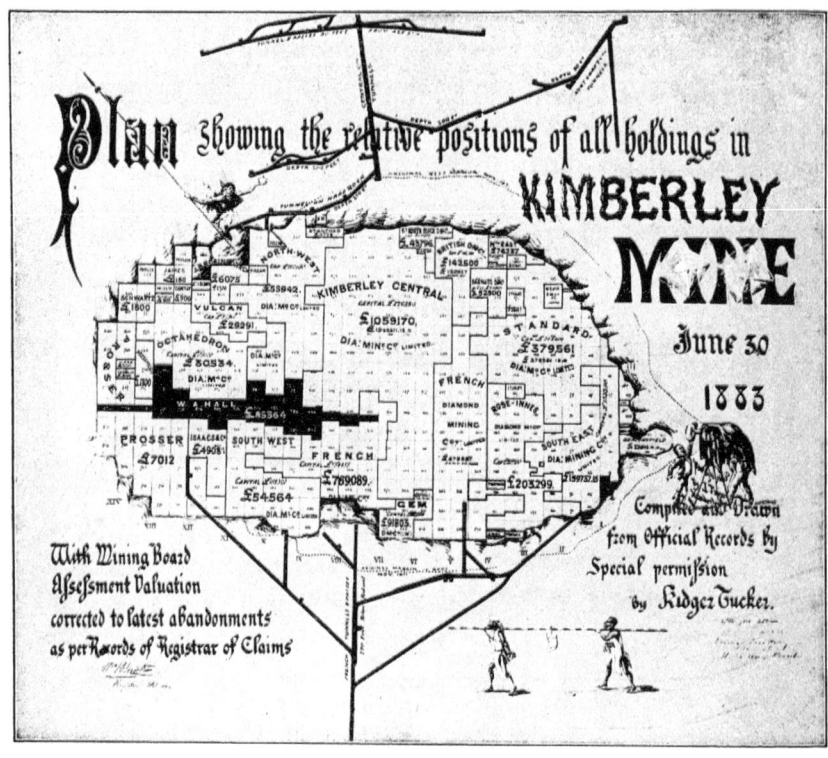

Before describing the subsequent application of engineering
science to underground mining, it is desirable to trace the progress of the other mines on the fields to the period in development reached by the leader.   The claim-owners in De Beers
mine profited greatly by the object lessons given in the opening
of the great pit of Kimberley.   For the first twelve years after
the discovery of the mines, the Kimberley mine ran far ahead of
the others from the superiority of its yield for some distance
below the surface.   The fatal error of the neglect of the claim-

KIMBERLEY MINE, 1886,

Showing Shafts at the Bottom of the Open Mine.

Reef Slips, Kimberley Mine, 1874.

Kimberley Mine, showing how the Ground cracked before Subsidence.

owners and Mining Board to cut back the mine walls was apparent in time to save many thousands of pounds to De Beers. This mine was also fortunate in the comparative hardness and stability of the basaltic rock stratum overlying the shale and forming the marginal top of its pit walls. By cutting back the reef in terraces, the De Beers Mining Board saved the mine from any serious rock falls for a considerable number of years. Only two hundred and fifteen thousand cubic yards of reef were removed in the five years ending with 1882, but this sufficed to protect the mine for the time. The cost of its removal was only £76,000, a slight burden compared with the charges at Kimberley mine, and showing a

The Central Company's Atkins Shaft.

cost per yard or per load of reef raised much less than the Kimberley average. This was a signal demonstration of the advantage of prudently cutting away the reef before it fell into the pit and buried prolific claims and increased the hauling charges.

This precaution, however, did not suffice to shield the mine from disaster when the pit was greatly deepened after the reef falls at Kimberley had diverted mining enterprise to De Beers. Over one hundred and forty thousand cubic yards of solid and broken reef were removed in 1883 and 1884, but reef slides were fast increasing, and it was judged necessary by the Mining Board to stop any further outlay for reef hauling when the mine bottom was 350 feet below the surface. The diamond-bearing ground had then been scooped out of the larger part of

The Last of Open Working, Kimberley Mine, 1889.

the funnel, but there was still a large area of yellow ground at the west end which had not yet been extracted because it contained so few diamonds compared with the other parts of the mine. The falls of reef had covered the eastern end of the mine, and early in 1885 the west end yellow ground caved in, and an enormous mass of nearly five million cubic feet fell in one day to the bottom of the mine, overlapping the fallen reef and burying the claims still open for work. This disastrous fall forced the stoppage of mining for six months until some part of the reef and yellow ground could be taken out, and mining was then resumed in a partial and half-hearted way in the open pit, though it was evident that further pit sinking in the face of such disasters was irrational mining.

R. D. Atkins.   (Manager of Kimberley Mine in the earlier days.)

The only possible resource was the introduction of a system of underground mining, and the first attempt in this direction was made in 1884 by the opening of a large circular shaft at a point 1000 feet from the north margin of the mine. This shaft was sunk vertically about 320 feet in the reef and then abandoned as too costly. In its place an incline was sunk, starting from a point about 150 feet from the west side of the claims, and entering the mine at the edge of the amygdaloidal trap underlying the basalt and shale, so as to avoid the expense of cutting through this hard rock. This work was begun none too soon, for before the end of the year 1887 further open pit working was proved to be utterly impracticable, and was wholly abandoned when the deepest open digging had been carried in three years only fifty feet farther than the depth of 350 feet reached in 1884.

Dutoitspan mine opening was practically the same as the course followed in Kimberley and De Beers. Owing to the comparative poorness of the diamond-bearing ground, pit sinking was not pushed as rapidly as it was at Kimberley, and, in 1874, most of the miners went over to Kimberley and were glad of the

No. 2 Incline Shaft, De Beers Mine.

chance of working over the "waste ground" which had been
cast away from the cradles and sieves of the early diggers. Two
years later, when improved methods of handling the ground
were coming into use, the miners flocked back to the abandoned
ground and took out fresh claims. Warned by the experience
of Kimberley, a circle of solid blue ground was left as a buttress
against slides and falls of the encasing reef of shale, and for ten
years this expedient served to shield the miners.

But this safeguard failed when the open working had reached
a few hundred feet in depth. Warning surface cracks had been

Eldorado Road, Dutoitspan Mine, 1874.

noticed on the northern margin of the mine, but the ardent diamond seekers kept on digging recklessly, until one day in March, 1886, when a huge mass of blue ground and reef broke away suddenly from the northern end of the mine and rolled over like the surge of a monstrous breaker, falling hundreds of feet with a fearful crash upon the doomed men at the bottom of

Claims in Dutoitspan Mine.

the pit.   The loss of life would have been frightful, but happily for the miners the fall was at the noon dinner hour, when the work of hoisting blue ground was stopped and blasting in the mine was begun.   Most of the workmen had left the mine, but eighteen poor fellows — eight white men and ten Kafirs — had taken shelter from the blasting in a pumping engine house in the pit.   The avalanche of rock fell on the house, and every one in it was fatally crushed or scalded by the escaping steam.   One hundred thousand cubic yards of shale and blue ground buried the claims on the pit bottom, and this fall was followed by others which ruined the open workings in 1887, when the mine had reached a depth of 400 feet.

In Bultfontein there was only another variation of the same tale of open pit working and final wreck.   The work of extracting the yellow and blue ground was well planned at the outset, under existing circumstances, by the cutting of inclined roadways over which the ground was hauled in bullock carts.   In

BULTFONTEIN MINE, 1878.

1880 effective hauling machinery was substituted for the carts, and the precious ground was extracted so rapidly that the depth of about five hundred feet was reached in the open working, a point probably beyond any attained in the other pits. Here, too, as at De Beers, there was an effort to protect the mine by cutting back the reef in terraces; but this safeguard was tried too late, and in any event it could only have deferred for a few years the fate of the mine. Before the close of the year 1889 almost the whole of the pit bottom was covered with fallen reef and only four engines were at work hauling blue ground.

The Extraction of the Diamonds

While the sinking of the pits was progressing with improved mining appliances, there had been a considerable advance in the methods of concentrating the diamond-bearing ground and win-

Bultfontein Mine, 1879.

ning the diamonds. For the first three years after the opening of the mines, the handling of precious ground was exceedingly crude and wasteful. The broken ground taken from the craters was crushed more or less finely by pounding with shovels and

mallets and clubs.   Then it was sifted in rocking troughs, fitted
with sieves like the placer miners' cradles, and the concentrate of
pebbles and crystals and coarse rock grains was spread on tables,
or sheets of iron and wood laid on the ground, where it was
scraped over by hand, and the gems picked out.   In this rough
process a third and perhaps a half of the smaller crystals were left
in the waste ground, and the losses from theft were enormous.

In 1874 there was a change for the better in the introduction of water in concentrating.   By building dams and sinking

The First Rotary Washing Machine.

wells the water supply of the camp was increased materially, and
it was possible to divert a portion for the diamond-washing
appliances.   Most of the early machines for this purpose were
simple cradles with riffles or ridges set at intervals on the
bottom, and a sieve at the end.   The pulverized ground was
dumped into a cradle with a sufficient flow of water to carry off
the slime, while the rocking shook the ground, and caused a
settling of the heavier mineral deposit at the bottom.   With one
of these rockers from six to thirteen cartloads of ground were
washed in a day.   Another device was a circular trough or pan,

fitted with a revolving set of iron teeth like a comb, that stirred
the ground and water and caused the settling of the concentrate.

Another Early Washing Machine, 1874.

This puddling trough would concentrate from twenty-five to
thirty-five cartloads in a working day and cost at first, about
£250, while the simpler cradle could be bought for £15, or
less.   There were other more elaborate devices, but their cost
put them out of the reach of the ordinary digger.   All were
based on one adaptation or another of the puddling principle,
and the fall and separation of minerals of different specific
gravities.

The sorting of the concentrate from the
puddling troughs was done by the same
method employed after the dry sifting,
but there was some improvement in the

Horse-power Washing Machine, 1875.

EARLY HORSE-POWER WASHING MACHINE, 1874.

THE FIRST WASHING MACHINE WITH ELEVATOR TO CARRY AWAY THE TAILINGS.

precautions against loss by theft.   The natives, who were commonly employed in scraping and picking over the mineral deposit, were more carefully watched.   Some were lodged in tents and sheds adjoining the stables belonging to claim-owners, and there was some oversight of them by night as well as by day. When the claim-owners combined in companies, their workmen were frequently kept together in enclosures called "compounds," where they were furnished with food and shelter at moderate charges deducted from their pay.   This separation and partial restriction was of undoubted service, not only in diminishing the opportunities for successful theft and disposal

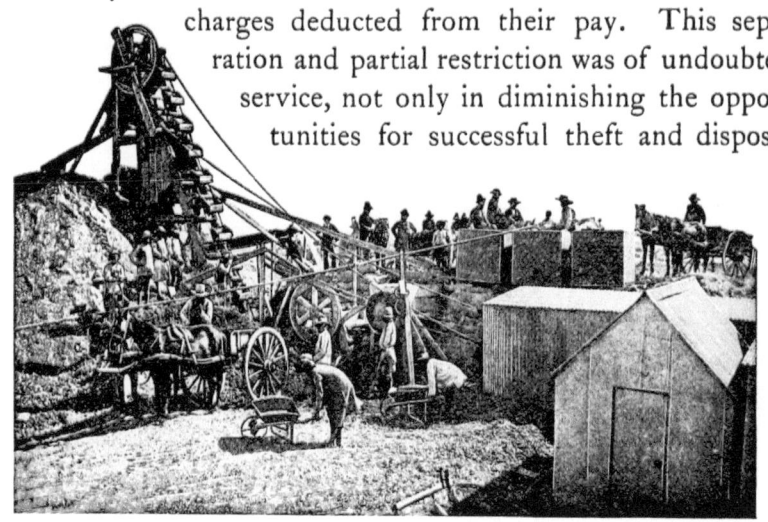

Washing Gear, Bultfontein Mine.

of stolen diamonds, but in checking the drunkenness of the black workmen and the outbreaks in the canteens and streets.

Progress was made, too, though much too slowly, in the more perfect pulverization of the blue ground.   It was soon observed that the broken ground would crumble upon exposure to the air, and after some weeks or months, according to its hardness, a mass of breccia, thinly spread out and raked over, would be very largely decomposed to fine sand fit for washing, without further treatment.   This natural pulverization was far cheaper and better than crushing with mallets; but the burden of accumulating and storing great quantities of ground was too heavy for the ordinary claim-holder, who was dependent upon quick returns: so only the larger companies maintained stores

of ground on their depositing places or "floors," and none of
these, even, were disposed to wait for the adequate pulverization
of the ground by the natural agencies of the sun, air, and rain.
Still the floors were gradually enlarged on the veld, and were
frequently fenced in with wire.   Year by year an increasing
proportion of blue ground was pulverized.   The average yield
of a truck load, or sixteen cubic feet of blue ground, from Kim-

Steam Washing Gear, Kimberley Mine.

berley mine, was computed to be one carat in diamonds, a valuation ranging from twenty-eight to thirty-six shillings, according
to prevailing market rates.

The mining camps changed, year by year, more completely
to the appearance of thriving mining towns.   De Beers fused
with De Beers New Rush in the town of Kimberley, while
the town of Dutoitspan rose on its camp site two miles away.
The connecting roadway was lined with straggling houses.
There was little available timber fit for building purposes, but
galvanized iron was very largely substituted for the canvas tents
during the first ten years, and, from 1880 on, many brick buildings were erected at Kimberley.   Outside of the main business
street there was little attempt at first,to lay out regular avenues,

and the diggers shifted their tents or "tin houses" to any vacant place that suited their fancy. The little galvanized iron buildings were so light and strongly riveted that they could be picked up and carried away by a few strong Kafirs. But with

Webb's Washing Machine, 1878.

the growth of the towns stands became more valuable, and title and possession were more sharply looked after. In 1876 the valuation of the town of Kimberley for assessment or taxation purposes was $5,151,500. Churches, schools, banks, hotels, theatres, concert rooms, and stores and offices of various kinds were erected to answer the demands of a prospering mining town. Sidewalks were laid along the principal streets, and after 1874 there was a regular appropriation for street watering. The houses grew in size and stability. Verandas and porticoes were added in place of the roof projections that gave a little shade to the early diggers, and many of the dwellings were set with a fringe of garden in front or on the sides, in which fruit trees and vines and choice flowers were planted.

With the advance of the diggings in depth, the combination

of claims, and the ending of widespread prospecting, the influx of whites to the camps fell off greatly. The shifting population of prospectors dropped to the number that could find employment in the mines or in the dependent towns. It was estimated in 1876 that the white population of Kimberley was about eight thousand, and the native from twelve to fifteen thousand. In Dutoitspan and Bultfontein there were perhaps six thousand more of whites and blacks.

The character of this population has been most absurdly decried. " The Diamond Fields of South Africa," writes one flighty reporter, " have been hot-beds of rowdyism. The libertines, forgers, bird-catchers, and other outcasts of Europe have found a refuge there as in Alsatia of old. The Houndsditch Jew and the London rough reign supreme." Thousands of witnesses might be summoned, if necessary, to refute this nonsense. Libertines and forgers drift elsewhere for prey than to hot, dusty

Cape of Good Hope Company's Washing Gear, 1878.

mining camps in the midst of the karoo ; though dainty folk might shrink from the roughness and grime of the diamond diggings, and weak nerves might be shaken by the boisterous exuberance of the bustling camp, the restless crowd tramping the streets, the uproarious canteens and music halls, and the capers of motley diggers and wild Africans.   Liquor drinking ran to excess, as it always does in a prosperous mining camp, and the natives especially were given to drunkenness; but the wildest sprees rarely threatened danger to life, for the hot spirits were blown off in yells, chants, and dances.   Every accurate record shows that murder and robbery and the more flagrant and brutal crimes were notably rare compared with the showing of the early American and Australian mining camps; and when the turbulence of the rush was over, and the bubbling camps simmered down to the comparative order and steadiness of the working

Washing Gear, Dutoitspan Mine.

mining towns, there was little disturbance from any outbreak of ruffianism.   In spite of all demoralizing influences, the conservative and civilizing agencies and public spirit that advance communities and exalt good citizenship gained in force year by year on the Diamond Fields.

Notable progress was made in the provisions for the health and security of the towns.   The most crying need, from the first, had been pure and abundant water.   The average rainfall of the mining field was only 17.5 inches, and the suffering from the lack of water in the dry season was scarcely endurable. Much was done to improve and increase the supply by the sinking of wells and extension of natural reservoirs and the more general introduction of filtering appliances. Dr. Morton noted in 1876 a marked advance in the health of the population on the Fields. The death rate at Kimberley, he said, was exceedingly small. The most sickly months of the year were August and January, marking the effect of the extremes of cold and heat. Outside of the ailments incident to the dust and exposure and sudden variations of temperature, there was little disease, and he particularly observed the complete immunity of the field from hydrophobia, though every man, woman, and child appeared to have a dog at their heels.

Washing Gear, Bultfontein Mine, 1878.

It was soon perceived, however, that a more certain and sufficient supply of water must be obtained to meet the growing demands of the towns and mines. This was secured through the enterprise of the men associated in the Kimberley Water Works Company, by the construction of a pumping station at Riverton on the Vaal River and the laying of a main sixteen miles in length to a reservoir on a ridge of the Bultfontein farm, near Kimberley. The water from the river was raised in three stages by powerful compound condensing engines, and carried to the large reservoir on the ridge, five hundred feet above

the river level.   From this reservoir it was distributed by a pipe
and hydrant system to the towns and the mines.   Since the
construction of this fine plant, the towns have been supplied with
filtered water at a cost of 1*s*. per 100 gallons ; and mines using
great quantities have a concession materially lowering this rate.
The amount of water sold to Kimberley annually has run as
high as 230,000,000 gallons and more than 300,000,000 have
been supplied to the mines.   The cost of the machinery and plant
was over £300,000.   Mr. E. A. Cowper, the consulting engineer of the Water Works Company, designed the machinery, and
Mr. George Buchanan, C.E., was the constructing engineer in
the erection of the plant.

The maintenance of peace and order on the Diamond Fields
was helped forward materially by the construction of "compounds," providing good lodging and food for the natives, checking their drunkenness, promoting steady industry, and enforcing
restrictions essential to the common security.   The police force
of the towns was from the start so small that the tolera-
tion of this condition attests the comparative rarity of brutal
crimes on the Fields.   Its very marked improvement with the
growth of the town, in later years, was rather due to the rising
demand for advance in every civic and social condition than to
any increase in disorderly conduct or the commission of crimes.

Diamond stealing and illicit diamond buying were, beyond
all question, the worst plague of the camps and towns.   Outside
of this line of operation there was practically no opening and no
temptation for the professional thief and receiver of stolen goods;
but the opportunities were unfortunately too apparent and easy
for filching and disposing of diamonds.   The sharpest oversight
could scarcely prevent nimble-fingered workers from slyly secret-
ing tiny crystals in picking over the concentrates on the sorting
tables or in handling the deposit in the rockers and puddling pans.
While the natives were allowed to rove about freely after their
day's work was done, they had little difficulty in transferring the
diamonds to the hands of the sharpers, who were always in wait
for the chance of buying stolen stones for little money.

Offices were opened by diamond buyers in the mining towns, either as independent merchants or as representatives of large, foreign wholesale dealers and diamond cutters, and besides these established purchasers, there were a number of traders who made regular rounds through the diggings, buying from claim-owners in their tents or houses or at the sorting table. These peripatetic dealers were familiarly known as "kopje wallopers," for kopjes were the sites of the chief surface digging. No doubt there were dishonest men among these dealers, small and large; for the frequent temptations were too strong for slight scruples, and it is certain also that many diamonds were bought under cover by saloon and shop keepers and other speculative traders who came into familiar contact with the diggers.

It is plain that it was impossible to trace or identify a stolen diamond, even when the theft was known, and great quantities of gems were secretly bought and carried to the coast towns for sale or forwarded stealthily to foreign markets. It has been estimated that fully fifty per cent of the diamonds taken from the diggings in the early years were secreted and sold speculatively. This is undoubtedly an extravagant reckoning, but there is no question that a large percentage were filched away.

To give some idea of the enormous quantity of diamonds that were stolen in the early days of the fields, and before the compound system was adopted, the following notice is reproduced : —

NOTICE

The undermentioned rough and uncut diamonds having from time to time been recovered by this Department, notice is hereby given to all whom it may concern, that unless proof of the bona fide right to the possession of such diamonds be given, or a proper permit for the same be produced within ten days from the date hereof, such diamonds will be sold and the proceeds of such sale carried to the account of the Government.

JOHN FRY,
Chief of Detective Department of Griqualand West.

MAY 24th, 1883.

Underneath the notice was a schedule showing —

The number of carats.   From whom recovered.   How acquired.

The number of carats ranged from half a carat to 6375 carats, which were found in the possession of one man. The total number reached 8443 carats, which were recovered from fifty persons. Two days later a similar notice appeared stating that 1573½ carats had been recovered, having been found in the possession of a well-known dealer in illicit diamonds. The total value of these two lots would amount to £30,000 or £40,000.

The practice of illicit diamond buying was so persistent and obnoxious that it was curtly styled I. D. B., and the strictest possible regulations were made to check it and punish offenders. A Special Court was established in 1880[1] to try cases of this kind, and a special police force formed with warrant to make the most rigorous search of suspected thieves and receivers. Under the Diamond Trade Act every parcel of diamonds taken from the Fields must be formally described and registered, and every transfer recorded from the date of discovery till the final shipment from the Cape Colony. No person was permitted to deal in diamonds unless he held a formal license, and his record books of purchase and sale were always open to police inspection. Thefts of diamonds and illicit purchasers were punished with all possible rigor.

---

[1] A Special Court was established under ordinance No. 8 of 1880. A barrister was appointed as special magistrate to act with the resident magistrate and the additional resident magistrate. Under Act No. 48 of 1882 the special court for mining offences consisted of three persons, of whom at least one was a judge of the Supreme Court. The other two were usually the resident magistrate and the civil commissioner. By proclamation No. 144, dated September 1, 1882, the districts of Kimberley, Herbert, Hay, and Barkly were within the jurisdiction of the Special Court. Act No. 34 of 1888 provided that the Special Court should consist of three members, two of whom must be judges of the Supreme Court. Persons convicted by the Special Court might appeal to the Supreme Court.

CHAPTER IX

THE MOVING MEN

N the rush of adventurers over the Diamond Fields the individual was inevitably merged in the mass. He might feel the pulse of latent powers, the unslaked thirst of ambition, but he must be for the time no more than a drop of water in the rapid, a locust in the swarm. He was one of a myriad which exulted in the enforced equality of living and opportunity.

There can scarcely be a purer democracy than an infant camp in such a field. Imperial sovereignty or feeble state assertion barely cast a shadow of authority over the stretch of " No Man's Land," the chrysalis of the Colony of Griqualand West. One man here was as good as another in his own mind, and free to maintain it. In the seething stream of humanity that poured into the Diamond Fields it mattered not whether one was to the manor born or cradled in a manger, the son of a peer or a beggar's brat. In the hot scramble for diamonds in the dirt, all ranks were levelled. The rough sailor jostled the captain, the university graduate swung his pick side by side with the navvy, and the last of the Vere de Veres snored in his sheepskin kaross back to back with a hopeless Japhet. The representative " Diggers' Committee" was merely the executive hand of the body of prospectors, the instrument of the will of the masses.

The distribution of the diamond beds from the start marked the strain for equality, the hostility to aggrandizement; and the relation of demand to supply compelled the division into little patches of holdings. It was years before the acquisition of more than two claims by one person was tolerated, and only imperious necessity forced the further consolidation of claims when the mines had reached a depth that made patch-working impracticable.

In this mass movement and equalizing of opportunity, the rise and display of strong individuality were necessarily subdued and slow to appear. In the years of the rush and the early advance of the mines, it is the life of the mass and not of the fractional unit that makes the history of the Fields. But with changing conditions, as the years rolled on, the way was opened for individual assertion, influence, and distinction. Then the men, hitherto unmarked, stood up preëminent. Then the brains that were capable of great conceptions and great performances found pressing occasion for all their foresight and energy. The history of the great mines that have explored the diamond-bearing craters so far beyond the pitfalls of the prospecting diggers is very essentially a story of remarkable men.

In July, 1873, a young Hebrew, Barnett Isaacs, took passage from England to Cape Town at the call of his brother from the

new Diamond Fields. His grandfather was a learned and honored rabbi, and the good standing of his family was marked by the marriage of his father, Isaac Isaacs, to a relative of Sir George Jessels, Master of the Rolls. But the son of the rabbi was only a small, plodding, frugal shopkeeper in London. His sons, Henry and Barnett, were trained in the excellent Hebrew Free School in Spitalfields, but both boys left school

Barnett Isaacs.

at the age of fourteen to help their father in his shop. Henry was drawn away in the current of the early rush to the Diamond Fields in 1871, and had such success as a kopje walloper that he wrote home to urge his brother to join him.

To the restless spirit and purely speculative mind of Barnett Isaacs there was magnetic attraction in such a field with its novel and gleaming opportunities. With instant decision he took the steamer for Cape Town, and made the tiresome trip over veld and karoo to Kimberley with unfailing pluck and good temper.

He was only twenty years old, and outwardly no more than a light-hearted boy, bubbling over with high spirits and comical conceits. But his fondness for athletic sports, theatrical extravaganzas, and practical jokes, and his contempt for conventional restraints, were merely the surface froth covering invincible energy and facile grasp of opportunities. He had an unshakable self-reliance, a quick perception, and a fertile resourcefulness that bore him up when feebler men sank. One could scarcely cast him in any society or any place on earth, where his nimble wits would not win him a living.

The impulse to go ahead was in his blood. " It has always been a superstition with me," he said, " never to turn back." He grew apace with the calls upon his powers. He did not profess to know more than he knew, but he was never content to know anything that interested him by report. " I must look into everything that concerns me for myself." This determination was a safeguard. He once boasted, in a rare fit of parade, that he had never made a mistake in his investment of money in his life. But his incessant activity was fatally wearing. He could not dawdle. He could hardly rest. For many years his extraordinary vitality and endurance kept him running. He had the precious faculty of dropping off to sleep at any moment of relaxation, and awaking after slumbering for a few moments. Nevertheless no creature of flesh and blood could endure the strain which he bore and recklessly courted. " Some day such a bundle of quivering nerves must snap, either life or brain must go," said one of his closest friends. But when young Barnett Isaacs wandered into Dutoitspan, " fit for anything," as he himself declared, after his long tramp and meals of porridge and biltong, nobody saw in him the raw material of one of the remarkable financiers of the century, or forecast, even dimly, the meteoric career of Barney Barnato.

His brother Henry had fancied and taken the name of Barnato, as a professional shift from his own family name, when he first came to the Diamond Fields and tried his luck first as a conjurer and vaudeville performer, relying upon the sleight of hand proficiency which he had gained in boyish practice to amuse his friends.   Henry soon turned his hand to the more profitable business of a diamond trader, but his stage name stuck to him, and passed naturally to his younger brother, who accepted it with easy indifference.   So young Barnett Isaacs became familiarly known as " Barney Barnato," and for the first year or two of his life on the Diamond Fields floated along in the current as " Harry Barnato's brother."   But his head never sank below the surface for a moment.   His first buoy was a cigar box.   He had money enough to buy sixty boxes of cigars after paying his way to Kimberley.   With this working capital he went into partnership with Louis Cohen, another newcomer, who had started as a kopje walloper.   The two young Hebrews picked out a shanty to their liking for an office.   It was a little tin shed, eight feet by six, owned by an Irishman who offered it for rent at a guinea a day.

"That is ridiculous," said Cohen.

"I don't know that," said Barnato.   "The situation is good, why not pay a guinea a day if you can make thirty shillings ? "

This keen measuring was typical.   Barney Barnato never counted cost alone if he wanted anything, but weighed it instantly against probable profit.   He was never a thoughtless or reckless buyer.   He did not shut his eyes to the risks of loss.   On the contrary, he reckoned risks with exceptional accuracy and precision of detail, but he reckoned profits with the same even-balanced judgment.   Hence he was not afraid to venture when others shrank back.   He was naturally sanguine.   He had faith in himself, and put all his working force into everything that he undertook.   So his high-pressure energy, persistently maintained, won success where a weaker and idler man would have failed.

There was no peculiar luck in his favor. Thousands around him had equal chances or better. He went to the front because he had the brains to choose aright and the working powers to make his choice profitable. He made mistakes as men of his sanguine temper must, but he did not make many mistakes, and no fatal or even greatly damaging ones.

There is no business without risks. The most prudent man cannot engage in mining or in trading in mineral products without risks. If hot-headed speculation has swamped fortunes in such a field, it is no less certain that overstrained caution has failed to win anything memorable. There is a happy and rare mean of sagacious judgment in mining operations, and Barney Barnato proved his possession of such judgment incontestably. His mind worked so quickly, and his mental calculations were so exact and minute, that it was often supposed that he jumped at conclusions. "Barnato's snap judgment," sneered a man whom he outbid in competition; "Barnato's sheer luck," growled the man who saw his judgment turn to gold.

The young partners, Barnato and Cohen, worked hard, early and late. Barnato's keen eye gained a valuable business connection in a way that suggests his kinship to Sherlock Holmes.

One of the most successful "kopje wallopers" (a name given to men who visited the various miners' huts for the purpose of buying diamonds) made regular rounds through the diamond fields on an old, lame, yellow pony, calling on men who had the best bargains in diamonds to offer. Barnato and Cohen tried repeatedly to follow him, but his track was soon lost in the labyrinth of tents, huts, and sand heaps. However, Barnato was able to see that the trader's pony had the habit of stopping at places where choice bargains were made, and when the broken-down beast was offered for sale one day by its owner, Barnato snapped at the chance to buy him for £27 10s., an enormous price for the old pony as a steed, but a great bargain for the keen diamond broker, for the walloper's business went with his pony, as he afterward saw to his chagrin.

Soon Barnato became known as a "walloping walloper,"

and in the third year of his push into the Fields he was able
to crown a new ambition by the purchase of a block of four
claims in one of the best-paying sections of the Kimberley mine.
His savings were then about £3000, and he put nearly every
pound he was worth into his purchase.   His seemingly risky
investment was quickly justified by the yield of his claims.
With the help of this great investment he came swiftly into
prominence.   Entering into partnership with his brother, he

established the firm of Barnato Brothers in 1880, as a
London and Kimberley firm
of diamond dealers and brokers in mining properties, and
crowned a further ambition
by combining his own claims
with adjoining holdings in his
first mining stock organization, " The Barnato Diamond
Mining Company."

He was one of many
quick-sighted and resourceful
men who perceived that the
day for any profitable working of individual claims had
passed, while the body of
miners was still struggling
along blindly in the great cav-

C. J. Rhodes, when a Student at Oxford.

ing chasms. He brought about a highly desirable amalgamation
of the claims which he controlled with those of the Standard
Company, one of the strongest organizations in the Kimberley
Mines, and later these claims were amalgamated with the holdings of the Kimberley Central Company, in which he became a
large shareholder.   It was at this stage in his fortunes that he
came into keen rivalry with the only competitor that could make
headway successfully against him, Cecil John Rhodes.

There was a singular likeness in some respects in the careers,

PORTRAIT OF CECIL JOHN RHODES.

conceptions, and calculations of these extraordinary men, although they were so markedly dissimilar in personal appearance and temperament. Cecil John Rhodes was the younger son of a Hertfordshire clergyman, and came as a sickly boy to South Africa in 1871, in the first flush of the diamond fever, to join his brother Herbert on a small plantation in Natal. The raw, dusty Diamond Fields were apparently one of the spots least likely to attract a youth whose health had broken down, and whose tastes were bent from early childhood toward a scholar's

Cape Town.

life in the cloisters of a university appealing to every high imagination in its memorials of every age since the dawn of letters in Britain. So indeed it seemed when young Rhodes turned his back on the fresh glitter of the new mines and entered his name on the rolls of Oriel College in 1872. But the same year saw his return, because of a lung fever that threatened his life, and made the shift from misty England to the mild clear air of the terraces of Natal an imperative prescription. Shortly after his return Herbert Rhodes slid into the current setting to the Diamond Fields, but Cecil stayed on the plantation until the following year, 1873, when his brother's report and his dawning

T

success as a claim-owner drew him, somewhat reluctantly, over the long sun-baked stretch to Kimberley.

So unknown to each other and blind to their future clash and union, Cecil John Rhodes, the clergyman's son, and Barney Barnato, the London shopboy, started abreast in the race for fortune on the same track. An ordinary observer of the two young men would probably have picked Barnato as the winner on such a track as the new Diamond Fields. Any one could see at a glance that the young Hebrew was unsinkable, and peculiarly fitted to make a good living in the stirring towns by his business training, quick wit, and racial genius for trade, while the English college student had no apparent fitting for success either as a digger or a business man. Kipling has told of the straining of the new ship, as a living thing, in the trial to find herself, and this fine conception has literal truth in the application to young manhood. So Cecil John Rhodes was forced to find himself, as he did, when he put away his books to plunge into the whirling life of the Great White Camps.

Tall, gaunt, shy, the stripling sat at the diamond sorting table, overseeing the Kafirs who scraped over the pebbles from his brother's claim, on a little "floor" near the edge of the big Kimberley pit. Roughly dressed, coated with dust, disdainful of any foppish touch, peculiarly self-contained, full of novel ideas and aspirations rising, turning, and shaping themselves in his mind, he was not one to mingle, like Barnato, in every stir of the froth in the camps, or ready to jump, like the London shopboy, into any gush of speculation, from a bet at cards to an auction sale. Externally the two young men could scarcely be more unlike than the little, chunky, bullet-headed, near-sighted, mercurial Hebrew, taking a hand in current sport or traffic, and the tall, thoughtful, young overseer, sitting moodily on a bucket, deaf to the chatter and rattle about him, and fixing his blue eyes intently on his work, or on some fabric of his brain.

Yet both were alike in their expanding ambition and power to grapple and mould in their distinctive ways the opportunities about them. Both had keen foresight, and extraordinary com-

Silver Trees.   (These trees grow only on the slopes of Table Mountain.)

prehension of great financial undertakings.   Both had, too, the essential poise and accuracy of judgment that shuns pitfalls and punctures illusions.   With variant motives they sought the same end of great riches : one for the sheer satisfaction of money making, of unfolding great schemes of production and flotation, of proving to the world that he was a master of finance; the other chiefly as a means to reach ends of Imperial scope, to throw the searchlights of civilization into every cranny of the Dark Continent, to lift the prodigious dead weight of unnumbered bygone ages of barbarism, to make the waste lands fruitful and open the arteries of traffic, to create a Greater Britain than the most daring fancy before him had conceived, and stretch the hand of his Queen over a realm transcending the farthest sweep of the Macedonian or the Roman.

Both realized very keenly the practical necessity of effecting combinations of the claims covering the diamond mines in order to provide a uniform and efficient development and to secure a scarcely less essential control of the diamond output. The patent collapse of the open pit mining forced the undertaking of underground works, and compelled the further consolidation

of holdings; but for too many years there was no common realization of the urgent need of the systematic development of the mines as a united property, and not as a complex collection of discordant parts. The

A Cape Cart.

working of the parts was at best cramped and conflicting. The prosecution of any well-designed plan was heavily handicapped by the lack of coöperation in adjoining properties.

This was sharply etched in by Barnato after Rhodes had successfully pressed the amalgamation of the variant interests. "I think I can prove to you, gentlemen," he said, in addressing a shareholders' meeting in 1889, "that in order to work the underground system, you must have the mines intact. You all remember the trouble and friction that took place when the De Beers mine was being worked by the De Beers Company, the Victoria, the Oriental, the Elma, the Gem, and others. Why was the underground system not a success in this case? Because one company was working against another; that is to say, if one

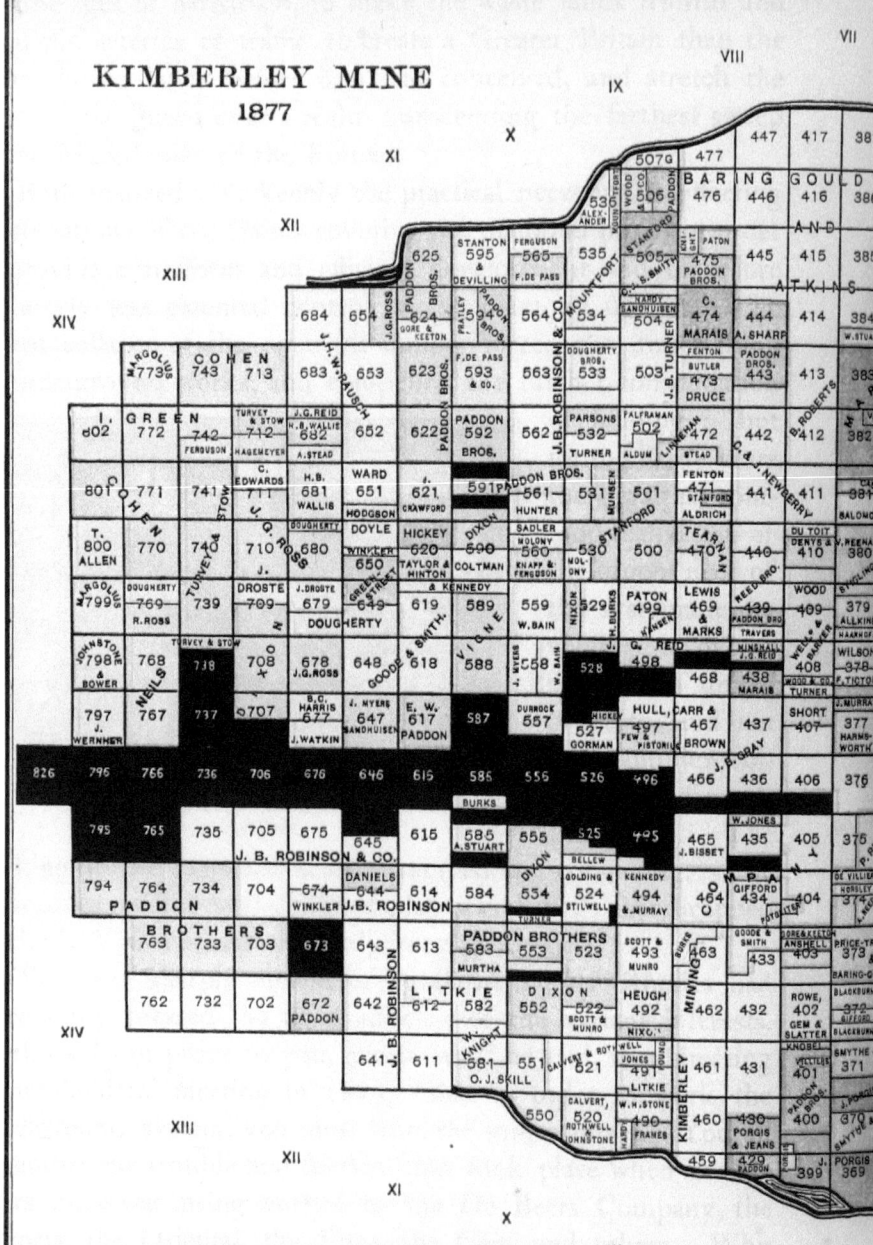

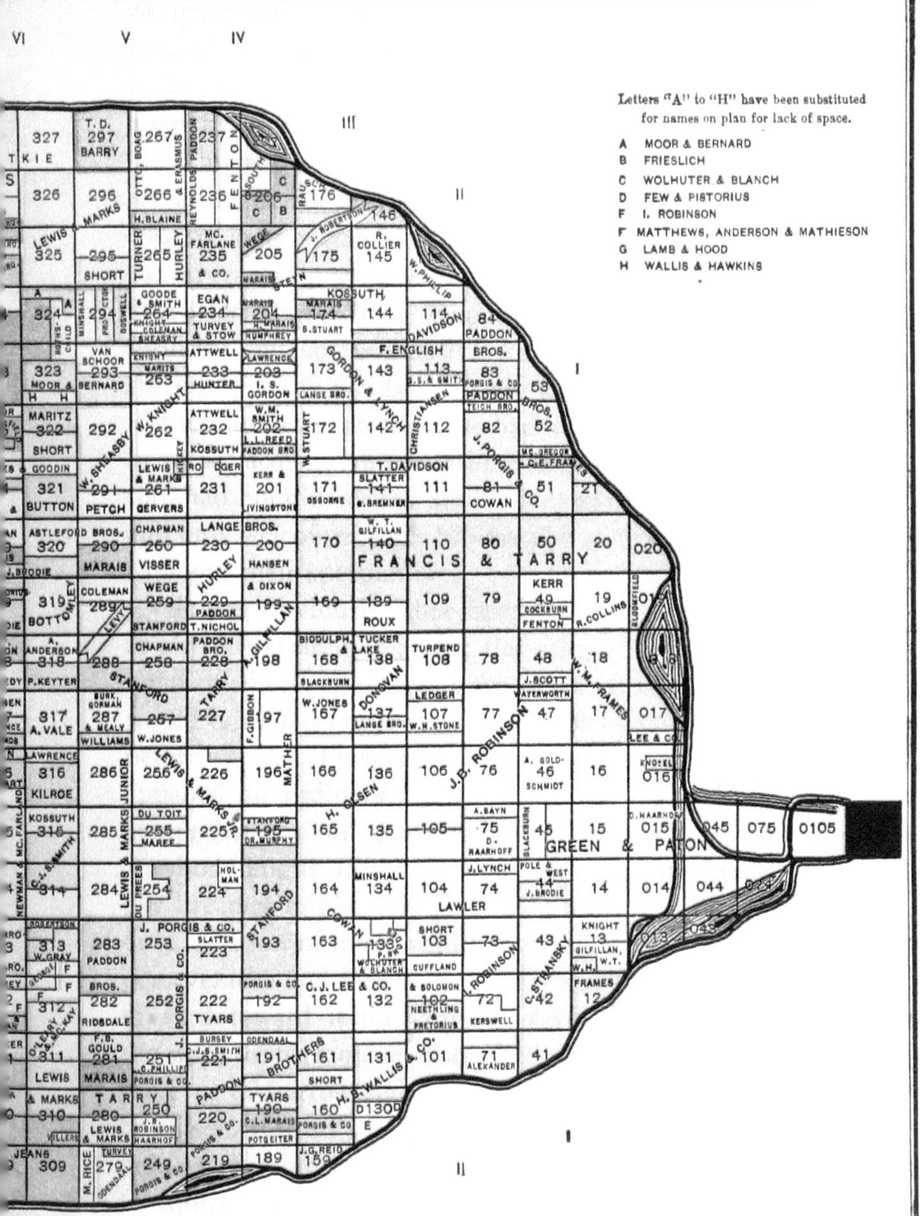

ENGRAVED BY BORMAY & CO., N.Y

Letters "A" to "H" have been substituted
for names on plan for lack of space.

A   MOOR & BERNARD
B   FRIESLICH
C   WOLHUTER & BLANCH
D   FEW & PISTORIUS
F   I. ROBINSON
F' MATTHEWS, ANDERSON & MATHIESON
G   LAMB & HOOD
H   WALLIS & HAWKINS

company was on the five hundred feet level, the opposing companies could go and eat into each other's boundary walls and pillars to such a dangerous extent that the entire mine was in a condition which threatened collapse at any moment."

This was so patently true, and more particularly in Kimberley mine, that it may seem surprising that the disastrous conflict was so long maintained. But it must be borne in mind that the average shareholder was not as quick to see and prompt to move for a remedy as Rhodes, and comparatively few had his intimate and comprehensive knowledge of the condition of all the mines in the Fields. A very large proportion of the investors in these mines were men who had never been on the Fields at all, or whose acquaintance was limited to a sightseer's visit. Many, too, had bought shares simply as a gamble in the stock market, and only welcomed such information or reports as were calculated to boom their speculations.

It was obviously labor lost to attempt to interest such men in any far-reaching plan for the union and systematic development of all the mining claims in the craters, and most of them would have sneered it away as a mere chimera if it had been laid before them. This was indeed a project which might well have appalled an ordinary man, even if he had the clear sight and comprehension of the position essential to a true judgment. Anybody might dream of such a gigantic combination, and some day-dreamer might babble about it to his gossips, but what man, or association of men, would have the foresight and patience, the perseverance and tact, the integrity and fulness of talent, to push forward toward it for years, to thrust aside or crush blocks in the way, to harmonize discordant and jealous interests, to open the eyes of narrow-sighted selfishness, to win the confidence of the distrustful, to design a scheme of union that would make all holders of good working claims common shareholders on a basis of equity and assured profit to all, and finally to provide the enormous capital necessary for the consummation of the scheme, and the development of the great diamond mines in a really great way?

Here was a task of such tremendous magnitude and difficulty that men of good ordinary judgment might well question its feasibility. What man in or out of the Fields would dare attempt it? Who could do it, if he dared to venture? There is a mighty fillip to the conceit of man, that in such great exigencies as these — in times when some prodigious undertaking is imperatively needed — the man or men who can carry it on to completion are almost always forthcoming. "Nothing is impossible nowadays," said the "Bonanza King," Flood, when doubts were raised of the practicability of piping water from the Sierra Nevada Mountains to the Comstock Silver Mines on the Virginia range; "the only question is, will it pay?" That seems, indeed, the only touchstone which men of such pith and temper are disposed to apply to any object. It was again made evident on South African Diamond Fields how far the possible stretches when men with Flood's touchstone are the adventurers. The moving men, who could comprehend the need for union and effect it, came irresistibly to the front in the Fields.

The undertaking to which they set their hands should be clearly set forth. In spite of the ruin of the open mine workings in the competing development scramble, and in spite of the continuing conflict and recurrent disasters in the underground mining so cogently enforcing the call for union, there were still, at the end of 1885, no less than ninety-eight separate holdings in the four mines. In Kimberley mine there were eleven companies and eight private holdings; in De Beers there were seven companies and three private holdings; in Dutoitspan, sixteen companies and twenty-one private holdings; in Bultfontein, eight companies and twenty-four private holdings. Thus the four mines were operated by a total of forty-two companies and fifty-six private firms or persons, all clashing within a surface area of 70 acres. The original location claims, aggregating 3600, had been united to this extent, merely, at the close of fourteen years of mining on the helter-skelter plan.

It is hardly just to credit Rhodes and Barney Barnato with an equal perception of the imperative call for the union of all

the discordant interests in the diamond mines. Each reached
the conclusion that it was no longer possible to continue to
work the mines divided into small
holdings which were controlled by
men antagonistic to one another.
Rhodes's interests were mostly in
De Beers mine, and Barnato's
largely in the Kimberley mine. In
the same year, 1880, in which Barnato floated successfully his first
diamond-mining corporation, "The
Barnato Mining Company," consisting of a few claims in a rich
section of the Kimberley mine,

Mr. C. D. Rudd.

Rhodes and others founded the De Beers Mining Company,
on the contiguous diamond-bearing crater.

It is of interest in this connection to trace the origin of De
Beers Mining Company through
the early years of De Beers mine.
In 1873 Rhodes united his claims
in De Beers mine with those of
C. D. Rudd, and they slowly increased their holdings. Robert
Graham joined them in 1874,
and later Runchman, Hoskyns
& Puzey took part with them in
the purchase of Baxter's holdings.
This combination, in addition to
mining their own ground, took
pumping contracts to drain the
mine. Besides the above combination there were other competi-
tors for the purchase of claims,
such as Dunsmure & Alderson,

Mr. Robert English.

Stow & English, and these three firms gradually acquired all the
best ground in De Beers mine except the Elma Company, owned

by Thomas Shiels and others, the Victoria Company in which J. Ferguson was then the leading spirit, and the United Diamond Mining Company.

The De Beers Mining Company was formed on the 1st of April, 1880, with a capital of £200,000, by the union of the three firms first mentioned. It progressed with extraordinary success, extending its range of ownership, absorbing step by step its floundering neighbors, and finally standing out preëminent in March, 1885, with a capital of £841,550, upon which dividends of $7\frac{1}{2}$ per cent had been paid during the last fiscal year, in spite of the heavy charges of development work and the unavoidable hampering of its mining operations. Mr. Rudd states that at one time Rhodes and he had the offer of the entire De Beers mine for £6,000, and they walked about a whole day talking it over, but finally decided they could not finance it. The licenses at that time were so costly that it was thought wise not to risk the purchase. Money was not very plentiful among these men in those days, as is shown by one of the first checks of the De Beers Mining Company, which was drawn by Rhodes in his own favor for £5, "as an advance against his salary as secretary."

It is possible that Barnato may have tried to bring about a further consolidation of some of the various interests in Kimberley mine, but there is nothing to show that he contemplated any broad scheme.

For nearly six years Rhodes concentrated his efforts in the Diamond Fields toward obtaining complete control of De Beers mine by himself and his chosen friends, and he brought about this consolidation of all the holdings in May, 1887. His master mind was steadfastly bent on the attainment of the control of the development and output of the four great diamond-producing mines of South Africa, and his work of first uniting all the interests in De Beers mine was but the beginning of his great dream. The range for amalgamation of the four mines was so great that no single man, however ambitious, could hope to cover it by any single-handed effort. The consolidation of

all the companies in De Beers mine was on the lines conceived
by Rhodes, and carried out by the support given him by the
leading men who were interested in the various companies.

Up to this time there was no rivalry between Rhodes and
Barnato, for no measures had been taken by Rhodes to obtain
a footing in Kimberley mine. The first steps taken in this
direction were to try to purchase the claims in the west end of
the Kimberley mine held by the Cape of Good Hope bank,
and known as W. A. Hall's claims. This was in the beginning
of May, 1887. Unfortunately, however, for Rhodes's scheme,
these claims had already been offered to a syndicate in London,
headed by Sir Donald Currie, and were purchased by that syndicate for £110,000. The plan which Rhodes had in his mind
was to purchase these claims, and also to purchase the claims
of the " Compagnie Française des Mines de Diamant du Cap
de Bon Espérance," known as the " French Company." The
" French Company " held a block of claims which ran nearly
across the mine from north to south, and divided the holdings
of the Central Company. It also held a block of claims adjoining those of W. A. Hall, but these were not connected with the
main body of their claims, being separated by the intervening
claims of the Central Company. These two companies were
so antagonistic to one another that neither would allow the
divided blocks of ground to be worked by means of tunnels
driven through the diamond-bearing ground of the opposing
company. The Central Company worked its claims by two
separate shafts sunk in the blue ground at the bottom of the
open mine, and the ground hoisted in the shafts was sent to the
surface by means of aerial trams, while the " French Company "
was compelled to drive tunnels into the walls of the mine adjoining the claims and connect them by a cross tunnel, as they were
working through one shaft only.

To create a powerful company in Kimberley mine was substantially all that the leading men in that mine had been working for, but this was far from satisfying Rhodes. Barnato viewed
the situation as a speculator and investor. Money making

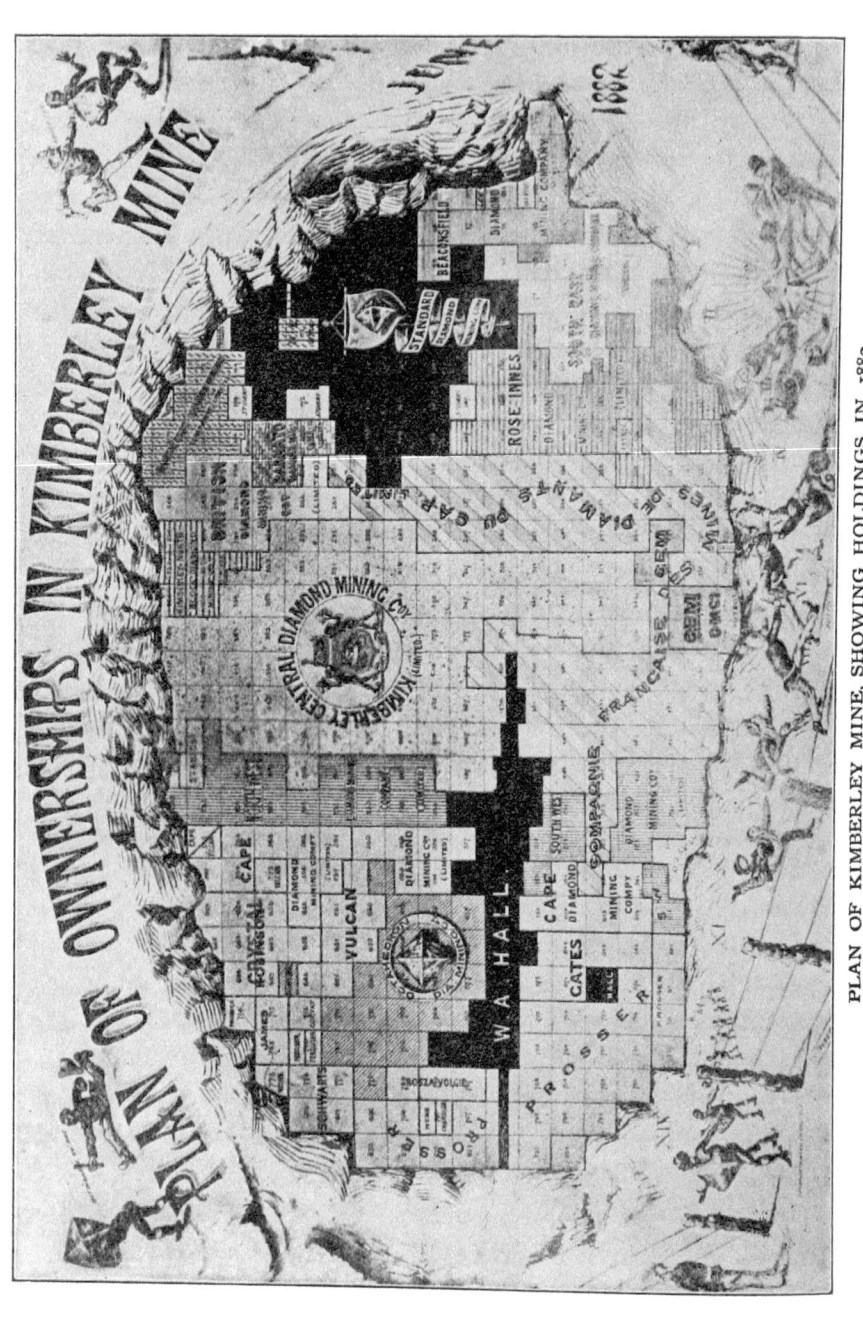

PLAN OF KIMBERLEY MINE, SHOWING HOLDINGS IN 1882.

through mining on a sound basis was avowedly the limit of his scheme, apart from a natural pride in figuring as the foremost operator in these marvellous Diamond Fields, and a rising star of the first magnitude on the London Stock Exchange. But the assurance of money making was, at most, a minor consideration with Rhodes. He, too, valued money highly, but not for the bare delight in piling it up or for the luxuries which it would purchase. Great wealth was to him the essential means for the furtherance of great plans. He wanted millions in hand, or the assured control of millions, to push his design for the lighting-up of the Dark Continent by the torchbearers of civilization, for the carrying of the flag of Greater Britain from the Cape to Cairo.

A man of kindred spirit, but of far more quixotic temper, the great soldier, General Gordon, once told him of the offer of a roomful of gold by the Chinese Government for his extraordinary services in subduing the Tai-Ping rebellion.

" What did you do ? " said Rhodes.

" Refused it, of course," said the disdainful Gordon. " What would you have done ? "

" Done," said Rhodes, " why, I would have taken it, and as many more roomfuls as the Chinese would give me. It is no use to us to have big ideas, if we have not got the money to carry them out."

The range of his plans and how he pursued them will be presented in detail in the chapter dealing with the far-reaching undertakings of the great Chartered Company which he conceived and brought into existence. It is sufficient to note at present that he pushed the development of his grand political aims apace with the means at his command, from the very beginning of his appearance as a prominent factor in the development of the diamond mines. He entered the Cape Parliament as a member for the district of Barkly West, almost coincidently with the formation of the De Beers Mining Company. From the day of his entrance into the political field, he worked unwaveringly for the extension of British dominion into the heart of Africa.

HOUSE OF PARLIAMENT, CAPE TOWN.

The northern boundary of the province of Griqualand West, formed by the inclusion of the new Diamond Fields, had not been determined by careful surveying, and the location of the line was disputed by the Batlapin chief, Mankoroane, who claimed control of the territory which is now Lower Bechuanaland. Rhodes prevailed on the Cape Government to form and send out a Delimitation Commission for the settlement of

Avenue of Oaks, Cape Town. House of Parliament at the Left.

the dispute, and his appointment as one of the commissioners was a natural recognition of his interest and competence. Shortly after he reached the frontier he was able to satisfy himself that the complaint of the chief was well founded. Some seventy farms belonging to Mankoroane's tribe had been included in error within the bounds of the British province, and justice demanded this acknowledgment. But instead of abandoning the ground, Rhodes saw that restitution might be made in a way to accord with his aim for the extension of

British sovereignty, and his cogent appeal persuaded the Batlapin chief to place all his territorial holdings, covering half Bechuanaland, under British protection by cession to the Cape Colony. To his mortification, however, the Colony declined the offered cession with its contingent obligations. Then Rhodes appealed to the Home Government, and finally succeeded in obtaining the establishment of a Protectorate over Lower Bechuanaland in 1884.

But it was only by the most pressing insistence that this advance was maintained. The Cape Colony was so stubborn in its refusal to bear the expense of any new acquisition, and the Imperial Government was so doubtful and sluggish in grasping its opportunities, that Rhodes was forced to the conclusion that the only assurance of the accomplishment of his aims must come from his own private enterprise, — through the forming of some great corporation, whose capital and interests might be engaged in his undertaking for the control and development of the resources of the vast barbaric interior of Africa. It was for this cause chiefly that he was so unflaggingly insistent upon the farthest possible stretch of amalgamation in the control of the diamond mines, though it must justly be observed that the thorough amalgamation of conflicting interests in the Fields was very highly desirable, if not absolutely essential, for its systematic development and the marketing of its output. A possible combination, with which Barnato would have rested content, would have wholly failed to accomplish the end which Rhodes had so deeply at heart.

In the year 1887, shortly after taking charge of the De Beers Mining Company, Mr. Rhodes requested me to write to two of my friends in London, Mr. Hamilton Smith and Mr. E. G. De Crano, who founded the Exploration Company of London, and who were intimately connected with the Messrs. Rothschild, and request them to ask Lord Rothschild if he would supply the funds for the purchase of the French Company in the Kimberley mine, provided Rhodes could come to some agreement with that Company for the purchase of the

property.   Before any answer could be received, even by cable,
Rhodes, who had gone from Kimberley to Cape Town to
attend the Session of Parliament, became very impatient about
securing this property, and wired me to join him, and we sailed
from Cape Town on the 6th of July.   In my letter to Messrs.
Smith and De Crano I put before them the plan which Rhodes
proposed to carry out, and the object he had in purchasing the
French Company's property, viz., to prevent the amalgamation
of all the interests in that mine, which might be set up as an
independent company in conflict
with the interests of De Beers.

On our arrival in London we
met Lord Rothschild, and Rhodes
discussed the plan with him.   In
the meantime, while we were on
the water, Rhodes's scheme had
been presented to the late Mr.
Tite and to Mr. Carl Meyer of
Messrs. N. M. Rothschild &
Sons, who were very favorably impressed with the business, and had
discussed it with Lord Rothschild.
Mr. De Crano had made several
trips to Paris, and had already

Mr. Carl Meyer.

paved the way for a conference between Rhodes and the directors
of the " French Company."

At the close of the interview, Lord Rothschild said, " Well,
Mr. Rhodes, you go to Paris and see what you can do in reference to the purchase of the French Company's property, and in
the meantime I will see if I can raise the £1,000,000 which you
desire."

On leaving the room Lord Rothschild stopped Mr. De
Crano for a moment, and said to him, " You may tell Mr.
Rhodes that if he can buy the French Company, I think I can
raise the million pounds sterling."

The same evening Rhodes, Mr. De Crano, Mr. Harry

Mosenthal, and myself left for Paris, and after several meetings with the French Company's directors, we settled upon the terms for the purchase of their property, which they valued at that time at about £1,400,000, including all their assets.   On returning to London Mr. Rhodes arranged with Lord Rothschild that he should furnish him with £750,000, which would be sufficient for the time being to complete the arrangements that he had made with the French Company.   In my letter of the 18th of June, it was mentioned to Messrs. Smith and De Crano that Rhodes would be willing to issue De Beers shares in payment of the loan at £1 less than the ruling market price of the shares at the date the money was paid, and would pay Messrs. Rothschild a handsome commission for transacting the business.

The final arrangement made for the payment of this money was the issue of 50,000 De Beers Mining Company's shares at £15 per share, and a syndicate was formed to take up these shares with the able assistance of Mr. Ludwig Lippert, of Hamburg.   It was agreed between Lord Rothschild and Rhodes that the profit on the rise of the shares between £16 and £20 during the next three months should be divided between the purchasing syndicate and the De Beers Company.   The shares rapidly rose, and, before the expiration of the time, had already reached £22 per share.   The De Beers Company received £100,000 as their portion of the profit on the rise of the shares. Shortly after the completion of this business Rhodes returned to the colony and awaited the result of the French Company's shareholders' meeting to confirm the sale which had been made to him by the directors of that company.   Barnato and others interested in the Kimberley Central Company, upon hearing of the transaction that had taken place, determined to use every effort to prevent the consummation of this sale, and threatened to offer the shareholders of the French Company at their general meeting £300,000 more than the amount for which the directors had pledged the company to Rhodes.

As a general of a great army is obliged to have the assistance and coöperation of competent lieutenants to carry out the plan of campaign which his superior mind has conceived, so Rhodes looked about for the strongest and ablest men to join him in repelling the vigorous attack which was being made against him. The first check which he gave his opponents seemed at first sight to be a complete surrender to them. Instead of allowing Barnato and his colleagues to bid against him for the purchase of the French Company, Rhodes arranged with them that he should complete the purchase upon the lines agreed upon with the directors of that company, and promised to unite the interests so purchased with the Kimberley Central Company, in which Mr. Francis Baring-Gould, who was the chairman, Barnato, and others held the controlling power, taking shares in the Central Company in payment.

Mr. Alfred Beit, while a Resident of Kimberley.

In this, as well as in subsequent transactions, Rhodes was most ably assisted by Mr. Alfred Beit, the Kimberley representative of Jules Porges & Co., who started business in Paris as diamond merchants in 1869. The men who from time to time have been connected with Mr. Porges and the successors to him, Messrs. Wernher, Beit & Co., took the keenest interest in Rhodes's scheme, and assisted him more than all others in bringing about the consolidation of the diamond interests. As early as 1871 Mr. Julius Wernher went out to Kimberley in the capacity of diamond buyer for Jules Porges & Co., and became partner in the firm in 1878. The firm grew in importance, and became owners in some of the largest companies in the four mines. They were the founders of the Griqualand West Diamond Mining Company in Kimberley mine, which was afterward re-formed into the " French Company." Mr. Alfred Beit came to the fields in 1875 as a diamond buyer for the firm of Lippert & Co., of Hamburg, and after a few years established himself in business as a diamond

u

buyer on his own account.   In the year 1882 he joined the firm of Jules Porges & Co., as their representative in South Africa, and became a partner in the firm in 1886.

In 1889 Mr. Porges retired from the firm, which was re-formed as Wernher, Beit & Co., Mr. Max Michaelis joining the firm.   Mr. Michaelis came to the Fields in 1878, and went into partnership with Mr. S. Neumann.   He organized the Cape Diamond Mining Company in Kimberley mine.   In 1880 he entered into an arrangement with Jules Porges & Co. to carry on his diamond business on joint account with them, which arrange-ment remained in force until Mr. Porges retired, when he became a partner in the new firm.   Mr. Michaelis assisted in bringing about the fusion of several of the large claim-holders in the Kimberley mine, such as Baring-Gould & Atkins, and Baring-Gould, Price & Tracy, with the Kimberley Central Company.

The great initiative and business capabilities of Mr. Beit were heartily recognized by Rhodes, and he was very largely instrumental in building up the diamond-mining industries, and bringing the dreams of Rhodes into practical shape and on business lines.

At a special general meeting of the shareholders of the De Beers Mining Company Limited, held at Kimberley on the 31st of March, 1888, for the purpose of considering ⎰ and con-firming an agreement entered into between the ⎱ respective

The Diamond Market, Kimberley, 1875.   (First Office of Mr. Alfred Beit at the Left.)

Boards for the amalgamation of the De Beers Mining Company with De Beers Consolidated Mines Limited, Mr. Rhodes gave his reasons for the necessity of acquiring either the control of the Kimberley mine or of entering into some arrangement with the directors of the Central Company, who controlled the mine, by which the output of both De Beers and Kimberley mines could be regulated. He saw that by skilful and systematic mining on the underground system, the output of the mines could be increased far beyond the world's requirements. It was clear, too, if these two mines were run in opposition to one another, it would result in the flooding of the market with diamonds, and a consequent depreciation of their value, with a fall in market prices almost ruinous to both companies. He saw that the output of diamond-bearing ground could be made almost unlimited, and in referring to this he said: "We had to face either an arrangement with the Kimberley Central Company, or obtain control of the Kimberley mine. We approached the Kimberley mine management in every possible way we could conceive. I valued the De Beers mine higher than they did, but I was willing to give way in order to obtain control. I was met simply with smiles and obdurate statements. I was met with the arguments of the gentleman at 'the corner,' who said the Kimberley mine was worth three times as much as De Beers. We had to choose between the ruin of the diamond industry or the control of the Kimberley mine. We saw this, that you could never deal with obstinate people until you got the whip hand of them, and that the only thing we had to do to secure the success of our industry was to get the control of the Kimberley mine."

As soon as Rhodes had bought the French Company and amalgamated his interests with the Kimberley Central Company, he found that the management of that Company was headstrong in its determination to run the Kimberley mine in rivalry with De Beers. This was diametrically opposed to his conviction that monopoly was the essence of success in diamond mining; for, as he said, "Our engineers had long ago shown us that, by underground working, Kimberley and De Beers mines could

THE RIGHT HONORABLE, CECIL, JOHN RHODES, AND ALFRED BEIT, ESQ. OCTOBER

produce far more diamonds than the world would take." By the purchase of the French Company, De Beers Company held one-fifth of the capital of the Central Company, and after many attempts to bring about a friendly union of the two mines, Rhodes came to the conclusion that the only feasible plan was to buy a sufficient number of shares in the Central Company to obtain control. To accomplish this would take at least £2,000,000 sterling. Fortunately Mr. Alfred Beit, whose command of capital for such great undertakings was unequalled in South Africa, stood fast by him in determined coöperation. In answer to Rhodes's natural question, " Where is the money to come from ? " Mr. Beit said pithily, " We will get the money if we can only get the shares."

Then ensued a most keen contest. Mr. Beit and Rhodes began buying all Central shares that could be secured with apparently limitless means. Both were leaders in the contest, but Mr. Beit furnished most of the money. Meanwhile Barney Barnato was bidding against them with unfailing pluck for the control of the Company. The price of shares mounted by jumps, but never too high for Barnato, who was persistent in his claim that the Kimberley mine was worth two of De Beers. Rhodes's version of the story of this struggle is that in his purchase of shares he had the support of the loyal directors and shareholders of his Company, while his principal opponent was handicapped by the fact that he was forced to buy out his own largest shareholders. There is little doubt that Barnato felt this apparent lack of loyalty keenly, but he was too strenuous a fighter to concede defeat. As a matter of fact, he came to terms with his antagonists only when the price of shares had been bulled to a figure that seemed out of reason even to his sanguine estimate, while the price of diamonds had been forced down unprofitably by unobstructed competition. After many and long conferences, Rhodes made Barnato one last offer, which he accepted. For his interest in the Kimberley Central Company he was paid with De Beers shares at the current rates of shares on the day of the sale. By this purchase De Beers' holding of Central shares was brought up to eleven thousand out of seventeen thousand shares. Under the trust deed of the Central no amalgamation could be made unless half the capital was present at a meeting called for the purpose, and no new resolution could be carried without a two-thirds majority of those present. The bargain with Barnato gave De Beers the control.

So having finally obtained the control of the Kimberley mine by purchase for £5,338,650, Rhodes turned his attention to what he called the poorer mines, Dutoitspan and Bultfontein. At a meeting of De Beers shareholders he said he was reminded of a story he had read about a certain mine, of which it was said "it was too rich to leave and too poor to pay," and he would thus describe the mines alluded to. " Nothing," he said, " was so extraordinary as the way in which the people would hold scrip from year to year that never pays, but it was always said, ' Oh, next year it will pay,' and so it went on from year to year." He wished to state " that so far as the amalgamation of the diamond mines was concerned, it would not help the poorer mines, but rather the other way. It was generally noticed in mining matters that following upon one success a number of unsuccessful ventures were floated. And this was why they had secured these interests in Dutoitspan and Bultfontein mines." He did not look upon the purchase of properties in these mines as a good investment, with diamonds at the price they were bringing at the time of the purchase; but as these two mines were large factors in the production of diamonds, their yield, even if mined at a loss, would affect in a very large degree the price which could be obtained for the product of the richer mines. Although Dutoitspan mine could not be worked at a profit at the market price of diamonds, and the mine had already begun to be troubled with reef falls burying the blue ground below, still he considered it necessary to get control of the principal companies in this mine. In Bultfontein mine, where the reef troubles had already begun, there was still a large portion of the mine in process of working, and he described it as being " on the margin of cultivation." If the reef remained standing, and

FAC-SIMILE OF CHECK GIVEN IN PAYMENT FOR THE PURCHASE OF THE KIMBERLEY MINE.

the price of diamonds was fair, the mine could be worked at a small profit.

Rhodes continued the purchase of the properties in both these mines until the whole of the two mines came into the possession of the corporation organized as De Beers Consolidated Mines. He showed the shareholders in the various companies that the fate of the poorer mines lay in his hands, because he could produce twice the amount of diamonds the world required from De Beers and Kimberley mines alone. Even at the low rate of fourteen shillings a carat, he made it clear that the richer mines could pay to the shareholders dividends which would satisfy them. "The poorer mines, 'on the margin of cultivation,' would have to accept our offers, or fight us on two grounds, larger output and lower rates."

In his speech at the annual meeting of the De Beers Mining Company, held at Kimberley on the 12th day of May, 1888, Rhodes bore tribute cordially to the essential coöperation of Mr. Beit in his great undertaking.

In moving a vote of thanks to the chairman, his former antagonist, Barnato, briefly referred to the struggle which was closed by the purchase of his shares in the Kimberley mine. He said "no person knew better than he did the labor Mr. Rhodes had to convert him into the De Beers Mining Company." He could say that day after day and night after night Mr. Rhodes was laboring to get him to take De Beers for Centrals. He gave way when he saw diamonds down to eighteen shillings a carat, and on those conditions he joined Mr. Rhodes. It is only just to Barnato to note in closing that he was as loyal in his later coöperation as he had been persistent in his antagonism. It is sad to recall how his brilliant and versatile mind gave way under the enormous strain brought upon him by the various obligations incurred through his numerous investments and flotations in the gold fields. His tragic death was a distressful close to his phenomenal career. On his way to England from the Cape, in June, 1897, he suddenly sprang overboard and was drowned.

BARNATO'S RESIDENCE, KIMBERLEY.

CHAPTER X

IT has been told why and how the conflicting interests on the Diamond Field were fused in one dominant organization. The signal services of this amalgamation are now too obvious for dispute. By the formation of De Beers Consolidated Mines Limited, it became practicable to design and conduct mining operations systematically and economically and to regulate the output to the market demand. It was soon apparent, too, that the organization of this extraordinary joint stock company was the creation of a power of yet unmeasured service for the development of the resources of South Africa and the push of civilization through the Dark Continent.

The only approaches to the far-reaching conception of this organization must be traced back to the old Dutch and English East India companies, or the visionary project of John Law, exploding in air as the Mississippi Bubble. At the outset, on the 12th of March, 1888, a seemingly unpretentious joint stock company was formed and established at Kimberley with a capital of £100,000 sterling, divided into twenty thousand shares of £5 each. Authority was granted, however, in the articles of the association, to the shareholders of the company to increase this small capital in general meeting, from time to time, for the acquisition of new property, by creating new shares to any extent, or, in the exact words of article 39, " such amount as may be deemed expedient." No provision for expansion and acquisition could be more liberal, and the particular specifications of the articles of the association show that " new property,"

in possible range at least, was not confinable to the Diamond Fields or Cape Colony, or even, perhaps, the scope of the whole Dark Continent.

It was remarked somewhat caustically at the time, but with undeniable keenness, that it was much easier to tell what this amazing Company could do than to determine what it should not do under its articles of association and trust deed incorporation under the limited liability laws of the Cape of Good Hope. It might shift its head office from Kimberley to any other place on earth. It might "acquire by purchase, amalgamation, grant, concession, lease, license, barter, or otherwise any houses, lands, farms, tracts of country, quarries, mines, mining or other claims, rights and privileges, water rights, waterworks or other works, privileges, rights and hereditaments, diamonds and other precious stones, gold and other minerals, ores, coals, earth, and any other valuable product or substance, machinery, plant, utensils, trade marks, patents for invention, licenses to use any patented invention, and other movable and immovable property of any description in Africa or elsewhere." Under this liberal license, the only apparent obstacle to its ownership of the face of the earth is the declination of other holders to sell or give it away.

It was further specifically authorized to carry on a mining and general trading business in any part of the globe, and to construct, maintain, and operate any tramways, railways, roads, tunnels, waterworks, canals, gas works, electric works, reservoirs, water-courses, furnaces, stamping works, smelting works, factories, and in general, "any other works and conveniences which the Company may think conducive to any of its objects." It might also become interested in, promote, and undertake the formation and establishment of such institutions or companies (trading, manufacturing, banking, or other) as may be considered to be conducive to the profit and interest of the Company, and to carry on any business, in short, "calculated directly or indirectly to render any of the Company's properties or rights for the time being profitable." There was also provision for the

A GROUP OF OFFICIALS,

De Beers Consolidated Mines Limited.

Standing.

From Left to Right.

Sitting.

Dr. R. Harris,
W. H. Craven.
G. W. Compton,
Major Goold Adams,
Gardner F. Williams,
Judge J. G. Lange,
Lieutenant Colonel Harris.

Richard Solomon,
D. J. Haarhoff.
Dr. L. S. Jameson,
Cecil J. Rhodes,
Mr. Newton.
Henry Robins,
E. E. Nind.

possible acquirement of any tract or tracts of country of any size in Africa or elsewhere, together with any rights that might be granted by the rulers or owners thereof, and the expenditure of any sums deemed requisite and advisable in the development and maintenance of order and good government in such acquisitions.

In view of the enjoyment by the shareholders of such privileges and liberties, it was only natural that the directors of the Company should not be grudgingly confined. This was, indeed, the case, and two specifications of powers, in particular, have proved to be highly serviceable in practice, for there has been no abuse of discretion. The directors were authorized " to purchase, hire, or otherwise acquire for the Company any share in any kind of joint stock company, property rights, or privileges which the Company is authorized to acquire, at such price and generally on such terms and conditions as they may think fit; also to sell, lease, abandon, or otherwise deal with any shares, property rights, or privileges to which the Company may be entitled, on such terms and conditions as they may see fit, and to amalgamate with any other company or companies having objects altogether or in part similar to the objects of this Company." They were further empowered " to found, promote, float, and acquire interest or shares in any companies, undertakings, or institutions, as they may deem advisable in the interests of the Company ; also to acquire interests in, promote, aid, or subsidize any useful industry or undertaking in any country where the Company may be carrying on business."

At the outset, De Beers Consolidated Mines Limited was preëminently what is termed a close corporation. Four men held all but twenty-five shares of its stock. These, in the order of their subscription to the articles of association, were Alfred Beit, holding 4439 shares ; Barnett I. Barnato, holding 6658 ; Cecil J. Rhodes, holding 4439 ; and Frederick S. P. Stow, holding the same number as Beit and Rhodes. These four, by the articles, were practically authorized as shareholders to create " five life governors or permanent directors of the Company, four of whom shall be Cecil John Rhodes, Barnett Isaac Barnato, Frederick Samuel Philipson Stow, and Alfred Beit." If "so minded," these four had the power by unanimous resolution of themselves or their survivors to appoint the fifth authorized "life governor," and to fill any vacancy occurring in their number by reason of death or otherwise. These four were further constituted the first directors of the Company, and had

A Group of Directors, De Beers Consolidated Mines Lt'd.   (Mr. Frederick Samuel Philipson Stow in the centre, holding a book.)

power to appoint other directors, if they so desired, to act in conjunction with them until the first ordinary general meeting of the Company, when the shareholders were called upon to determine how many directors there should be besides the life governors, and to elect "such number as they determine to be necessary."

From the point of view of ordinary investors in ordinary stock companies the unlimited sweep of this unique organization and the powers confided to its controlling directors may be

summed up in the familiar outcry of Dominie Sampson.   They
are indeed " prodigious," but the phenomenal success of this
combination is a stubborn fact that must be faced in any contention that its scope and method of conduct were unwarrantable
and unadvisable.   Its base of operation was not Lombard Street,
but the heart of South Africa, in a field so unique, in a situation so perplexing, in unavoidable touch with such far-rang-
ing and conflicting interests, that ordinary limitations, hampering
freedom of expansion and action, would have been crippling and
possibly disastrous handicaps.   The powers of the directors are
great, but who can justly deny that they have been greatly used
for the reconciliation of jarring interests, the comprehensive and
rational development of the diamond mines, the safety and comfort of the miners, the profit of the shareholders, the promotion
of allied industries, and the general welfare of South Africa ?
The possible range of expansion of the interests of the corporation is a bugbear to some good people, who would prefer the
harmlessness of the deserted village to the risk that civilization
might " git forrid sometimes upon a powder cart."   But what is
there to show, to-day, of the actual stretch and exercise of the
corporate powers beyond the judicious limits of profitable investment, sagacious development of tributary resources, and discharge of patriotic obligations ?

The expansion of the original corporate foundation was
rapidly pushed.   The plan in detail was presented by the chairman of the corporation, Mr. Rhodes, on March 31, 1888, at
the special general meeting of the shareholders of the De Beers
Mining Company.   The programme thus presented was unani-
mously endorsed by the shareholders of the Company, accepting
it without alteration as the best feasible proposition for the consolidation of the diamond-mining interests.

At this general meeting of the shareholders the De Beers
Mining Company was formally merged in the new corporation.
The shareholders of the old Company received two fully paid
£5 shares in De Beers Consolidated Mines Limited for every
fully paid £10 share in the old Company.   Having effected

this acquisition, transferring the whole of De Beers mine, and the interests of the late De Beers Mining Company in all outside mining properties, the Consolidated Mines pushed forward steadily their undertaking of a comprehensive consolidation. The first and most important step was the securing of the whole of Kimberley mine, the greatest producing factor next to De Beers. The method by which the property of the Kimberley Central Diamond Mining Company was finally turned over to the Consolidated Mines has been described in the preceding chapter.

In the acquisition of Dutoitspan and Bultfontein mines a different plan was adopted. These mines, at the time, were not profitable producing properties, and it was practically certain that they could not be operated to advantage in view of the output from the greater and richer mines. For several years each of these mines had produced diamonds to the value of over half a million carats annually; but this production was rapidly declining, owing to the unresisted falls of reef. Among the assets taken over from the old De Beers Company were a number of shares in the Griqualand West Company of Dutoitspan and in the Bultfontein Consolidated Company. By the influence secured through this acquisition, it was not difficult to effect permanent working agreements with De Beers Consolidated Mines, by which the new corporation attained complete possession of both mining properties in consideration of the payment of a fixed annual dividend. During the second year after the incorporation, the Consolidated Mines purchased the property of the Anglo-African Mining Company, the Compagnie Générale (including its interest in the Conivieras mines in the Brazils), the Sultan Diamond Mining Company, and the United Diamond Mining Company, representing nearly all the properties of material consequence and extent in Dutoitspan mine except the Gordon Company's holdings. During the same period the Consolidated Mines bought in the Bultfontein Mining Company, the Spes Bona Diamond Mining Company, and the South African Diamond Mining Company, comprising a considerable part of the properties in the Bultfontein mine.

GROUP OF LIFE GOVERNORS, DIRECTORS, GENERAL MANAGER, AND SECRETARY, DE BEERS CONSOLIDATED MINES LTD.

The actual cost of the properties thus acquired by the Consolidated Mines was approximately £14,500,000. There would have been no difficulty in expanding the capital of the corporation by the issuing of shares to an amount sufficient to cover this immense acquisition, but a more conservative course was adopted. It was decided not to increase the capital of the corporation beyond £3,950,000. The purchases in excess of this issue were provided for by the issue of debentures. The adoption of this plan necessitated a provision for covering very heavy fixed charges in the early years of the operations of the Consolidated Mines; but this obligation was undertaken with confidence in view of the assurance of the control of the diamond market, brought about through the consolidation, and the actual return in the rapidly increasing output of the mines with systematic and scientific development.

During the financial year following the completion of consolidation, De Beers produced 2,195,112 carats of diamonds. This product, including the proceeds of diamonds from débris washing, realized in the market £3,287,728. In that year the total weight of diamonds produced by all the mines in the Kimberley division was 2,415,655 carats. Thus approximately ninety per cent of the total production was then furnished by the Consolidated Mines. The net profit of the operations for the year exceeded £1,000,000 sterling, and two half-yearly dividends of ten per cent each were paid to the shareholders. The actual cost of winning over 2,000,000 carats of diamonds, including all expenses at the mines and office charges, was a little over a million sterling, or roughly 10s. per carat. The difference between the estimated net profit and the costs of operation was expended in the payment of interest on debentures and obligations and in provision for their redemption, and in the setting aside of an exceedingly liberal provision of over £500,000 as an offset for depreciation of plant, etc.

The directors of the De Beers Consolidated Mines could point with high satisfaction to this profitable showing in contrast with the records of disastrous competition and conflicting mine operations. No exact statistics are obtainable of the production in the early years, when no official returns from the mines were made. The late Barney Barnato, who made a special study of the probable rate of production, estimated the product from 1873 to 1880 as ranging annually from a million to a million and a half carats. After 1880 there was a considerable increase, and in 1883, when official returns were first rendered, the quantity of diamonds produced was 2,319,234 carats. The average value of this product was reckoned at 20s. 4¾d. giving a total of £2,359,466. In 1884 the product was 2,264,786 carats, valued at £2,562,623, showing an average of 23s. 2¾d. per carat. This was the top notch in market value, for in the following year, 1885, the diamonds produced amounted to 2,287,261 carats, with an average value of only 19s. 5¾d. per carat. In 1886 the production reached the high total of 3,047,639¾ carats, but the demand increased in proportion, so that the average selling price was fully 2s. higher per carat than during the previous year. In 1887 and 1888, through the increased facilities for production in De Beers and Kimberley mines, the total output rose to 3,646,889 carats, and 3,565,780¾ carats successively. The average price during these two years ranged from 21s. 6d. per carat to 22s. 1½d. but the market was flooded, and prices were falling perilously close to the cost of production even in the richer mines. There was no assurance of any far-sighted regulation of the output and market prices, and, lacking this, diamond mining properties were commonly reckoned as little better than gambling ventures. It has been clearly shown how this disastrous condition was at once changed to stable assurance and prosperity by the control of the new organization.

To the shareholders in the mines, after this reorganization was effected, the returns were unprecedented. This profit was largely due to the complete control of production, systematic operation, and regulation of the output; but the comparative showing was also greatly enhanced by the shrewdness of the financiering in the organization, and the withdrawal of inflation from the stocks of the various mining properties included in the new incorporation and its leased holdings. The capital of De Beers Mines before consolidation was £2,009,000. The capital of the Central Company was £1,779,650. De Beers stock at the time of consolidation was selling at £40 a share, representing a capital of £8,036,000. The stock of the Central Company, controlling the Kimberley mine, was selling at £50 for each £10 share, making a total valuation of £8,898,250 for the mine. At this market estimate the valuation of the two great mines was £17,934,250. The capital of the Dutoitspan was approximately £3,500,000, and of Bultfontein, £2,000,000 nominally, making a gross valuation for the four mines of £23,434,250. By consolidation the capital stock was compressed to £3,950,000, and almost absolute control of the mining in all four of these great properties was secured at an annual charge of about £320,000 for interest on debentures and for leases of two companies, one in Dutoitspan mine and one in Bultfontein mine. The business of the Company grew so rapidly that it was necessary to establish transfer as well as general business offices in London.

Mr. E. R. Tymms, Secretary of the London Board, De Beers Consolidated Mines Lt'd.

CHAPTER XI

UST acknowledgment has been made in a former chapter of the essential service rendered to the diamond mine owners by the device of Mr. Edward Jones for underground work beneath the fallen reef covering the bottom of the open pits. This was, however, confessedly only a temporary makeshift, enabling the claim-holders to defray the heavy costs of sinking shafts through the hard rock outside the craters, and pursuing some systematic plan for the extraction of the diamond-bearing breccia by underground workings. Deepshaft sinking was undertaken with renewed heart by several companies owning claims in Kimberley and De Beers mines, but for some years there was an obvious lack of essential coöperation and unity of method. Eight shafts were sunk, or were under way, in 1885, within and without the craters, for opening De Beers and Kimberley mines, and through these shafts the blue ground was extracted by four different methods of stoping, none of which was satisfactory. The system instituted by the Central Company, the largest operator in Kimberley mine, illustrates sufficiently the inherent defects in all. Here galleries fifteen feet wide were driven to the right and left of a main tunnel, with pillars fifteen feet thick between them. Passages or winzes for broken ground were sunk at short intervals to a tunnel below. The ground was stoped to the height of fifteen feet above the main tunnel, and then below it until the stope reached the next level. The passes became filled frequently with large pieces of ground, and had to be cleared. Under this system the mine was assuming the shape

of a section of a gigantic honeycomb cut in two longitudinally, the spaces for the honey representing the worked-out part of the mine, and the comb, the support for the superincumbent mass of débris. After a short period of working, the pillars began to show signs of crushing, and the mine was considered too dangerous to allow the men to remain in it. They were withdrawn just in time to prevent a disaster, for the whole underground

The Last of Open Mining, Kimberley Mine.

works collapsed shortly after the last man had left the mine. Fortunately no one was killed. The mine had to be reopened from top to bottom, for every underground excavation was filled up at the close of the year 1888.

The errors in engineering were further accentuated, during the early stages of underground mining, by the jealous bickering of rival owners, which was constantly impeding the progress of the workings, and it was seemingly impracticable to agree upon any plan securing concert of operation and expert opening of the

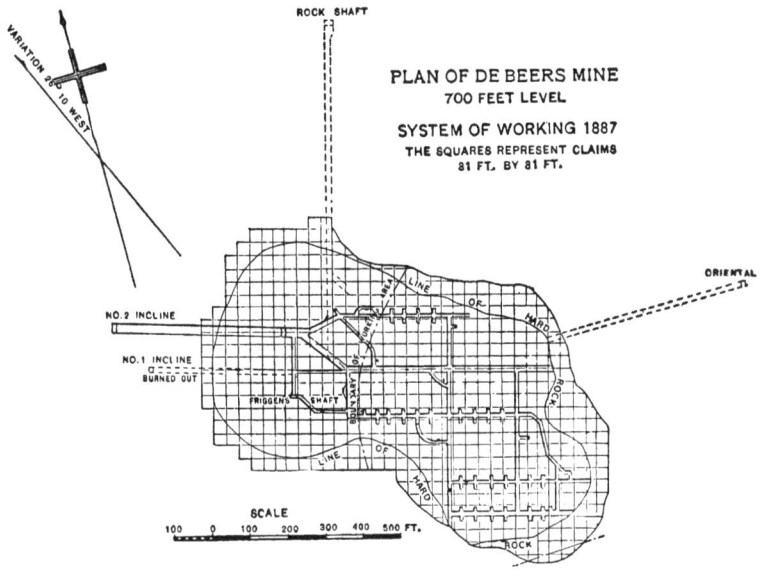

PLAN OF DE BEERS MINE
700 FEET LEVEL

SYSTEM OF WORKING 1887
THE SQUARES REPRESENT CLAIMS
81 FT. BY 81 FT.

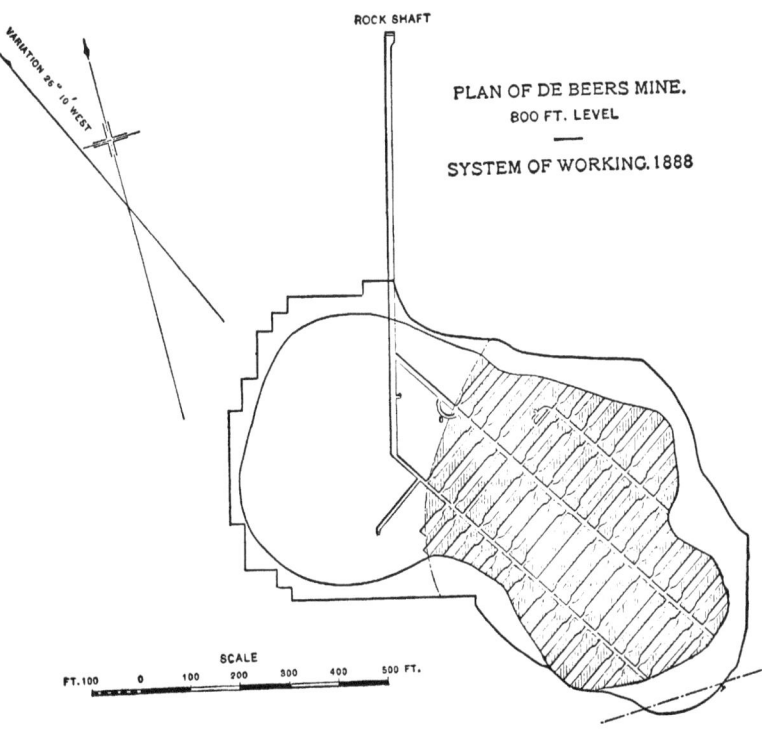

PLAN OF DE BEERS MINE.
800 FT. LEVEL

SYSTEM OF WORKING. 1888

mines. At the end of the year 1885, although the need of amalgamation of claims was obvious and imperative, there were still, as has been noted, ninety-eight separate holdings in the four mines. Prior to the consolidation of the holdings in De Beers and Kimberley mines, the underground workings were prosecuted with the general design of withstanding pressure and sliding of the reef by leaving sufficient solid blue ground, in the form of "floors" or "roofs," between the series of levels, supported by buttresses and pillars of blue ground. Costly experience by frequent collapses of the roofs and crushing of the pillars

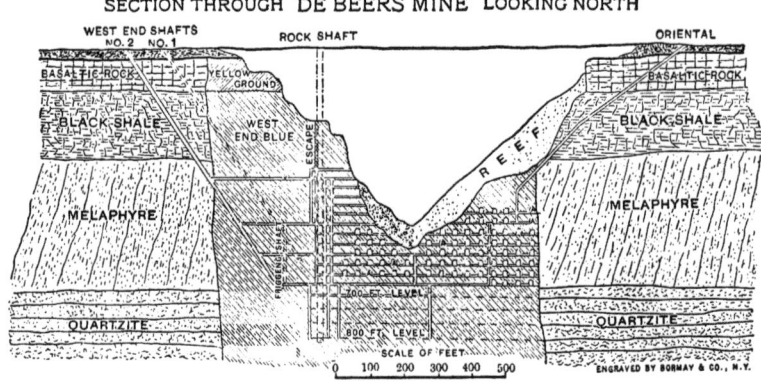

SECTION THROUGH DE BEERS MINE LOOKING NORTH

proved that the levels were too near one another, and that galleries driven full size from the offsets were difficult to maintain and unsafe for the workmen.

The heavy expense of sinking vertical shafts and driving tunnels through the hard rock surrounding the mine had led to the adoption of inclined shafts in order to reach the blue ground more quickly; but, for several reasons, these inclines were not adapted for the prosecution of deep underground works. The chief defects may be briefly summarized. They were difficult to maintain, as they were sunk obliquely through the horizontal strata of the shale, which frequently gave way and crushed the shaft timbers. Secondly, being inclined to the horizon (De Beers 56°, and Kimberley Standard Shaft 32°) and situated not far from the margins of the mines, they soon reached blue ground,

SECTION THROUGH DE BEERS MINE
LOOKING WEST.

and were continued down in this breccia, which must sooner or later be mined. Some of these shafts, as at De Beers, had a uniform slope from top to bottom, while others, as at Kimberley mine, changed to a steeper slope in depth and in one case to a vertical shaft. De Beers No. 2 worked well to the depth of 800 feet, and the Standard shaft, Kimberley mine, was fairly serviceable to the depth of 845 feet. The shafts were not sunk with the view of putting in proper pumps, and when steam was taken into the mines through them, for pumping purposes, the

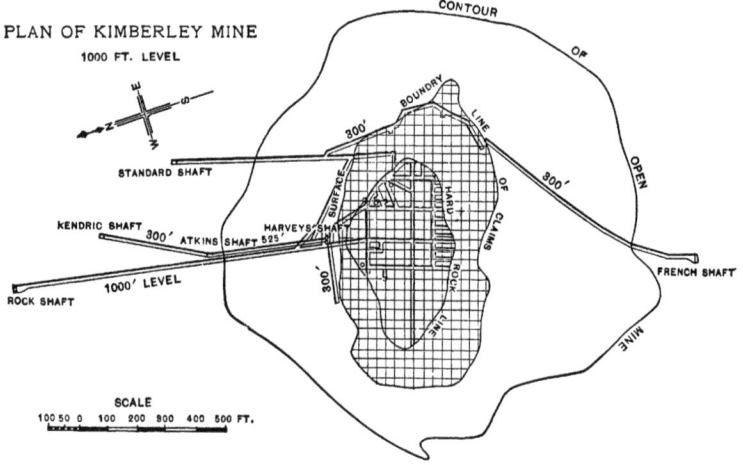

natives had to pass up and down the same shafts by means of ladders. As all the inclined shafts were upcasts, the heat was insufferable.

When I took charge of De Beers mine, in the year 1887, it was worked under what was then known as the Gouldie system, which had been copied from the hematite mines of Cumberland, and first introduced in the Kimberley mine by Mr. Joseph Gouldie, then manager for W. A. Hall, and afterwards mine manager of De Beers Mining Company. At De Beers mine an inclined shaft had been sunk to the 500-foot level, with intermediate levels 30 feet apart between the 380-foot and 500-foot levels.

SECTION OF KIMBERLEY MINE
LOOKING NORTH
100 50 0 100 200 300 400 500 600 FT.
SCALE

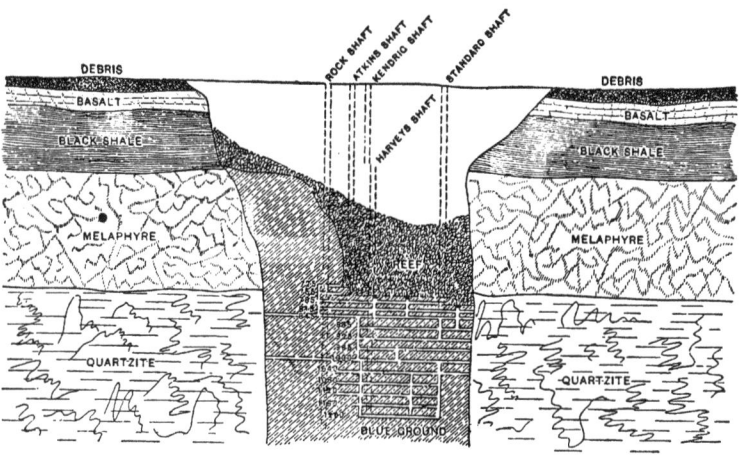

SECTION OF KIMBERLEY MINE
LOOKING EAST
100 50 0 100 200 300 400 500 FT.
SCALE

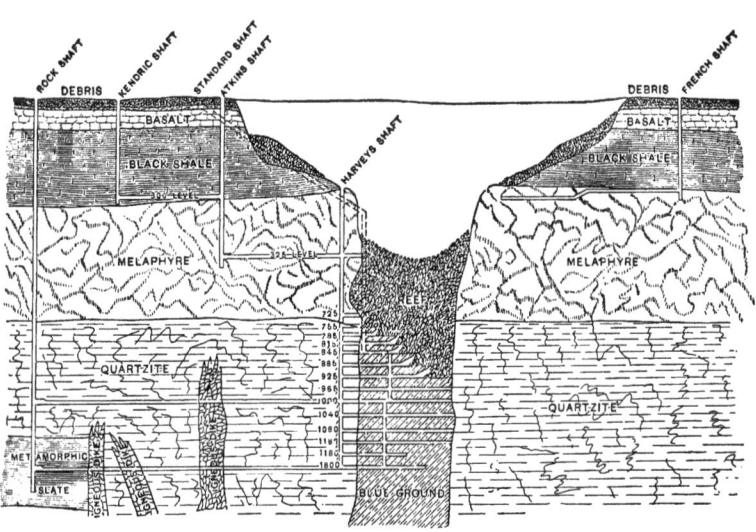

The plans on other pages illustrate the manner in which the various levels were laid off. Tunnels were driven across the crater at De Beers mine from west to east, about 120 feet apart, and galleries 18 feet wide and 18 feet high were opened every 36 feet along the main tunnels, and were worked up to within 12 feet of the loose ground in the top levels. Pillars of solid blue ground 18 feet thick were left between the galleries, but later on first the roof and then the pillars were taken out.

This method of mining was fairly successful for a time; but, as already stated, as depth was attained, the roofs of the galleries or rooms became unsafe before the galleries were opened through to those on the next level above, and they frequently gave way, thus making the extraction of the blue ground exceedingly difficult. This system was both expensive and dangerous. No timber was used except in the main tunnels or drifts, the nature of the blue ground being such that the roofs and sides of the excavations stood fairly well for a short time, provided they were well ventilated.

In other parts of De Beers mine various companies were working or trying to work underground; but as no regular system of mining could be carried on owing to the irregular shape of their holdings, and the more or less temporary methods adopted, it was clearly impracticable to devise and carry into effect any comprehensive system of operation for the rapid and economical handling of the diamond-bearing breccia in the craters, until the union of all the claims through the formation of one controlling company permitted the installation of a single uniform system of mining.

It has already been narrated how this was effected for De Beers mine during the year 1887, by the combination of all the holdings in the mine into one company, and the organization of De Beers Consolidated Mines Limited, in March, 1888. Kimberley mine came formally into the possession of this great corporation on the 1st of June, 1889, and controlling interests in the other two mines, Dutoitspan and Bultfontein, were also secured. The assured control of all the mines and their operation by De Beers Consolidated Mines Limited enabled its directors to institute and conduct successfully a single broadly comprehensive plan for extracting the diamond-bearing rock and for disposing to the best advantage the total product of their mines.

This system of mining was devised and applied by me shortly after my appointment as general manager of the De Beers Consolidated Mines Limited, and was based essentially on a method suggested by the miners themselves and without reference to any other system. Instead of attempting to withstand, even for a time, the pressure of the superincumbent mass of broken reef, the new system contemplated was a caving in and a filling of the excavations, after the precious blue ground had been extracted.

In order to make the output of diamond-bearing ground as great as possible, the levels in De Beers mine were at first opened up in the new system according to the following plan : —

When the numerous small tunnels had been driven to the margin of the mine, *i.e.* to the point where they reached the sides of the crater, the blue ground was stoped on both sides of and above each tunnel until a chamber was formed extending along the face of the rock for 100 or more feet, with an average width of about 20 feet, and about 20 feet high. The roof of the chamber or gallery was then blasted down or allowed to break down by the pressure of the overlying mass of broken diamond-bearing ground or débris. I mention diamond-bearing ground here, for in the early stages of underground mining there was an enormous amount of this ground which had been left behind when open mining was discontinued, and had been crushed either by the moving sides of the immense opening or by the collapse of the underground pillars when mined by the old system. It happened frequently, after breaking through to the loose ground above, that clean diamond-bearing ground would run down as fast as it was removed for weeks or months at a time. The galleries would at times become blocked with large pieces of blue ground, which had to be blasted, and then a

further run of blue ground would follow. When the blue
ground was worked back toward the centre of the crater, larger
boulders or fragments of basalt, which had come down through
the loose reef from the surface, would be met with. This system of working would be continued until reef alone came
down, the waste or reef removed being sent to the surface by
itself and dumped on the reef tips ; it formed, however, only an
inconsiderable proportion (one to four per cent) of the total
output. It will be remembered that, when the roof caved in,
the gallery was nearly full of blue ground. By the work which
followed, only a part of this ground was removed by the men
working on that level, the miners preferring to take it out on the
next level below. This process of mining was repeated from
level to level until finally there was no more loose ground to be
recovered. The cost of extracting blue ground while loose
ground existed was very low.

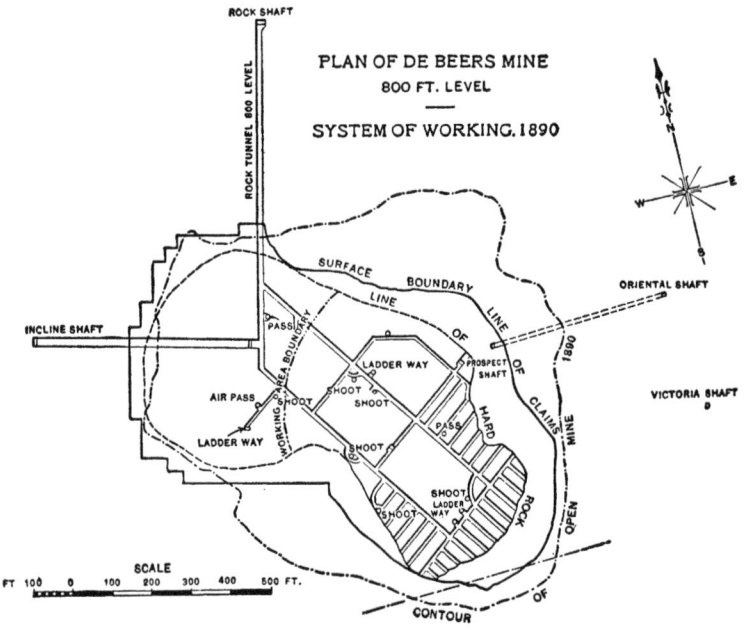

Now all this has changed, and the plan of opening up new
levels has altered somewhat, but the system remains the same.
By referring to the plan, given above, it will be seen that the

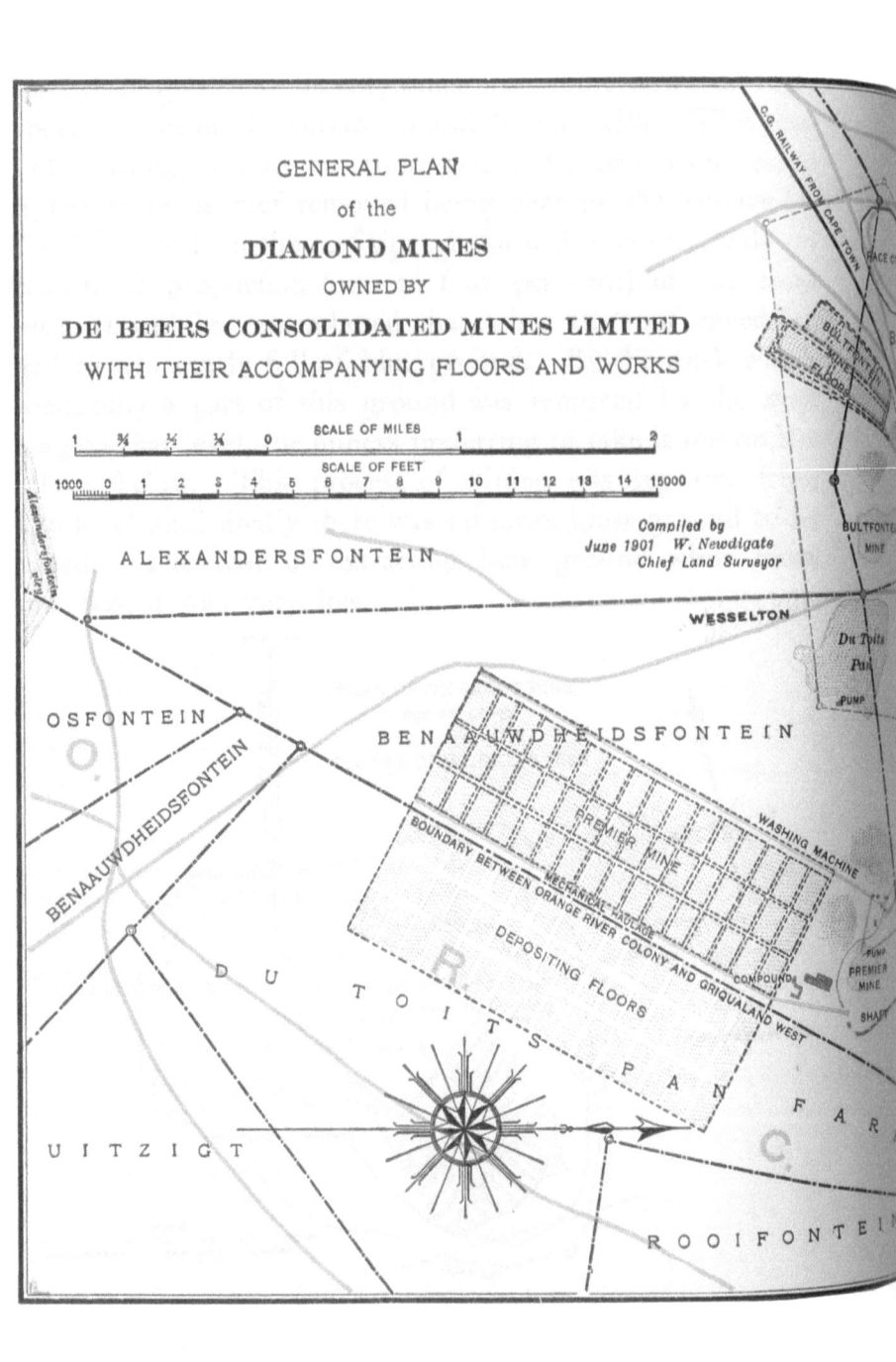

GENERAL PLAN

of the

DIAMOND MINES

OWNED BY

DE BEERS CONSOLIDATED MINES LIMITED

WITH THEIR ACCOMPANYING FLOORS AND WORKS

SCALE OF MILES

1   ¾   ½   ¼   0   1   2

SCALE OF FEET

1000 0 1 2 3 4 5 6 7 8 9 10 11 12 13 14 15000

Compiled by
June 1901   W. Newdigate
Chief Land Surveyor

ALEXANDERSFONTEIN

WESSELTON

OSFONTEIN

BENAAUWDHEIDSFONTEIN

BENAAUWDHEIDSFONTEIN

PREMIER MINE

WASHING MACHINE

BOUNDARY BETWEEN ORANGE RIVER COLONY AND GRIQUALAND WEST

MECHANICAL HAULAGE

DEPOSITING FLOORS

DU TOITS PAN

FARM

UITZIGT

ROOIFONTEIN

BULTFONTEIN
MINE

Du Toits
Pan

PUMP

COMPOUNDS

PUMP
PREMIER
MINE

SHAFT

C.G. RAILWAY FROM CAPE TOWN

RACE COURSE

BULTFONTEIN
MINE
FLOORS

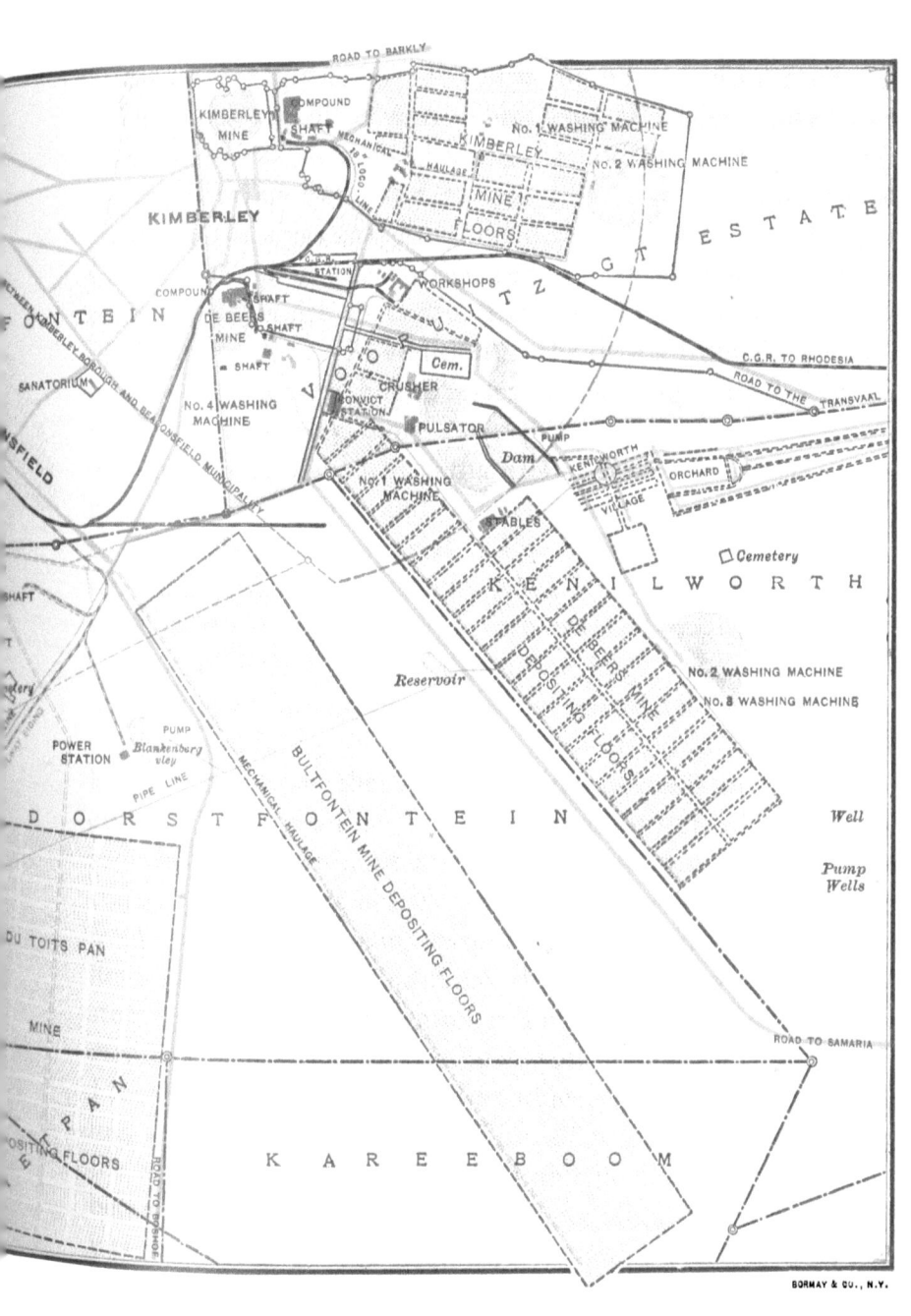

ROAD TO BARKLY

KIMBERLEY MINE

COMPOUND SHAFT

No. 1 WASHING MACHINE

KIMBERLEY

No. 2 WASHING MACHINE

MECHANICAL HAULAGE TO LOCO. LINE

KIMBERLEY

MINE

FLOORS

G T Z

E S T A T E

COMPOUND SHAFT

STATION

WORKSHOPS

DE BEERS MINE SHAFT

SHAFT

C.G.R. TO RHODESIA

ROAD TO THE TRANSVAAL

SANATORIUM

Cem.

CRUSHER

No. 4 WASHING MACHINE

CONVICT STATION

PULSATOR

PUMP

Dam

KENILWORTH

ORCHARD

No. 1 WASHING MACHINE

KENILWORTH VILLAGE

STABLES

□ Cemetery

K E N I L W O R T H

Reservoir

DE BEERS MINE DEPOSITING FLOORS

No. 2 WASHING MACHINE

No. 3 WASHING MACHINE

POWER STATION

PUMP Blankenburg vley

PIPE LINE

MECHANICAL HAULAGE

BULTFONTEIN MINE DEPOSITING FLOORS

D O R S T F O N T E I N

Well

Pump Wells

DU TOITS PAN

MINE

ROAD TO SAMARIA

K A R E E B O O M

OSITING FLOORS

SORMAY & CO., N.Y.

levels were opened around the east end of the mine. When the underground works had reached the depth of 800 feet or more, a new danger appeared. It will be borne in mind that the huge open mines are filled with débris from the sides, caused by the removal of the diamond-bearing ground by open quarrying to depths varying from 200 to 500 feet. As the supports were removed, the sides caved and filled the open mine. This débris was composed of the surface red soil, decomposed basalt, and friable shale, which extended from the surface down to a depth of about 300 feet. In addition to the débris from the surrounding rocks there were huge masses of " floating shale," resembling indurated blue clay more than shale. Large heaps of yellow ground and tailings, which the early diggers deposited near the margin of the mines, and west-end yellow ground contributed to the mud-making material. The black shale which surrounds the mines disintegrates rapidly when it falls into them. It contains a small percentage of carbonaceous matter, and a large amount of iron pyrites. When the huge masses of shale fell into the open mine, they frequently ignited, either by friction or, more probably, by spontaneous combustion, as they have been known to do on the reef tips, and burned for months and even for years at a time. These masses of burned shale become soft clay and form a part of the mixture which fills the open crater. This débris moves down as the blue ground is mined from underneath it, and becomes mixed with the water which flows into the open mine from the surrounding rock and with storm water, and forms mud. This overlying mud became a menace and danger to the men working in the levels below. Frequent mud rushes occurred suddenly, without the least warning, and filled up hundreds of feet of tunnels in a few minutes, the workmen being sometimes caught in the moving mass. It became evident that the method of working shown on the plan was dangerous in case a mud rush took place, the men being sometimes either shut in or buried in the mud coming from the opposite side of the mine. It was decided, therefore, to work the mines from one side only, and to have the offsets to the rock connected one with the other at as few points as would still allow the ventilation of the working faces. The plan illustrated

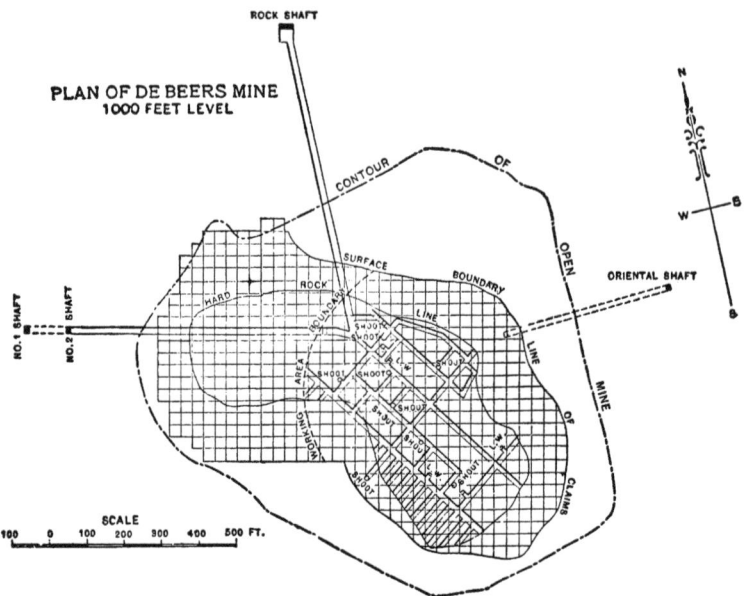

in the above figure shows the method which was then adopted and is still in use. Kimberley mine is worked on about the same general system.

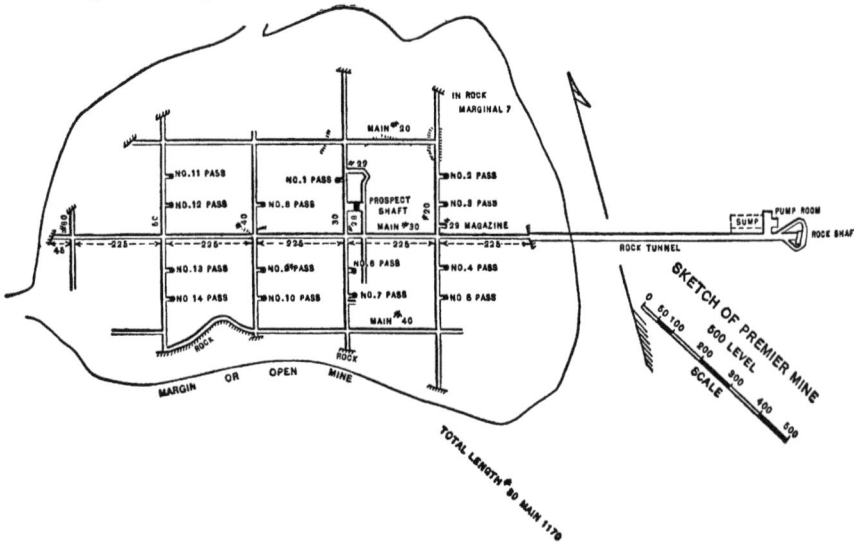

The method of laying out the workings is also here shown. Main tunnels are driven across the crater upon its longer axis,

Stoping.

and, at right angles from these, small tunnels are driven out every 30 feet until they reach the hard rock on the south side of the mine.   These tunnels are widened, first along the rock until they connect one with another, and, at the same time, the roofs, or " backs," are stoped up until they are within a few feet of the

SKETCH SHOWING METHOD OF STOPING

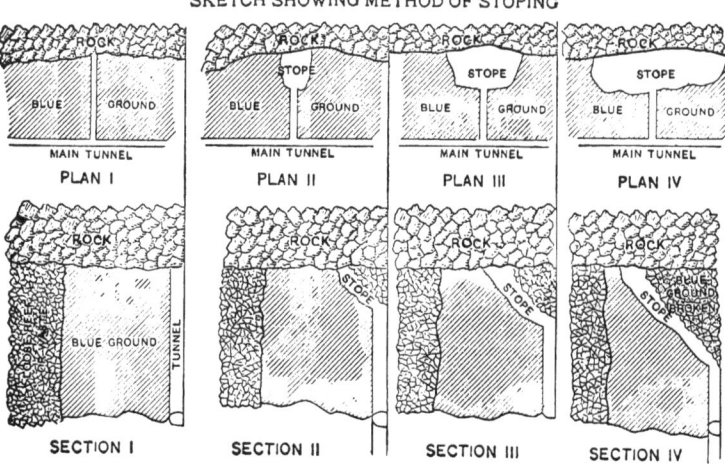

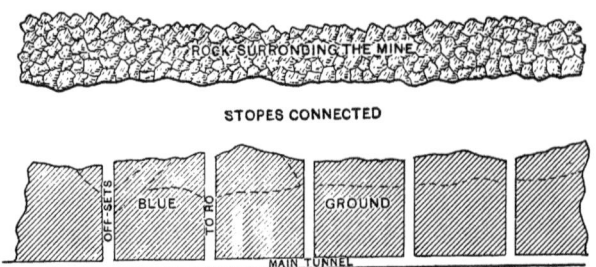

loose ground above, thus forming long galleries, filled more or less with blue ground, upon which the men stand when drilling holes in the backs. The working levels were at first 30 feet apart vertically, but, for greater economy,

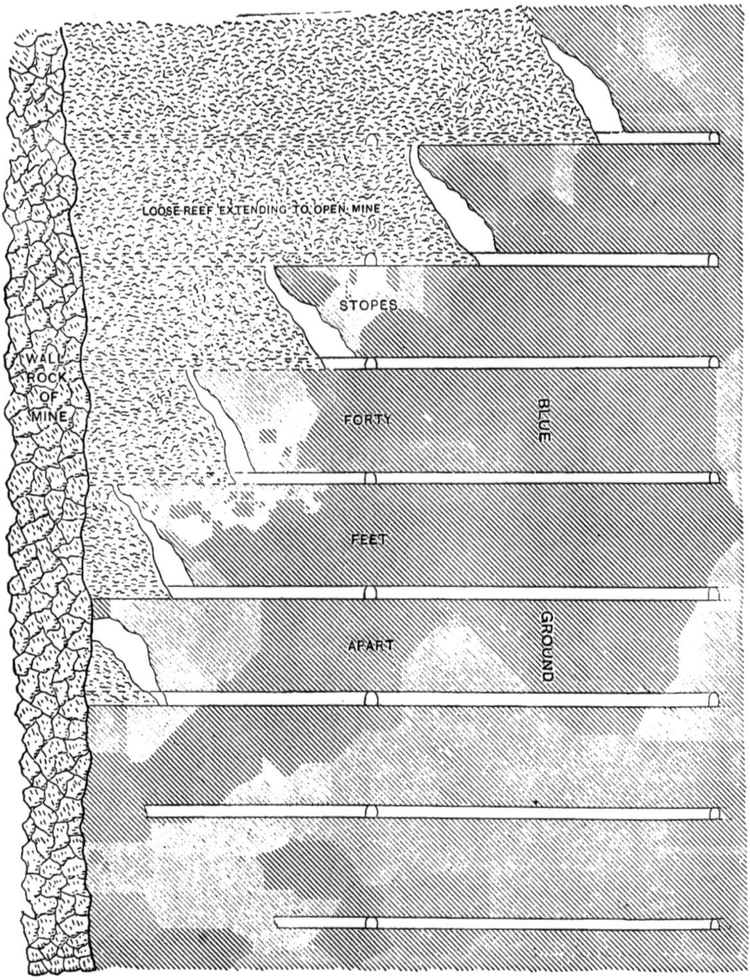

Method of Stoping, Vertical Section.

the distance was soon changed to 40 feet. The broken blue
ground lying in the galleries is taken out, as a rule, before there
are any signs of the roof giving way. At times this is impossible, and the roofs cave upon the broken ground, and the blue
ground is covered with reef. As the roofs cave or are blasted
down, the blue ground is removed, and the loose reef lying above
it comes down and fills the gallery. Tunnels are often driven

Timbering Tunnels.

through this loose reef, and the blue ground, which has been cut
off and buried by débris, is taken out; but it is sometimes left
for those working the next level below to extract.

After the first " cut " near the rock is worked out, another
cut is made, and in this manner the various levels are worked
back, the upper level in advance of the one below, forming terraces as shown in section on page 320. In De Beers mine there
are now eleven levels on which work is progressing, commencing
at the depth of 800 feet and extending down to the 1200-foot

Y

Timbered Tunnels at the 1000-foot Level, Kimberley Mine.

level. At Kimberley mine nine levels are being worked, from the 1200-foot to the 1520-foot level inclusive. The galleries are not supported in any way by timbers, but all tunnels in soft blue ground are timbered with sets of two props and a cap of round timber, and are covered with inch and a half lagging.

Natives drilling, De Beers Mine.

Soft blue ground is drilled with long jumper drills sharpened at both ends. In hard blue ground, drills and single-hand hammers are used. The native workers become very skilful in both methods of drilling, and do quite as much work as white men would do under similar conditions.

Winding Shafts

The grand winding shafts and plant by which the enormous output of diamond-bearing ground is brought to the surface are illustrated in accompanying figures. The present working shafts are all vertical. De Beers rock shaft was the first large vertical shaft of any importance, from the present mining point of view, which was sunk in any of the mines. It is 20 feet by 6 feet in size inside timbers, and contains four compartments, two for skips lifting blue ground, one for a cage for taking men and material up and down, and one for pumps and ladderway. A

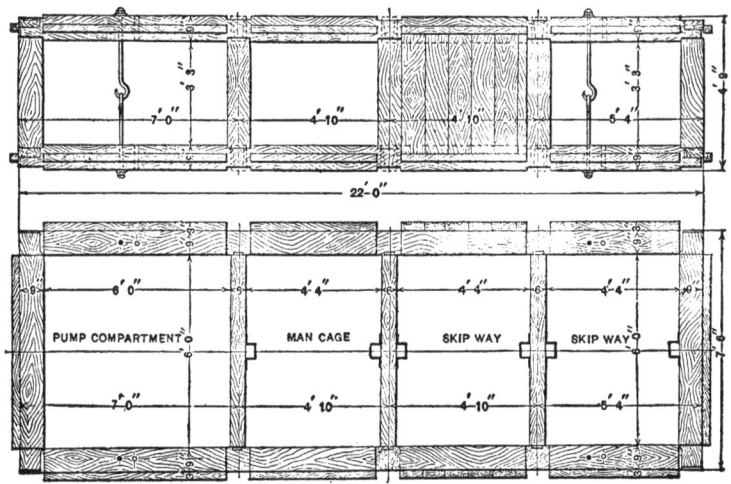

Detail of Sets for Rock Shaft.

balance weight for the cage runs in the pump compartment, which is also the downcast shaft through which the whole mine is ventilated.

No. 1 is the upcast shaft. It has two compartments for skips, two for cages, one for pipes, etc., and a double ladderway.

At Kimberley mine the rock shaft is a duplicate of De Beers rock shaft, except that the pump compartment is larger.

At De Beers, tunnels 11 feet wide by 8 feet high have been driven from the rock shaft at the 800, 1000, 1200, 1400, and 1720-foot levels, and from No. 1 shaft at the 380, 800, and 1400-foot levels.

At Kimberley mine the rock shaft is connected by similar tunnels with the mine on the 1000, 1200, 1520, 1840, and 2160-foot levels. The present depths (1902) of De Beers shafts are 1720 feet and 1400 feet respectively, and of Kimberley rock shaft 2160 feet.

Trucks holding 16 cubic feet transport the blue ground in the mines from the loading places to the main chutes or passes, and from these to the shaft. The trucks are hauled by an

A Shaft Station.

endless chain which rests upon V-shaped clips fastened to the trucks, the motive power being supplied by engines driven by compressed air, carried through pipes from the surface. At the shaft there is a large station cut out of the solid rock, some 30 feet wide, and extending back toward the mine for a distance of 70 feet to the point where the tunnel (8 by 11 feet) commences. There is an extension on one end of the shaft for a small cage-way to bring up any ground that may spill over the skips while being loaded. This prevents delays in the skip hoisting. The shaft is also lengthened for a few feet at the pump end, where a set of pumps is put in.

Loading the Trucks.

As one descends the shaft in the cage in pitch darkness and suddenly comes to a large opening brightly lighted with numerous electric lamps, the scene is weird and confusing. A score of natives, half dressed, each vying with the other in shouting his own comments upon the visitors as they come forth from the cage ; the whirl of heavy iron trucks as they go to and fro ; the banging of the tippers as they turn over and deposit the contents of the enclosed truck into a chute below, — all present a picture unique in itself and only to be seen in passing through the shafts at De Beers and Kimberley mines. Those who have travelled through the native centres, or have seen the negroes loitering about the towns, and have thought them lazy, indolent, beer-drinking beings, should visit the diamond mines, and especially the scene upon the " flat sheet " as described above, and they will get a new impression of the working capacity of these despised black men. The natives working in the diamond mines, if they are old hands in the service, are uniformly active and industrious men, while natives fresh from the kraals are soon taught their duties, which they learn to perform with nearly as much skill as most European miners.

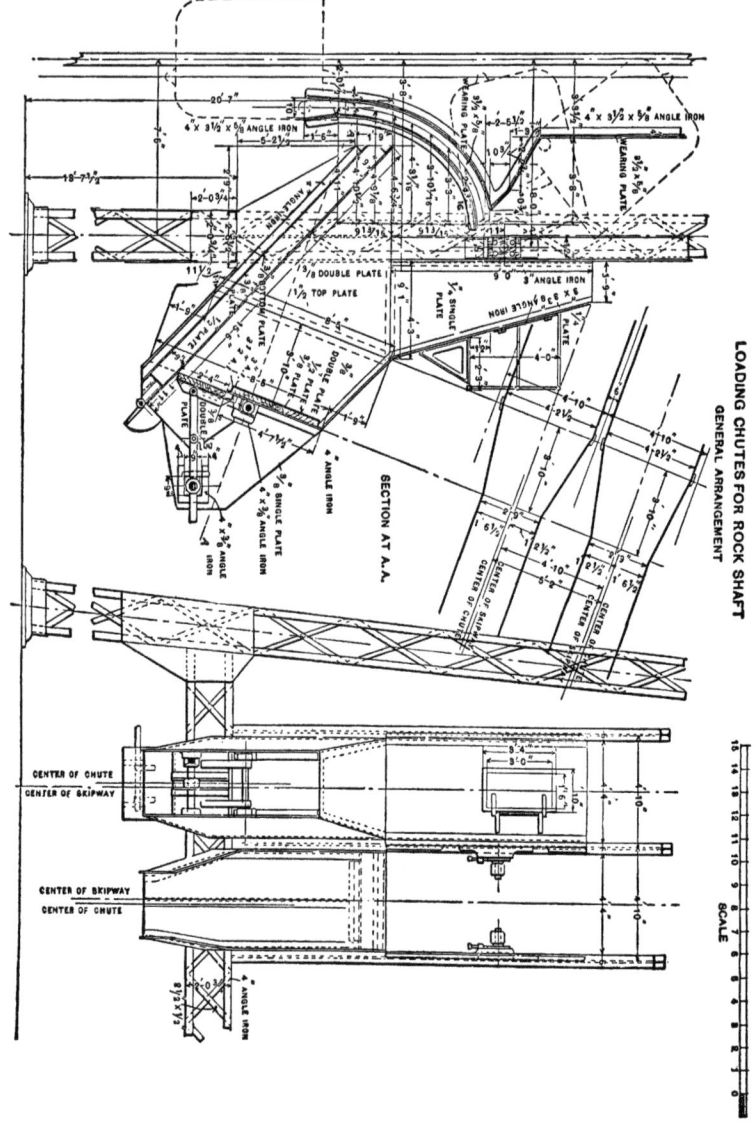

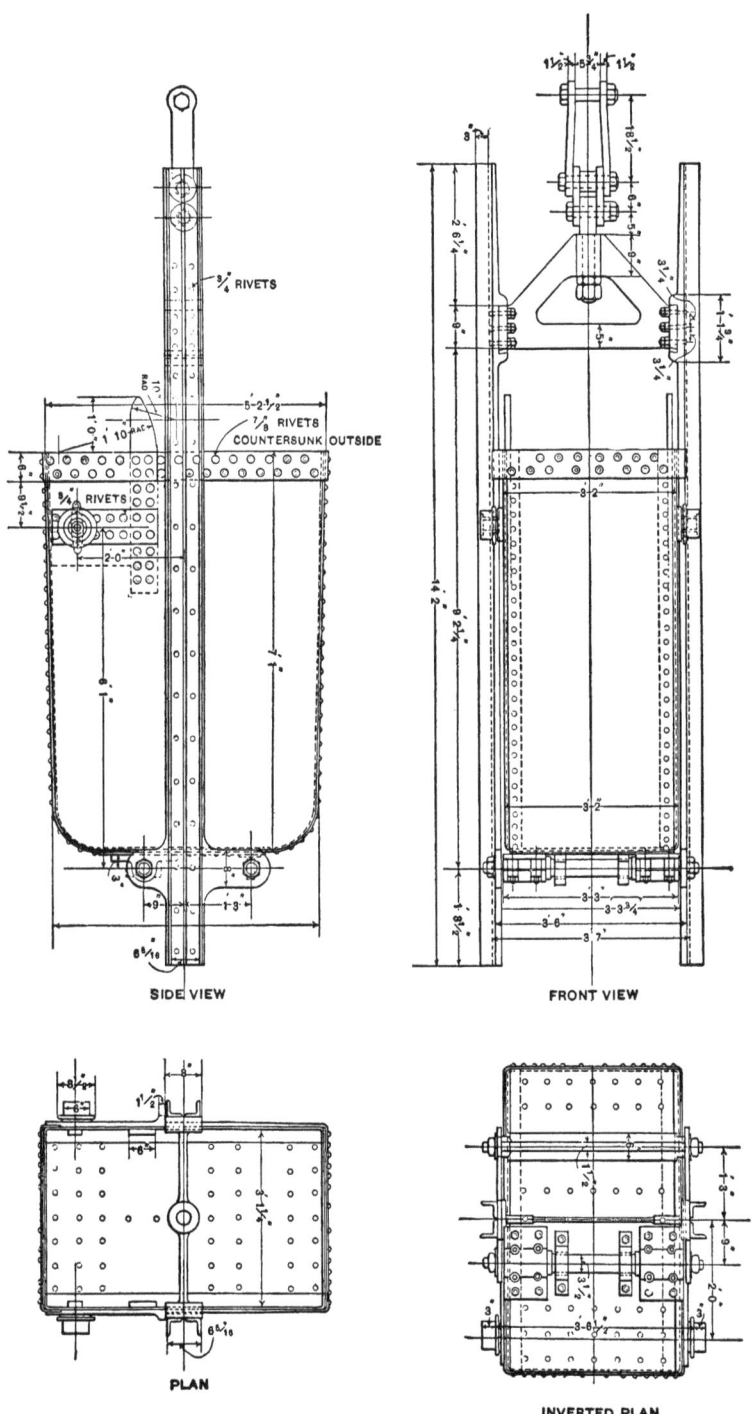

SIDE VIEW

FRONT VIEW

PLAN

INVERTED PLAN

PLAN OF SKIP FOR SIX LOADS.

Main Shaft, Kimberley Mine.

No more rapid handling and extraction of the blue ground seems possible than is effected by the aid of these alert workers and the perfected mechanical devices.   As soon as the loaded trucks reach the shaft, they are tipped into loading chutes holding six truck-loads (96 cubic feet).   As the skip reaches the bottom a door is opened, and the contents of the chute run into the skip and are hoisted to the surface.   Experience has shown that the best results are obtained by sending up loaded skips from one level at a time.   The simple and efficient device early adopted in the mines for tipping the loads from the trucks into the skip at No. 2 incline of De Beers consisted of an iron chute.   Four end-tipping trucks were placed close against the edge of the chute and the catches loosened.   As soon as an empty skip was lowered past the chute the trucks were tipped and the loads ran into the chutes so rapidly that the engine-driver frequently received a signal to hoist before his engine had been stopped.   The skip in this incline held 64 cubic feet, or four truck loads weighing 1600 lbs. each.

The time of the journey through the shaft now varies only a little with depth, being from thirty-five to forty seconds from the 1200 or 1520-foot levels.   On reaching the surface, the blue ground is tipped automatically from the skips into loading boxes. The "self-dumping" skips in present use were introduced by me in 1888, and were made from drawings supplied by the Union Iron Works of San Francisco, and are similar to the skips used in the mining districts of the Pacific Coast.   (On page 327 are shown the plans for the skip and the manner of tipping into the surface chutes.) From these chutes the blue ground is loaded into side - tipping trucks holding 20 cubic feet each. The average weight of the blue ground in a surface truck is 2000 pounds. The trucks used underground hold 16 cubic feet, and are end-tipping in the inter-mediate levels where the ground is dumped into passes, but have solid ends on the main levels where revolving tippers are used.

The Rock Shaft, De Beers Mine.

From the depositing surface boxes at the winding shafts, the ground is taken by means of an endless wire rope haulage to the " floors," where it is treated as described in another chapter.

Record Hoisting

With alert and orderly handling of the blue ground in the mines, the rapidity of extraction has advanced to extraordinary record points. During the month of July, 1889, 142,567 loads were hoisted through a single shaft in No. 2 incline, De Beers mine.   The best day's work of 24 hours was 6222 loads of 16 cubic feet, or 4977 short tons. For an hour at a time hoisting was carried on at this shaft at the rate of five skip-loads every three minutes, or 400 truck-loads an hour,

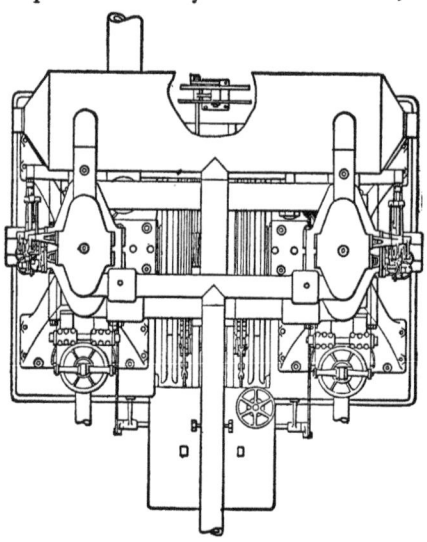

lifted from the 700-foot level, a distance of 840 feet through the inclined shaft. The total amount of blue ground hoisted during the fiscal year from April 1, 1889, to March 31, 1890, was 1,355,089 loads, aggregating 1,084,071 tons of 2000 pounds. This remarkable record was made under unfavorable conditions, because the hoisting engine was small, nominally of 70 horse-power, and not designed for such rapid service.

VERTICAL TANDEM COMPOUND CONDENSING WINDING ENGINES

With the construction of new shafts and the setting up of engines and fittings of the best and latest designs, the efficiency of operation was greatly increased. Two types of winding engines have been erected, and it is interesting to follow the changes which have been made in this portion of the plant. The first large engine erected by the De Beers Company was the one at De Beers rock shaft. Its cylinders were 24 inches in diameter, with a stroke of 5 feet. It had two drums, each 4 feet 4½ inches in width and 10 feet 6 inches in diam-

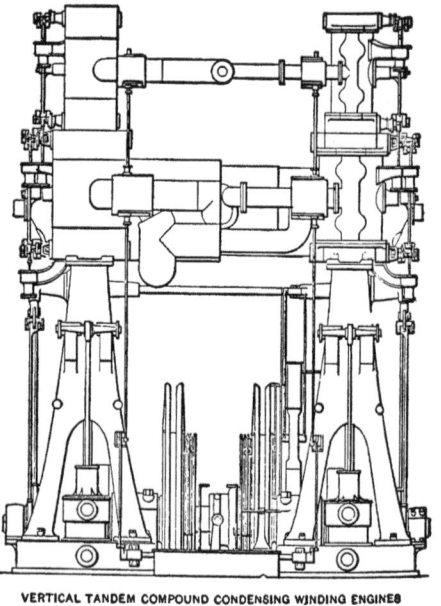

VERTICAL TANDEM COMPOUND CONDENSING WINDING ENGINES

eter, with a grooved tread to prevent friction on the rope. This engine was built by well-known makers of winding engines, whose works are too near the cheap coal centres of England. The engine was what is called in America a "sawmill engine." In the timber districts of America, the boilers are fired with slabs cut from the round logs in squaring them. Enormous quantities of these slabs accumulate about the mills, where they must be consumed in some way, or carted away at a considerable expense. To get rid of the slabs, engines that consume the greatest amount of steam are those most sought after.

VERTICAL TANDEM COMPOUND CONDENSING WINDING ENGINES

In South Africa, on the contrary, the extraordinary consumption of steam was a heavy drawback. Welsh steam coal then cost £8 10s. ($41.25) per ton of 2000 pounds, delivered at Kimberley, so this "sawmill" engine was converted from two high-pressure cylinders to a cross compound, with cylinders of 26 inches and 40 inches diameter, and the consumption of fuel dropped more than 30 per cent. After several years of constant service, the engine was stopped June 11, 1896, the old drum and crank shaft, weighing 32 tons, were taken out bodily, and a new set, weighing 40 tons, substituted and made ready for service in less than 48 hours. (See illustrations.) With this new outfit there was soon a series of record-breaking performances, which are given below.

At the Kimberley mine, the main or rock shaft was started on the north side of the mine in March, 1889. In the first year this shaft was sunk to the depth of 699 feet, and, in the following year, it was pushed to the depth of nearly 1300 feet. The driving of the tunnel to the mine from this shaft on the 1000-foot level showed how exactly vertical was the wall of the crater,

Winding Engine, Kimberley Mine.

for the tunnel, at this depth, entered the blue ground 1134 feet from the shaft, corresponding almost precisely to the distance from the mouth of the shaft to the edge of the melaphyre at the

Mr. Louis I. Seymour.

depth of 300 feet. For hoisting service at this shaft, a winding-engine plant was especially designed by the late Mr. Louis I. Seymour, mechanical engineer for De Beers Consolidated Mines Limited, and constructed by James Simpson & Co., of London, England. This plant consisted of a pair of vertical tandem compound engines driving two reels. These engines were designed to hoist six truck-loads in one skip, from the 1000-foot level, in 45 seconds, including filling, starting, stopping, and discharging;

but in practice they pulled up the skip carrying this load from the 1000-foot level in from 30 to 35 seconds. Flat ropes were used, at first, on the reels, but when the shaft was sunk some hundreds of feet deeper, round ropes were substituted by the adoption of the " Whiting system," first used by Mr. S. B. Whiting, general manager of the Calumet and Hecla Copper Company of Michigan. The dimensions and description of the engines are given in Appendix II.

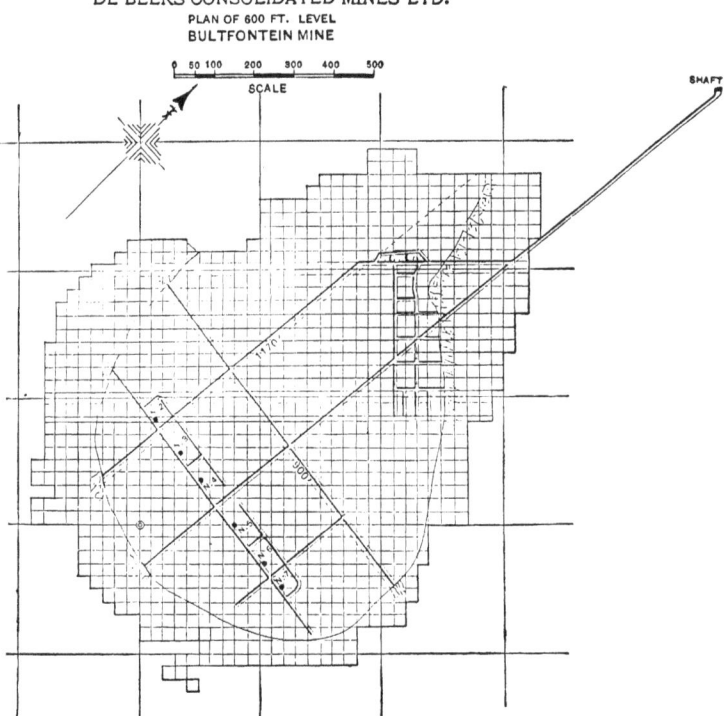

DE BEERS CONSOLIDATED MINES LTD.
PLAN OF 600 FT. LEVEL
BULTFONTEIN MINE

The only excuse I can offer for having adopted flat ropes for winding is that I was persuaded to do so against my own judgment by a number of American engineers, and experience proved that I erred in so doing. Leaving all other disadvantages aside, and they are many, the extra cost of ropes per load is sufficient to condemn the flat rope. The average cost per load for flat ropes was .6 of a penny against .076 of a penny with the present Whiting system, the saving amounting to more than £2000 per annum. This system as modified in the diamond mines is as follows : The round winding rope, made of the best plough steel, extends from the skip over the sheave on the pit-head frame down to the reel on the crank shaft of the engine, thence four times around this reel and a corresponding reel on a lay shaft (centres of shafts being 12 feet apart) ; thence the rope passes around an idler sheave, the shaft of which runs on bearings set upon a movable frame, which is attached at each end to a carriage by means of trunnions. The carriage in this case runs upon a track 50 feet long. From the idler or tension sheave the rope passes around a second reel which is loose upon the crank shaft, the centre of which is in line with the second sheave upon the pit-head frame.

By the completion of the new plant the output of blue ground from the Kimberley mine was greatly increased. During the fiscal year ending June 30, 1893, 1,453,152 loads were taken from the mine as against 1,310,994 loads, the output for fifteen months previous, an increase almost wholly due to the new hoisting facilities, for fully three-fourths of the yield was drawn through the main shaft. The product of De Beers mine for the same year was still greater. The total quantity hoisted was 1,637,031 loads, of which 1,403,060 loads were drawn through the main or rock shaft, and only 233,971 loads through the No. 2 or west end incline shaft.

Drainage

Thorough drainage is of manifest importance in the operation of any mine, but it is peculiarly essential in these diamond mines. At the commencement of underground mining the inflowing water was removed by steam pumps. The use of such pumps was an error, for the resultant heat and moisture caused the blue ground to crumble, and made the ladderways so hot that they were at times impassable.

As soon as the vertical shafts were completed at De Beers

and Kimberley mines, Cornish pumping plants were put in, by means of which all the water is now pumped from the mines. The average quantity of water taken from De Beers mine is 4350 gallons per diem, and from Kimberley, 8385 gallons. Nearly half of the latter influx comes from a crevice at the junction of the quartzite with an intrusive dike of igneous rock which was struck while driving the 1200-foot tunnel at a distance of 600 feet from the mine. While no water is found in the blue ground or mine itself, that which flows into the mine from the surrounding rock mixes, as before described, with the débris which has fallen into the worked-out portion of the De Beers and Kimberley mines, and makes mud. Enormous quantities of this mixture are from time to time forced suddenly into the working parts of the mine, which are connected by tunnels with the loose débris. At times hundreds of feet of tunnels were filled in a few minutes. Mud rushes became so frequent that the working of the mines was seriously interfered with, and the loss of life was very great.

At Kimberley mine, large springs of water flowed into the open works at the junction of the melaphyre with the shale. Only a small part of the melaphyre was then exposed to view, and the position of the other part was unknown. A tunnel was started from the Standard shaft, and driven to the south around the mine. Another tunnel was started from the Harvey shaft and driven to the west end around the mine in the opposite direction until the two tunnels met. The total length was 2097 feet. Through these tunnels all the surface water and all water coming into the mine above the melaphyre was taken up and led to the pumps by means of pipes. All water which enters the mine in the deeper workings is taken down in passes, sunk in the rock outside of the mine. By these precautions mud rushes have been completely stopped in Kimberley mine, and none have occurred for many years past.

De Beers mine has not been so fortunate, and mud rushes are of frequent occurrence, although the quantity of water in

336   THE DIAMOND MINES OF SOUTH AFRICA

this mine is only about one-half that of the Kimberley mine. The following work is being done with the view of preventing them. A tunnel is being driven around the mine at the hard rock (melaphyre) level, about 380 feet from the surface, in order to take up the water which flows into the open mine below the shale. Tunnels are also being driven on the 1000-foot level on the south and east sides of the mine, which will be continued until they meet. Diamond drills are at work making holes between levels, with the view of tapping the water. Everything feasible will be done to free De Beers mine from the plague of water as perfectly as it has been done at Kimberley mine. The problems are not the same, however, for in Kimberley mine the débris had followed down as the blue ground was extracted, and had left the hard rock more or less exposed to view, and one could see in places where the streams of water flowed into the open mine; but in De Beers no hard rock has been exposed until lately, and one must grope in the dark, as it were, to find out where the water enters the open or worked-out portions of the mine.

The pumping plants for freeing the mines from water have kept pace fully with the advance in the hoisting plants. For the service of De Beers mine, a new pumping engine was erected at the rock shaft in 1889. This is a compound surface-condensing engine made by James Simpson & Company, of London. Its high-pressure cylinder is $14\frac{3}{64}$ inches diameter, and its low pressure, 21 inches, with a stroke of 30 inches, It is capable of developing 120 horse-power. With this engine an average of nearly 6000 gallons an hour was readily drained from the mine from the start, and no difficulty was experienced in lifting over 8000 gallons an hour at times. The cost of pumping is largely offset by using the water drained from the mine for washing the pulverized blue ground. By combining this supply with that obtained from surface reservoirs, enough water was obtained for the use of the concentrating plants, except in very dry seasons. For the Kimberley mine a Cornish pumping plant of 400 horse-power, from designs by the late Mr. L. I.

Seymour, was erected in 1891. This is a vertical triple-expansion condensing engine, with cylinders $15\frac{1}{2}$ inches, $23\frac{1}{4}$ inches, and 37 inches in diameter, and a stroke of 36 inches. The gears for this engine were made by Fraser & Chalmers, of Chicago, Illinois, and the crank shafts by Sir J. Whitworth, of Manchester, England, but the main constructors were James Simpson & Co. Ltd., of London. (See Appendix III.) With this plant an average of over 12,000 gallons a day was readily pumped from the mine in the first year after its erection, and since then there has been no further difficulty in handling the influx of water into the workings.

Compressed Air

For all underground service in the mines, in driving sinking engines, mechanical haulages, rock drills, and any other machinery where power is necessary, steam has been supplanted by compressed air. Electricity has also been used for some of these purposes, and is the cheaper and better power for many of the uses for which steam and compressed air have been used.

Lighting

For lighting, the application of electricity has already proved to be almost indispensable. All tunnels and ladderways throughout the mines are lighted by electricity. In the stopes and other working faces candles are used. Electric lights have been found to be of the greatest assistance in enabling the men to get away from rushes of mud. These occur at times when some of the galleries are "hung up" (to use a miner's expression), which means when the tops of some of the galleries are choked with huge pieces of blue ground. The roof suddenly gives way from the pressure of mud above, and all open lights, such as candles, are put out by the force of the concussion of the air, and, were it not for the electric lights, the tunnels in the vicinity of the mud rush would be in total darkness.

z

Other Electric Service

Electric bells are in use throughout the mines, and have very greatly promoted the rapidity of hoisting through the shafts. Owing to instantaneous communication between the man in charge of loading the skips and the engine-driver, hundreds of loads more are sent to the surface daily than could be forwarded under the old "pull bell" system. For instant additional communication between the surface and the underground work telephones have been installed, and the same rapid communication extends to the depositing floors, concentration works, and offices of the company.

Natural Ventilation

The Kimberley mine is ventilated in a somewhat peculiar manner. The rock shafts at both De Beers and Kimberley mines are downcast, i.e., the air for ventilation goes down these shafts, along the bottom tunnels and thence up through the various levels, and it is fortunate for the men working in the mine that it is so, for the cool air comes in at the bottom and ventilates the mine much better than if the rock shaft drew the heated air down through all the lower workings. The upcast in Kimberley mine is through the Harvey shaft, the top of which is 300 feet below the top of the rock shaft. This shaft, with which the various levels of the mine are connected, extends down to the 1200-foot level, and a similar shaft or winze situated near it extends from the 1200-foot to the levels below. As the top levels in the mine are the hottest, the current of air ascends through the Harvey shaft. The usual direction of air currents in mines with two shafts and natural ventilation is down the shorter shaft and up the shaft the mouth of which is situated at the greater height on the surface. The reverse is the case at Kimberley mine. The quantity of air which passes down De Beers rock shaft was 33,300 cubic feet per minute until 1898, when the enlargement of the upcast shaft was completed, and the air current was increased to 45,000 cubic feet per minute. In the Kimberley mine the influx of air per minute is 25,500 cubic feet.

Temperatures

At De Beers, with temperature of the air on the surface 79° F., the temperature ranges from 75° to 77° in the tunnels leading to the mine on the 1000, 1200, and 1400-foot levels. The temperature of the air as it leaves the mine on the 800-foot level is 84°. The temperature of the mud after a mud rush was on one occasion 85° F. Temperatures at Kimberley mine in the 1200-foot tunnel were, for the air, 71°.5 ; for the rock, 72°.1 ; for the large spring of water 78°.9 F. The quantity of water flowing from this spring, which is about 600 feet from the crater, is 3500 gallons an hour. The temperature in the working galleries on this level is 87°. Springs of water on the 1520 and 1840-foot levels gave 83°.8 and 81°.9 respectively, the water in the lower level being the cooler.

The Output of Blue Ground

The table of statistics (Appendix V) gives the amount of blue ground produced from De Beers and Kimberley mines since the formation of the De Beers Consolidated Mines Limited. In the same table of statistics will be found the average cost of production per load. It will be seen that the lowest cost was for the year ending June 30, 1894, — 6s. 6.8d. per load. This includes all charges from the mining of the ground to the delivery of the diamonds to the valuators. All mine charges, including shaft-sinking, tunnelling, etc., are charged to current expenses. It is interesting to note in the same table the greatest output of each mine through a single shaft for various periods of time. The maximum amount of blue ground produced in one year was 1,746,240 loads from De Beers mine for the year ending December 31, 1897. This ground was raised from a depth of 1000 feet from the beginning of the year until June 14, when winding commenced from the 1200-foot level.

At Kimberley mine, the maximum output from one shaft was 1,600,422 loads for the year ending December 31, 1893, hoisted from a depth of 1000 feet. These figures do not include the waste or "reef" which is taken out, amounting to 100,651 loads from De Beers mine and 64,799 loads from Kimberley mine during the year.

During the month of November, 1898, 208,013 loads of blue ground equal to 166,410 tons of 2000 pounds, were hoisted through the two skip compartments of De Beers rock shaft and from a depth of 1200 feet. The winding stops from Saturday night at eleven o'clock until Monday morning at six. The average number of loads of blue ground hoisted per hour was 349. The average daily output for a full day's work was 8376 loads, and for Saturdays 5933. The best day's record was 9790 loads, the best week's record was 50,450. In the above records no account has been taken of stoppages during working hours nor is the quantity of waste, which was 11,992 loads during the month, taken into account. Previous to this the best month's production was from De Beers mine, in November, 1897, a total of 197,173 loads from the 1200-foot level. In Kimberley mine, the best records for a month were in November, 1893, when 157,847 loads were taken from the 1000-foot level, working three shifts of eight hours each per day, and 108,627 loads from the 1200-foot level in May, 1895, working twelve hours per day.

The best week's record from Kimberley mine, winding by day only, was 27,418 loads in sixty-nine hours from the 1520-foot level for the week ending September 22, 1897. No account has been taken of any lost time.

From the above figures it will be observed that all records have been broken for winding ground through a single shaft with two skip compartments.

Labor and Wages

The following table shows the average number of men employed in and about the mines worked by De Beers Consolidated Mines Limited, during the year ending June 30, 1897 : —

EMPLOYÉS

The average number of persons daily employed is as follows :—

	DE BEERS.		KIMBERLEY.		PREMIER.		WORKSHOPS.		ON THE ESTATES AND ELSEWHERE.	
	Whites.	Blacks.	Whites.	Blacks.	Whites	Blacks.	Whites.	Blacks.	Whites.	Blacks.
Above ground	477	1851	187	925	46	423	388	211	28	118
Underground	212	2001	183	1322	105	489	—	—	—	—

		TOTAL.	
		Whites.	Blacks.
Above ground	. . . .	1126	3528
Underground	. . . .	500	3812
Grand total .	. . . .	1626	7340

The average number of white men employed has increased to over 2000 and the number of natives to over 11,000.

NATIONALITIES OF WHITE EMPLOYÉS
PERCENTAGES

JANUARY.	IN AND ABOUT THE MINES AND FLOORS.		AT THE WORKSHOPS.	
	1894.	1897.	1894.	1897.
English	52.2	46.5	41.5	37.1
Scotch	6.2	7.2	23.3	20.6
Irish	4.8	5.6	2.4	2.8
South Africans	33.1	36.8	27.1	33.3
European	1.8	1.5	4.2	4.7
Other Nations	1.9	2.4	1.5	1.5
	100	100	100	100

WAGES

The following figures give about the average wages paid for various kinds of labor at the mines : mechanics, £5 to £7 per week ; miners, from £5 to £6 per week ; guards and tallymen, from £4 to £5 per week ; engine-drivers, £6 to £7 per week ; natives in the underground works, from 3s. to 5s. a day.

Overseers, from £3 12s. to £4 2s.; machine men and assorters, from £5 to £6 ; natives (ordinary laborers), 17s. 6d. to 21s. per week ; drivers, from 25s. to 27s. 6d. per week. Every employé has a percentage on the value of diamonds found by himself. On the floors the white employés receive 1s. 6d. and the natives 3d. per carat. Nearly double these amounts are paid for stones found in the mines.

Dutoitspan and Bultfontein Mines

If operations were not pushed with the like energy and lib-
erality of outlay in Dutoitspan and Bultfontein mines, it was
simply because of sound economic considerations, and impediments unreasonably placed in the way of projected developments.
Heavy falls of reef had very greatly damaged the open workings in Dutoitspan mine before it came into possession of the

Mount Ararat before Blasting.   (Removal of a piece of " Floating Reef," Bultfontein Mine, 1901.
It was 150 feet long, 50 feet wide, and 120 feet high.   180 holes were drilled in it and charged
with 1050 pounds of No. 1 dynamite.)

De Beers Consolidated Mines.   In spite of this obstacle, work
was carried on actively for a time, until it became certain that no
profit could be made by working this mine and the continuance
of operations would have caused great loss directly to the controlling corporation.   If diamonds were like gold and there was
an unlimited demand for the product, Dutoitspan mine would
assuredly have been worked as long as it paid expenses and the
barest margin of profit.   But, seeing that the demand for dia-

monds, or any other precious stones, is practically limited to the amount marketable without breaking down the prices disastrously to the producer as well as to every dealer and cutter, work in Dutoitspan mine was suspended at the close of 1889. The mine is still idle, but a large shaft was started in 1901 for the purpose of working it at some future date.

Shots Fired.

Bultfontein mine might have proved more profitable, but in 1889 an immense fall of reef, covering nearly the whole bottom of the mine, made open work impossible, except over a very small area. In face of this situation shafts would have been started and underground work on a systematic plan prosecuted, had it not been for obstacles set in the way by the lessors, the London and South African Exploration Company. It was not anticipated that there would be any profit in instituting these costly underground works at that time, but the directors of De Beers mines desired to furnish employment to miners out of work, and the mine would have been opened and explored on a return of bare expenses, if the lessors had seen fit to make reasonable terms. As their demands were considered exorbitant, work in this mine was also stopped in 1889, and was only commenced

A Second after Firing.

again in 1900. Plan on p. 333 shows how the mine is being opened. There are nearly 13,000,000 loads of blue ground in sight above the 600-foot level.

Premier Mine

In December, 1891, the farm, Benaauwdheidsfontein, adjoining Kimberley, and lying on the border line between Griqualand West and the Orange Free State, was purchased in full by the De

Premier Mine looking from Workings up through Incline where the Blue Ground is hauled.

Beers Consolidated Mines.   On this property the Wesselton or
Premier mine, situate about four miles from the town of Kimberley (plan at pages 316–317 gives its position relative to the other
mines), had been discovered in September, 1890, by a Dutchman,
Fabricius, who was prospecting for an old resident of the diamond fields, Mr. Henry A. Ward, who had a bond on the Wessels' estate, or an option to purchase the property for £175,000
within a stated period of time.   When a man has no money,

The Mine Filled with Smoke.

and Ward had little or none at that time, it matters very little
to him what amount he has to pay for such a property, for he
does not want the farm unless he finds a payable diamond mine,
and if he does find a mine, some one else supplies the funds.   In
this case the mine was found, but it was one chance in a million.
Only a small portion of Wessels' farm was in the Cape Colony,
and it was upon this portion that the mine was discovered.

Scores of sanitary pits had been sunk within a stone's throw
of the mine before the prospector Fabricius sunk a hole at ran-

dom, without any apparent reason, through ten feet of limestone and found yellow ground. It was soon noised about, and the mine was rushed and jumped by a crowd from Kimberley and Beaconsfield, consisting to a great extent of members of the Knights of Labor. Hundreds of claims (30 feet by 30 feet) were pegged off, and holes averaging 3 feet by 6 feet were sunk all over the place, looking far more like open graves than prospectors' shafts; in fact, they proved to be the graves of the

After the Smoke has cleared away.

hopes of the reckless jumpers of private property. Many of the holes were sunk outside of the area of the present mine.

Ward had the sole right of prospecting for minerals upon this farm, which was held under his agreement with Wessels; but for some time the jumpers held their ground regardless of its legal ownership, and their contest was the more bumptious from the fact that the mine was only a few hundred yards from the boundary line between the Colony and the Free State. Title to Wessels' farm was originally granted by the Free State.

By the laws of this State all minerals belong to the owners of the farms upon which they are found. In the settlement of the boundary line between the Free State and Griqualand West it was agreed that the farmers who had held titles to their farms under the laws of the Free State should retain the right to any minerals that might be found upon them. After months of wrangling, Ward's claim was established beyond dispute. Ward was without means to continue prospecting, and parted with half his rights for £3,000. When the mine was discovered, De Beers Consolidated Mines bought the interest which Ward had sold, for which they paid £120,000. Ward disputed De Beers ownership to an undivided one-half interest in the property. The case came to trial in the Supreme Court of the Cape Colony, the mine having been discovered in that part of the farm lying within the Colony. Judgment was given in favor of De Beers, and that Company became joint owner with Ward in the property, now called the Premier Mine, named by Ward in honor of Rhodes, who was at the time Premier of the Colony, and with whom he had conducted most of the negotiation in relation to the purchase of the mine and the final disposition of his interest.

In the meantime Ward had obtained an extension of his option for an additional sum of £125,000. The directors of De Beers mines were in no way consulted in this matter. The time for taking up the option was drawing to a close, and as Ward did not have the money to pay for his half, it was mutually agreed that De Beers should pay the purchase price of £300,000, Ward becoming responsible for the repayment of his half. After considerable negotiation Ward agreed to cede his interest in the mine on the following conditions: that he should take over the mine for a period of five years, during which time he had the right to take out 5,000,000 loads, equal to 4,000,000 tons of diamond-bearing ground. *Diamond-bearing* and *blue* are not synonymous terms here, for Ward took out *yellow* ground to the depth of about 60 feet. The mine was surveyed as accurately as possible. An allowance of 8 feet in depth was made for the surface limestone which covered the mine and

PREMIER MINE, OPEN WORKINGS.

which was supposed to be *non*-diamond-bearing. It was also agreed that a load of ground in place should be 9.6 cubic feet, but this was afterward increased to 10.6, as it was found that 9.6 cubic feet of yellow ground would not make a load when broken. From the preliminary washing of ground taken from various parts of the mine, it was estimated that the mine would yield about 16 carats per hundred loads washed. Ward took possession of the mine, and through contractors erected a large washing plant capable of washing 4,000 to 5,000 loads daily. During the five years Ward mined and washed the 5,000,000 loads to which he was entitled. The yield was about 20 carats per 100 loads by means of the first sortings, and possibly two or three carats more were obtained by subsequent sorting, so that the total number of carats obtained reached about 1,100,000. As to the price realized for these diamonds and the cost of producing them, I have no knowledge, but one may assume that the average value of the diamonds was about 18*s*. per carat, and that the cost of mining and washing did not exceed 1*s*. 6*d*. per carat, if it reached that figure. The first 60 feet were easily mined, as the ground was decomposed and could be sent direct to the washing machines from the mine. At the present time, under De Beers management, blue ground is mined and deposited, harrowed and watered, and then loaded and sent to the washing machines for a cost of about 2*s*. 2*d*. a load.

From the year 1871, when the four mines at Kimberley and the Jagersfontein mine were discovered, a period of twenty-one years elapsed during which no paying diamond mine was found, although continuous prospecting was carried on. The Premier mine was covered for an average depth of eight feet with lime, which for the most part was diamond-bearing. The formation of the lime seems to have been the result of the evaporation of water highly impregnated with lime, or possibly springs existed in the localities, whose waters were highly impregnated with carbonate of lime, which was deposited by the evaporation of the water. Water, in many of the lime-covered districts, is found very near the surface. On the Wesselton estate

PREMIER MINE,

PREMIER MINE.

it shows itself in numerous "fonteins" or springs. Below the lime coverlet the diamond-bearing yellow ground extended to a depth of sixty feet, where it changed to blue ground. The work which had been done proved the area of the mine, and it was found to contain about 1162 claims of diamond-bearing ground, equal to about 24 acres.

Under Ward's administration the diamond-bearing ground was removed by means of trucks drawn by an endless chain haulage, which delivered them at the top of a large washing plant, where it was at once treated.

In January, 1896, Ward's lease expired, and from that time work in this mine has been constantly carried on by the De Beers Consolidated Mines. An incline with a grade of one foot in five was constructed in 1896; the drainage water from all parts of the mine was concentrated in a sump, and a pumping plant erected capable of handling the great influx of water, averaging 42,726 gallons an hour in 1896, or about 7,178,000 gallons a week. At the end of that year the new works were so far advanced that 271,777 loads of blue ground had been raised. For the economic working of this mine, a complete mining and washing plant, with compounds, machine shops, stores, and other necessary buildings, was installed soon after the mine was turned over to the Company by the lessee.

The incline mentioned above was made through the marginal reef, and down to a depth of 185 feet. At present the diamond-bearing blue ground is hauled from the mine by means of an endless wire rope haulage (see illustration opposite) driven by an engine on the surface. The mine is being worked in sections of 50 feet in depth. The ground is broken by drilling deep holes (12 feet) with jumper drills and blasting with dynamite. The average number of loads broken per case of dynamite (50 lbs. net) is 416, equal to 333 tons. The breaking of the ground was formerly done by contract, and cost $5\frac{1}{2}d.$ per load delivered upon the "flat-sheets" near the mine end of the wire rope haulage. This mining is now done by the Company. Loading is done in the mine upon the contract system, by paying the

PREMIER MINE,

Showing Mine being worked in the Open. The water on the right represents about one million gallons daily, which finds its way into the mine, and has to be pumped out.

Premier Mine, looking up the Incline.

natives 15s. for 100 loads. The cost of hauling and depositing is about 6d. per load. In open mining the natives are paid 15s. per 100 loads (80 tons) for loading and delivering to a flat-sheet from 100 to 150 feet from the place of loading. On the floors, after the ground is pulverized, 12s. per 100 loads is paid for reloading. The ground is treated in the same manner as at De Beers and Kimberley mines, which will be described in the following chapter.

There is a large body of floating reef in the mine, which measured about 350 feet by 200 feet on the surface, but, at the depth of 500 feet, it has been nearly displaced by diamond-bearing ground. As already mentioned, these large blocks of floating reef are portions of the country rock which have broken loose

2 A

during the time the craters were being filled, and were not incorporated in the breccia of which the blue ground is composed. In some instances the "floating reef," or "islands," is

One of the Early Washing Machines.

the same as the amygdaloid rock or melaphyre, which surrounds the mines at a depth varying from 300 to 400 feet, but, as a rule, somewhat altered. It will be noticed that the aerial gear is not used at the Premier mine, and the reason is that for shallow depths, or for depths down to 200 feet, inclines, either open cuts or shafts inclined to as great an angle as it is practicable for a wire rope haulage to work, are more economical.

Up to the present time the difficulties of falling reefs, which caused so much trouble in the other mines, have not yet arisen. A belt of blue ground some seventy feet in thickness has been left standing in places to support the friable decomposed basalt and shale with which the mine is surrounded. This is but a temporary remedy, and one which does not recommend itself to

One of the Present Washing Plants.

the engineer, owing to the value of the ground which is being temporarily sacrificed. It is my intention to combine the open with the underground system, and to remove the blue ground which lies adjacent to the reef in the same manner as it is now done in De Beers and Kimberley mines, and at the same time to work the remaining portion of the mine in the open, as at present, so long as open mining can be safely and economically carried on. Owing to the enormous flow of water from the reef into the mine (the blue ground itself contains no water), it will be necessary to sink a shaft, and to drive tunnels to tap these large springs, and lead the water away from the mine. The

NO. 1 WASHING PLANT, DE BEERS FLOORS.

average quantity of water pumped from the mine is about 40,000 gallons per hour, or more than three times the quantity which is pumped from De Beers and Kimberley mines combined. In order to make use of this water, it is pumped to De Beers floors for washing the blue ground, and to the village of Kenilworth for irrigation purposes.

The average yield of diamonds for several years past under De Beers management has been three-tenths of a carat per load.

No. 2 Washing Plant, De Beers Floors.

The value of the Premier mine diamonds as compared with those from De Beers and Kimberley mines is about twenty per cent less, owing to the greater proportion of boart and small diamonds. The diamonds from this mine show distinctive characteristics, and a parcel of them can be easily distinguished from those produced from other mines. It is estimated that the production of this mine could be raised to 1,000,000 carats per annum. The mine is being developed for the commencement of underground mining. Plan on page 318 shows the shape and size of the mine on the 500-foot level. It is estimated that there are 13,000,000 loads, equal to 10,400,000 tons, of blue ground in sight above this level. The Premier mine may, therefore, be looked upon as a mine of very great value, and one which will play an important part in the future history of the diamond-mining industry.

Jagersfontein

Mention has been made previously of the Jagersfontein mine. It was the first of the so-called "dry mines" discovered. The mine is very large, containing 1124 claims. The average yield of the ground is about twelve carats per one hundred loads. The quality of the diamonds far surpasses the yield of any other crater. The mine is noted for its large blue-white diamonds, and, now and again, an exceptionally large stone is found. One stone cut as a brilliant weighs 239 carats and is without a flaw.

Two full-size reproductions are here given of the largest diamond found in the mine, its weight being 969½ carats. For

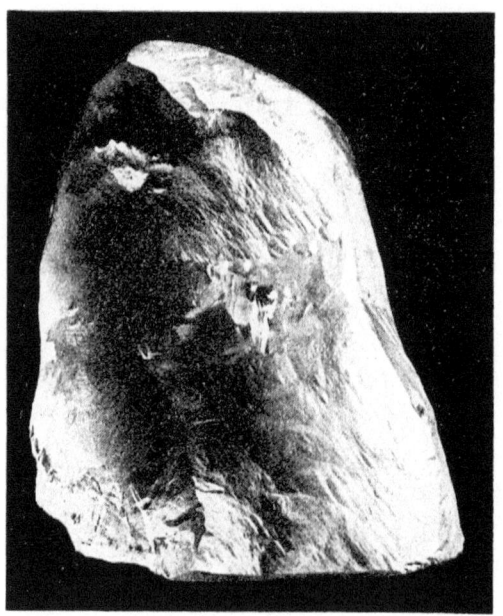

Excelsior Diamond, 969½ Carats. (Found in Jagersfontein Mine. Actual Size.)

many years after their discovery, the richer mines of Kimberley offered greater inducements to the digger as well as to the investor, but the fever for consolidation attacked the directors of some of the principal companies in this mine, and the New Jagersfontein Mining and Exploration Company Limited was incorporated in 1888, about the same time as De Beers, and the various interests were gradually absorbed.

The mine is still worked in the open, and during the last few years has had some difficulty with falls of reef, which is quartzite from the surface down. In 1900 the mine had reached a depth of 450 feet in the lowest workings.

The output of blue ground for the year ending 31st March, 1899 (the last full year's work before the war) was 2,600,000 loads, and the diamonds produced amounted to 289,000 carats, which realised about £500,000. The dividends paid during the same period amounted to £150,000, equal to fifteen per cent on the capital.

Another View of the Excelsior Diamond.

CHAPTER XII

T has been shown how resourceful engineering mastered the problem of the extraction of the diamond-bearing deposits swiftly and systematically, without injury to the mines. It was no less essential to advance and perfect the process of the winning of the diamond from the mass of extracted blue ground with corresponding speed and efficiency. For the handling of the mammoth bulk of breccia, through which the tiny, precious crystals were sprinkled in a proportion so infinitesimal, there was a practical call for every feasible stretch of invention in transportation, concentration, assorting of sizes, and final separation of the gems. The indispensable reconciliation of thoroughness in extraction with rapidity in working over the ground made the task greatly perplexing. It was only through years of experimenting and progressing from imperfect to improved designs that the present great diamond-winning plant of the mines was evolved. If this is still short of ideal suitability to the work, it is simply fair to observe how vast is the stride that has actually been made in a few recent years in diamond-winning methods, from the primitive Indian wooden shovels and drying mats, and the water holes and shaking plates of the Brazilian.

As fast as the blue ground is dumped automatically from the skips into the ore bins, it is carried away in trucks by an endless wire rope haulage, driven by steam, to the depositing floors. These floors are made by removing the bush and grass from fairly level stretches of ground. After clearing the face of the ground, it is hardened and smoothed with heavy rollers until it is fit for use. Receiving grounds are laid out separately for each of the diamond mines on the four farms, and cover an area of several thousand acres. The most extensive of any are the De Beers floors, which are laid off in rectangular sections, six hundred yards long and two hundred yards wide, on the farm, Kenilworth, adjoining the mine. They begin about a mile from the mines and extend for three miles in the easterly direction and a mile to the west.

The main tramway line from the mine is three miles in length, with two branches, one mile and three quarters of a mile

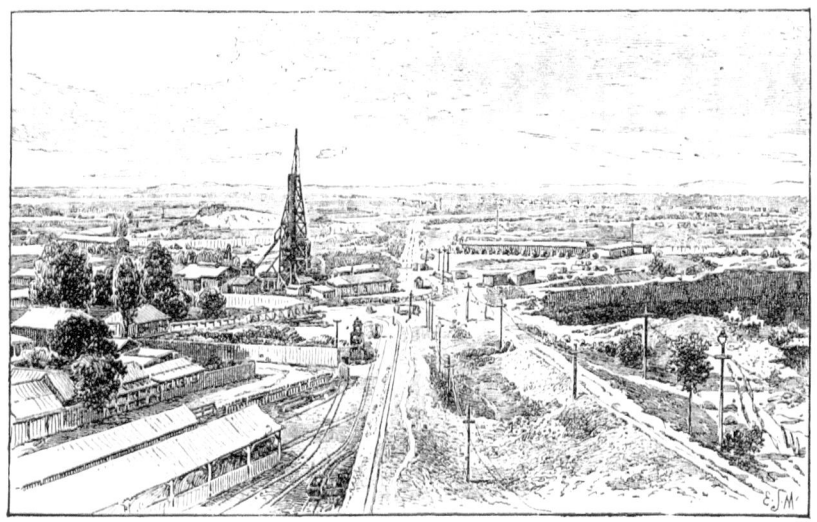

De Beers Mine and Floors. (Showing Haulage from Mine to Floors.)

in length respectively. The speed of the running trucks ranges from 2.5 to 4 miles an hour, and they are counted and greased automatically as they are sent on to the floors. There is a slight down grade from De Beers and Kimberley mines which is of material service in lightening the drag of the loaded trucks. When the trucks reach the floors, they are drawn by horses or mules over auxiliary tram lines at right angles to the main haulage line to any desired point of deposit. A full truck contains about 16 cubic feet of blue ground, weighing 1600 lbs. approximately; but it was found more convenient to supplant these end-tipping trucks by 20 cubic feet side-tipping trucks. The

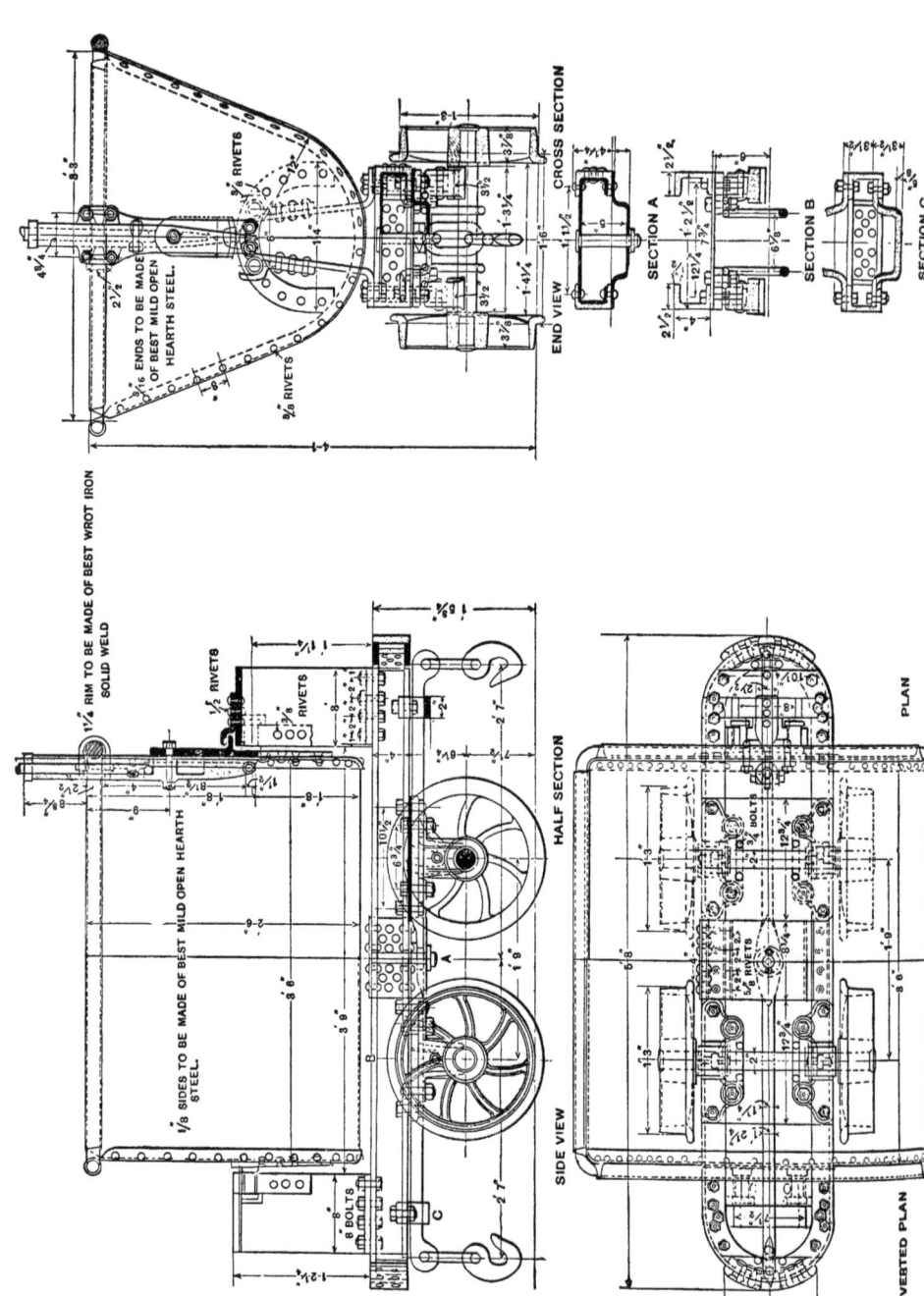

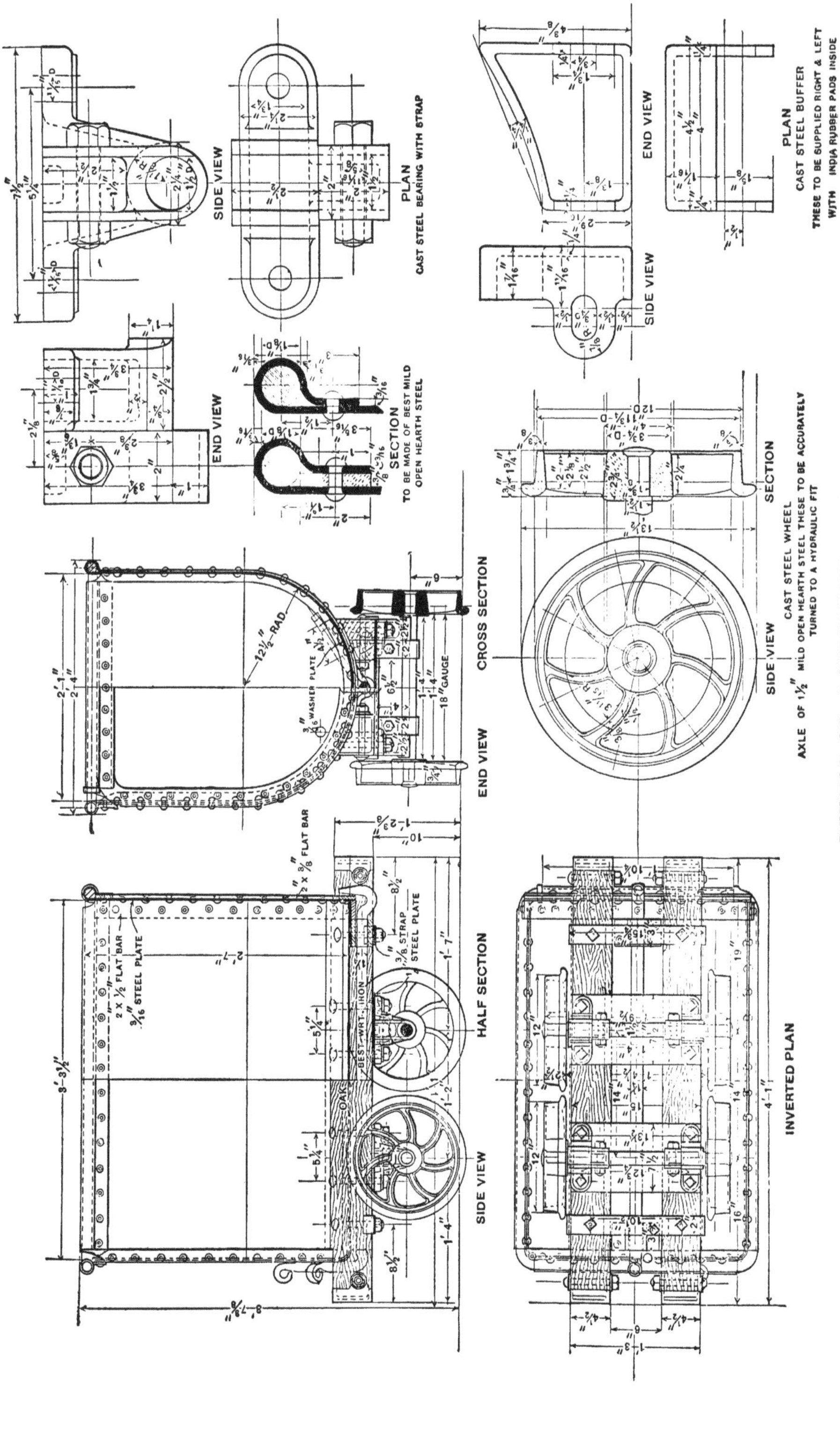
SIDE VIEW
PLAN
CAST STEEL BEARING WITH STRAP
END VIEW
SECTION
TO BE MADE OF BEST MILD OPEN HEARTH STEEL
END VIEW
SIDE VIEW
PLAN
CAST STEEL BUFFER
THESE TO BE SUPPLIED RIGHT & LEFT WITH INDIA RUBBER PADS INSIDE
CROSS SECTION
END VIEW
12½" RAD.
WASHER PLATE
18" GAUGE
SECTION
SIDE VIEW
CAST STEEL WHEEL
AXLE OF 1½" MILD OPEN HEARTH STEEL THESE TO BE ACCURATELY TURNED TO A HYDRAULIC FIT
2 X ⅜ FLAT BAR
2 X ½ FLAT BAR
3/16" STEEL PLATE
⅜" STRAP STEEL PLATE
BEST-WRT-IRON
OAK
HALF SECTION
SIDE VIEW
INVERTED PLAN
DETAILS OF 16-CUBIC-FEET TRUCK.

De Beers Floors.

old unit of measurement, 16 cubic feet, has been retained, and the automatic counters are so geared that every time four 20-cubic-feet trucks pass them, five truck-loads are registered. Each of the rectangular sections of the De Beers floors holds about 50,000 loads. The Kimberley floors are nearly as large, and substantially the same method is employed in covering them. On the depositing ground a truck-load is spread out to

Mechanical Haulage, Kimberley Mine.

cover about 21 square feet. So over the miles of floor surface is outstretched an enormous carpet of "blue" somewhat less than a foot in thickness, and sprinkled with invisible diamonds. It may appear to the reader that the word "invisible" is used to convey the idea that the diamonds are very small, but such is not the case, for many of the diamonds lying buried are as big as filberts, and it is not unusual to find them as large as walnuts. What is meant is that the diamonds contained in the blue ground are invisible to one walking casually over the floors even

De Beers Floors.

after the ground has pulverized. During the fifteen years of my charge of De Beers mines I have never found a diamond on the floors.

It will be seen that no pains have been spared to hasten and cheapen the flow of ground to the floors. After the blue ground has been spread out, it is necessary to wait patiently until the sun and the rain have contributed their service in disintegrating the breccia. The effect of the exposure of this curious compound to heat and moisture is very remarkable. Large pieces of blue, which are as hard as sandstone when freshly taken from the mine, soon begin to crumble on the depositing floors. To hasten the disintegration, the bed of blue is harrowed several

times to turn up the bigger lumps and expose fresh faces of
the ground to the sun. Spans of mules were originally used to

De Beers Floors.

drag the light harrows used in those days, but steam traction
engines are now employed to draw wheeled harrows with huge

Kimberley Floors.

teeth to and fro across the floors. So the great spread of the
floors looks like some vast ploughed farm where the laborers
are preparing the soil for seed.

The length of time required to effect the desired degree of
pulverization depends on the season of the year and the amount

Harrowing Blue Ground with Steam Traction Engines.

of rainfall. It is curious to note, also, that there is a marked
difference in the rapidity of disintegration of the blue ground in
each of the four mines. The blue from Kimberley mine
becomes well pulverized in three months with heavy rains in
the summer season, while the De Beers blue requires double
that time. The longer the ex-
posure, the more complete the
pulverization, and the better
for washing. The long continuance of droughts, which
are of frequent occurrence,
causes very costly delay.
During a period of more
than eight months in 1897
there was not sufficient rain
to wet the blue ground.
The lack of rain water was

Harrowing Blue Ground.

offset, in a measure, by artificial means; but as the blue ground
upon De Beers and Kimberley floors covers 2000 acres of land,

the difficulty of any approach to complete watering may be readily imagined. Under normal conditions soft blue ground becomes sufficiently pulverized in from four to six months, but it is better to expose it for a longer period, even for a whole year.

Traction Engine for Harrowing Blue Ground.

A certain percentage of the blue ground is not affected by exposure on the floors. This intractable ground, which is called hard blue, makes up about 5 per cent of the product of De Beers mine. The large pieces of hard blue are removed from the floors to be crushed in rock breakers and rolls, and large, worthless boulders and stones embedded in the blue, as well as large pieces of basalt and shale which fill the open mines, and have become mixed with the blue ground during the process of mining, are picked out to be thrown away. Then the well-

Loading Pulverized Blue Ground, De Beers Floors.

disintegrated blue ground is taken from the floors in trucks by endless rope haulages to the washing machines and put through the first stage of concentration.

The ground is dumped from the trucks into hoppers, at the bottom of which are small revolving tables upon which the ground is divided and fed automatically into two revolving cylinders. This automatic feeder, which was devised by Mr. Robeson, late mechanical engineer to De Beers Company, not

Loading Pulverized Blue Ground.

only divides the ground equally between two rotating washing machines, but delivers it regularly, so that the machines cannot become overcharged, which would result in loss of diamonds.

2 B

Washing Machine, De Beers Mine.

After leaving the automatic feeders, the ground is mixed with puddle (the name applied to the thick muddy water which flows out of the washing pans) and a quantity of clear water is added.

No. 2 Washing Plant, De Beers Floors.

This mixture serves to bring the fresh supply of blue ground in the pans to the proper consistency for washing, for experience proves that diamonds and the heavy minerals with them separate from the mass of lighter material much better in a fairly thick puddle than in comparatively clear water. From the chutes below the feeders the mixture flows into a revolving cylinder

De Beers Crushing Mill.

covered with perforated steel plates with holes $1\frac{1}{4}$ inches in diameter. All lumps larger than the holes pass out of the end of the cylinder, and are carried by a pan conveyor to crushing rolls for further treatment. Worthless stones carried in the ground are picked out by hand as the lumps move along on the conveyor.

The pulverized ground which passes through the screen holes of the cylinders is fed into shallow circular pans, divided so as to form an annular space, four feet in diameter, between the outer and the inner rim (see figures on pages 372–373). Here the ground is swept around by revolving arms attached to a vertical shaft, and carrying wedge-shaped teeth (see figure). These teeth are set to form a spiral which forces the diamonds and other heavy minerals to the outer side of the pan, while the lighter material flows out of the discharge situated upon the inner rim. Fifty per cent of De Beers ground, when well pulver-

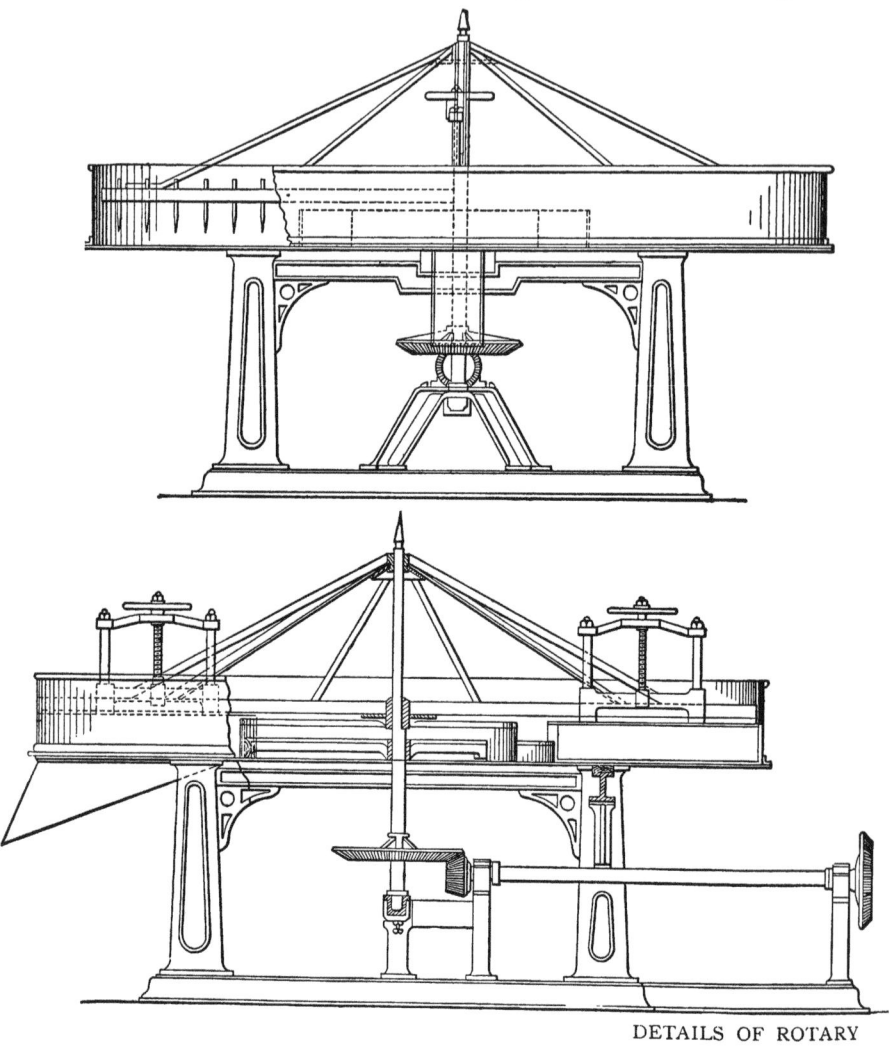

DETAILS OF ROTARY

ized, will pass through a screen with holes $\frac{1}{16}$ of an inch square, and 66 per cent of Kimberley ground will pass through the same screen. The big pieces of hard rock, which were brought out of the mines only a few months before, have crumbled almost to dust, which, during every working day in the year, passes through the pans in a flowing stream for ten hours a day, leaving its treasure behind. When the bare statement is made that nearly five million truck-loads, or more than four million tons of blue ground, have been washed in a year, the mind only faintly conceives the prodigious size of the mass that is annually drawn from the old craters and laboriously washed and sorted for the sake of a few bucketfuls of diamonds. It would form a cube of more than 430 feet, or a block larger than any cathedral in the world, and overtopping the spire of St. Paul's,

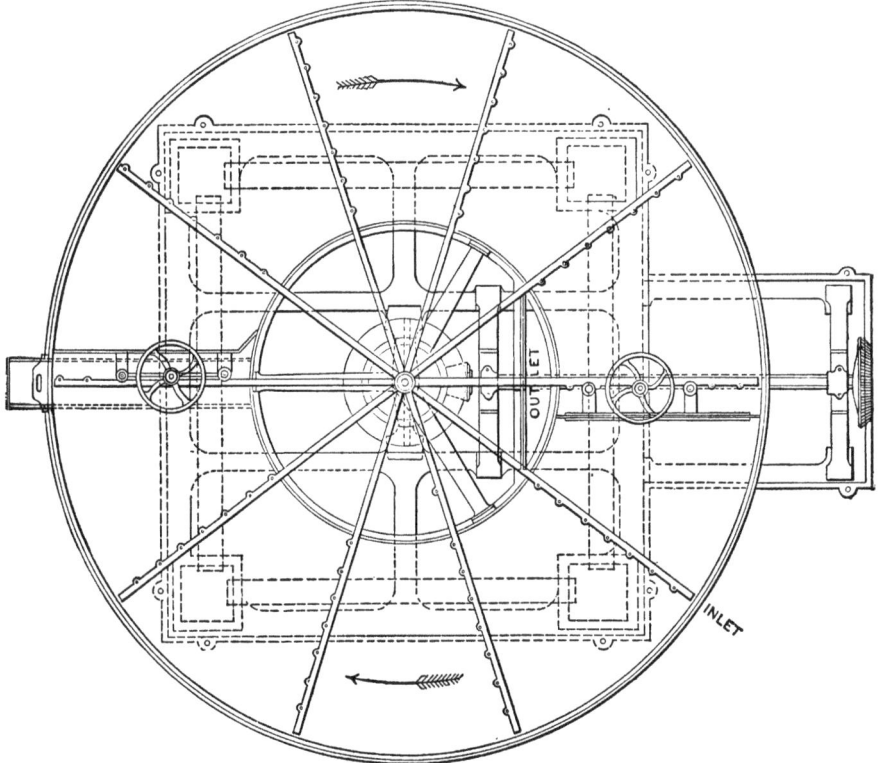

WASHING MACHINE.

while a box with sides measuring two feet nine inches would hold the gems.

When the day's work is completed, the pans, through each of which three hundred loads have passed, are emptied or "cleaned up," and the concentrated deposits of diamonds, mingled with the other heavy but valueless minerals, are then

sent to the Pulsator in trucks with locked covers, where they are
sized by passing through a cylinder covered with steel sieving with
holes from one-sixteenth to five-eighths of an inch in diameter.
The five sizes which pass through the cylinder flow upon a combination of jigs, termed at the mines the pulsators, and the name
Pulsator, which originally applied to the one set of jigs only that
did all the work for the De Beers Mining Company in 1886, is
still applied to the large concentrating plant and machinery where
the final concentration is done and the diamonds sorted from the
worthless minerals with which they are associated.

De Beers Crushing Mill, Back View.

Before tracing the diamonds through the Pulsator, it is desirable for the sake of clearness to sketch the treatment of the hard
blue ground taken direct from the depositing floors. For the
handling of this portion of the product of the mines an elaborate
and costly plant was erected on one of the old tailing heaps.
The driving power of the crushing mill is a compound vertical
engine of 1100 horse-power. The whole plant is divided into four
sections, and provided with friction clutches so that any portion
of the machinery may be stopped without interfering with the
running of the rest of the mill.

Washing Machine with 18 Pans. Capacity, 6000 Loads, equal to 4800 Tons, in Ten Hours.

An endless wire rope haulage carries all the refractory ground to the mill, where it is put through a series of crushing machinery. The first or "comet" crushers reduce the ground so that the largest pieces will pass through a two-inch ring. From these crushers the ground passes through revolving screens which separate the finely crushed from the coarse pieces. The fine size is conveyed to the washing pan, and the coarser ground passes from the end of the screen to revolving picking tables, where diamonds of the larger size may be seen and removed without risk of crushing by further pulverization. From the picking tables the ground is scraped automatically into two sets of rolls, and the pulverized product screened again and graded into three sizes. The finest size, passing a half-inch screen, goes to the washing pans, and the two coarser sizes to jigs. Large diamonds which have been separated from their envelope of blue are retained in the jig. The ground still holding the smaller diamonds passes out of the end of the jig and then through a series of rolls, screens, and jigs until the finished product is drawn from the bottom jigs into locked trucks running on tramways to the pulsator for further concentration and sorting.

From beginning to end of this process the crushed ground is carried by water, and the plant requires a flow of 400,000 gallons an hour. After leaving the last jig the water is separated from the fine ground by a revolving screen and the tailings are taken away in trucks to the tailing heap. Within the past three years the ordinary rotary pans have supplanted the jigs, and are found to be more economical.

The coarse ground, which passes out of the end of the revolving cylinders of the washing plants, is called "lumps." As the lumps leave the end of the cylinders they fall upon a conveyor and are taken to the end of the washing machines, where they are reduced by a similar, though smaller, crushing plant, with the exception that pans only are used for saving the diamonds.

Thus the screened and sized product from the washing pans and the crushing machines reaches the final stage of concentration in the Pulsator. This is a combination of jigs with station-

The Pulsator.

ary bottoms covered with screens with square meshes. The meshes are a little coarser than the perforated plates of the cylinders that size the concentrate for the jigs. Upon the jig screens a layer of leaden bullets for the finer sizes and iron for the coarser sizes is spread, forming a bed that prevents the deposit from passing through the screen too rapidly. The heaviest part of the deposit with the diamonds passes through the screens into pointed boxes, from which the deposit is drawn off and taken to the sorting tables. The lighter material or refuse flows over the ends of the jigs into trucks, which are hauled away and dumped on the tailing heap.

Only one per cent of the total amount of ground washed, or one in a hundred loads, goes to the Pulsator in the form of concentrate. Eight and a half per cent of this passes through the screens below the five-eighth inch size, thirty-three and a half per cent is above that size, and the balance, fifty-eight per cent, flows over the jigs as waste. Formerly, for every hundred loads washed, five-twelfths of a load passed over the sorting tables,

ordinary wooden tables covered with steel plates. Here the diamonds were picked out by hand, first by white men while the
deposit was wet, and later, when dry, by native convicts. The
concentrate was worked over as long as the cost of handling was
repaid by the gleaning of diamonds. The size of the stones
which reached the sorting tables ranged from one-sixteenth of an
inch to one and one-eighth inches.

Sorting Gravel for Diamonds.

Mixed with the diamonds in the concentrates are a number
of other minerals of high specific gravity, and some of notable
beauty though they have no marketable value. Among these
are the rich red pyrope, the flesh-colored zircon, the blue
disthene, bright green chrome diopside, pale green rhombic
pyroxene, and olivine occasionally in large, polished pebbles.
Some of the garnets are of fine quality, and one was recently
cut which resembled a pigeon-blood ruby, and attracted an
offer of £25. The complete list of minerals found on the
sorting tables includes: (1) pyrope, having a specific gravity

of 3.7 and containing from 1.4 to 3 per cent of oxide of chrome; (2) zircon (specific gravity 4.41 to 4.7), in flesh-colored grains and fragments, but no crystals — this mineral is commonly known on the Diamond Fields as Dutch boart; (3) disthene, or cyanite (specific gravity 3.45 to 3.7), discernible by its blue color and perfect cleavage; (4) chrome diopside (specific gravity 3.25 to 3.5), in fragments bright green in color and containing, according to Knopp, over two per cent oxide of chrome; (5) enstatite or bronzite with pale green rhombic pyroxene (specific gravity 3.1 to 3.3); (6) mica (specific gravity 2.7 to 3.1); (7) magnetite (specific gravity 4.49 to 5.2), occasionally found in octahedron crystals; (8) non-magnetic iron ore (specific gravity 4.5) containing chrome and titanium in varying quantities; that is to say, sometimes it is chrome iron, and sometimes titanium iron ore : according to analysis by Knopp, it contains from 13 to 61 per cent of oxide of chrome and from 3 to 68 per cent of titanic acid; (9) hornblende (specific gravity 2.9 to 3.4); (10) barite (specific gravity 4.29 to 4.3); (11) calcite (specific gravity 2.7); (12) pyrite (specific gravity 4.83 to 5.2); (13) olivine (specific gravity 3.3).

The work of picking out the diamonds by hand from the concentrate on the sorting tables was, of course, necessarily slow and tedious. It was the only division of diamond mining and winning which seemed beyond the application of blind and unconscious machinery. But men to-day are not inclined to admit that anything greatly worth doing is impossible.

A series of experiments was initiated by me with the object of separating the diamonds from the heavy valueless concentrates with which they are associated. An ordinary shaking or percussion table was constructed, and every known means of separation was tried without success. One of the employés of De Beers, Mr. Fred Kirsten, was in charge of the experimenting, under the supervision of the late Mr. George Labram, the manager of the large crushing plant, and afterward mechanical engineer to the company. Notwithstanding the fact that the specific gravity of the diamond (3.52) was less than that of several of the minerals associated with it, so that its separation would seem a simple matter, it was found in practice to be impossible owing to the slippery nature of the diamond. The heavy concentrates carried diamonds, and diamonds flowed away from the percussion table with the tailings. When it seemed that every resource to do away with hand sorting had been exhausted, Kirsten asked to be allowed to try to catch the diamonds by placing a coat of thick grease on the surface of the percussion table with which the

Automatic Diamond Sorter, called the Greaser.

other experiments had been made. Kirsten had noticed that oily substances, such as axle grease and white or red lead, adhered to diamonds when they chanced to come into contact, and he argued to himself, if these substances adhered to diamonds and not to the other minerals in the concentrates, why should not diamonds adhere to grease on the table and the other minerals flow away? In this way the remarkable discovery was made that diamonds alone of all minerals contained in the blue ground will adhere to grease, and that all others will flow away as tailings over the end of the percussion table with the water. After this was determined by thorough experiments, more suitable shaking tables were constructed at the Company's workshops. These were from time to time improved upon, until now all the sorting (except for the very coarse size) is done by these machines, whose power of distinction is far superior to the keenest eye of the native. Since the discovery of the affinity of grease for diamonds, experiments have been made with rubies and sapphires from Burma, and it was found that grease caught these gems with the same certainty that it catches diamonds.

After a thorough trial a number of these unique diamond-catching tables (see cut, p. 379) were constructed, and are now working on De Beers concentrates. Each shaking table is made of corrugated cast-iron plates in five sections, with a drop of about an inch from one division to another. Thick grease is spread on the plates to cover them to the top of the corrugations.

The concentrates are conveyed from the jigs upon a conveyor belt and deposited into hoppers, where the load is elevated to revolving cylinders covered with perforated steel plates. Through the graded screens of these cylinders the concentrates pass into small hoppers, one above each table, fitted with automatic feeders, — cast-iron cylinders with grooves corresponding to the graded sizes of the concentrates, — and are distributed evenly across the upper portion of the shaking tables, and carried down by a flow of water from a trough fixed behind the feeders. During the time the table is working it is rapidly shaken from side to side by an eccentric placed on a shaft under the table.

Strange to relate, the descending diamonds stick on the face of the grease while all other minerals pass over it. Only about one-third of one per cent of diamonds is lost by the first table, and these are recovered almost to a stone when the concentrates are passed over the second table. The discrimination of this sorter is surely marvellous. Native workers, although experienced in the handling of diamonds, often pick out small crystals of zircon, or Dutch boart, by mistake, but the senseless machine is practically unerring. It will catch rubies, sapphires, and emer-

alds as well as diamonds, but so far as it has been tested, it will
not cling to anything but a precious stone. The grease which
is used loses its power to catch diamonds after a few hours' work,
owing to its becoming more or less mixed with particles of
water. It is then scraped off the tables, together with the diamonds adhering to it, placed in a kettle made of finely perforated steel plates, and steamed. The grease passes away to tanks
of water, where it is cooled and is again fit for use. The diamonds, together with small bits of iron pyrites, brass nails from

The Manager of the Pulsator, Mr. James Stewart, through whose Hands £3,000,000 to
£4,000,000 Worth of Diamonds pass every Year.

the miners' boots, pieces of copper from the detonator used in
blasting, which remain on the tables owing to their high specific
gravity, and a very small admixture of worthless deposit which
has become mechanically mixed with the grease, are then boiled
in a solution containing caustic soda, where they are freed from
all grease. The quantity of deposit, from the size of five-eighths
of an inch downwards, which now reaches the sorting table, does
not exceed one cubic foot for every 12,000 loads (192,000 cubic
feet) of blue ground washed. As already stated, one-twelfth of
one per cent of the whole mass of blue formerly passed to the

£60,000 PARCEL OF DIAMONDS.

sorting tables; or, from 12,000 loads, which is about the daily average of the quantity washed at De Beers and Kimberley mines, 160 cubic feet had to be assorted by hand.

The first question usually asked by visitors is, What is the cause of this amazing discrimination? This is a very difficult question to answer with positive assurance. It is possible that the secret of the affinity may lie in the fact that water adheres to or enters into all minerals composing the concentrate except precious stones. These present comparatively dry faces to the grease and quickly adhere to it, while the wet stones flow over the table. The grease has no affinity for a piece of glass, which, when dropped on the table, flows away in the tailings.

From the sorting tables the diamonds are taken daily to the general office under an armed escort and delivered to the valuators in charge of the diamond department. These experts clean the diamonds of any extraneous matter, such as small particles of adhering blue ground, by boiling them in a mixture of nitric and hydrochloric acids (aqua regia), or, still better, in fluoric acid. When the stones are cleaned, they are carefully assorted with reference to size, color, and purity, and made up in parcels for sale, formerly to local buyers, who represented the leading diamond merchants of the world. For several years past De Beers Company has sold in advance its annual production to a syndicate of London diamond merchants who have representatives residing in Kimberley.

A Day's Diamond Wash.

CHAPTER XIII

OBSTACLES AND PERILS

N the open workings the imminent hazard of maiming and death by reef slides was ever hanging over the heads of the miners. In view of the rashness with which the pit sinking was pressed, it was a marvel, indeed, that the actual loss of life was, on the whole, so small. No complete or accurate records were ever kept of the men injured or killed in prosecuting the work before the advent of systematic mining.

In the journals of the Diamond Fields the most noteworthy casualties were recorded, and it is seen that in the years immediately following the undertaking of underground mining, the principal loss of life occurred from the falls of loosened pieces of blue ground or reef. This is expressly noted in the report of the Inspector of Mines at Kimberley to the Assistant Commissioner of Crown Lands on August 27, 1885. Underground mining operations in Kimberley and De Beers mines were then, he observed, becoming very hazardous. In both mines, but especially in the Kimberley mine, "some of the underground working places in diamantiferous ground are huge caverns of from 25 to 52 feet in height and 20 to 30 feet in width. The roofs of these workings, from exposure to atmosphere, shocks of blasting, and inherent weakness of the blue or diamantiferous ground, are becoming extremely unsafe; occasionally pieces of the ground or rock fall from the high roof or sides, to the imminent danger of persons working on the floors. During the last and current months there have been three deaths in underground working places directly due to the dangerous operations in the mines," and in view of this danger and loss of life, the inspector urgently recommended the limitation by the government of the height and width of the underground workings.

" Main tunnels to be used only for traffic not to exceed 8 feet in width and 8 feet in height.

" Working chambers or stalls from which the blue or diamantiferous ground is excavated in bulk, not to exceed 18 feet in width by 20 feet in height to the highest point.

" Partitions or pillars not to be of less thickness than half the width of the contiguous chambers or stalls.

" The roof of ceiling between one level and the next above to be not less than 20 feet in thickness at the highest point of the lower workings."

This recommendation had in view obviously the precautions enforced in the working of coal mines, and would doubtless have afforded an increased measure of protection, but the method of working proposed was not well suited to the development of the diamond-bearing ground, as was later conclusively determined. The slaking and crumbling of the diamond-bearing breccia upon exposure to air and moisture make roof falls and slips from the sides especially frequent and disastrous. The ground is full of soapy seams, and pieces of considerable size drop without a moment's warning, so that it is necessary, in places, to keep the tunnels timbered as near the working face as possible. Risk from this cause cannot be wholly obviated in such mining, but the introduction of the new system adopted for the working of the mines, shortly after they came under my management, has greatly diminished this peril, and the resultant loss of life or injury to the workmen. By the new system the levels are worked back from the surrounding hard rock or reef in sections, formerly 30 feet, now 40 feet apart, as before particularly described, in a series of terraces, extracting the ground from the uppermost level downward in succession. This method did away with any danger of collapse in the underground works, and by successively robbing out the roof and sides of the tunnels on each descending terrace, the caving of

2 C

the unstable ground was systematically anticipated and restricted. No feasible care in the direction of men working in such shifting ground can entirely do away with casualties. Some are scarcely to be avoided, but most are attributable, more or less, to the miners' heedless disregard of the warnings of overseers and proper precautions.

There was another serious risk in mining in the upper levels of the mines, where shale is heavily impregnated with bituminous matter, and no device could wholly prevent the gathering of carburetted hydrogen, which, mingling with air, forms the "fire damp" that has been so deadly a peril to miners. When sinking shafts or driving tunnels in the shale, miners are prohibited by the strictest injunction of the management, and the formal regulations of the Government Inspector of Mines, from carrying any lighted candle into passages where there is any possibility of this gas having gathered; but no prohibition has ever been able to prevent an occasional stretch of reckless-ness on the part of some careless miner. Locked safety lamps are provided abundantly for testing the atmosphere in such parts of the mine workings, but neglect of this precaution has caused startling explosions, scorching and striking men down, and in a few cases causing death. In 1883 there was a slight explosion of accumulated gas in the reef workings of the French Company, Kimberley mine. Here thin bands of coal had been struck in the black shale, and in an upward drive to meet a pass, some gas had collected in the interval from Saturday to Monday. A naked flame set fire to this gas and caused the explosion. Prior to this time two other cases were on record, in both of which workmen were severely injured. Perhaps the most notable instance of the gathering of this gas was in a heading of the workings of the Gem Company in De Beers mine in July, 1885. One of the workmen had his face and hands badly scorched by an explosion at the end of the heading, and a second explosion occurred shortly afterward, when the managing director and an overseer attempted to exam-ine the heading, taking candles to light their way. The director,

Mr. George McFarland, was severely burned by this blast of gas, which was described as a "fizz" almost noiseless. Since the workings have been carried down below the level of the shale, there has been no danger from fire damp, and the accidents from this cause have ceased to occur in the deeper mines.

The strictest precautions are enjoined in the storing and handling of explosives used in the diamond mines, and the need of such stringency was signally emphasized in the destructive explosion that wrecked a dozen magazines near the compound of the Victoria Mining Company on October 31, 1884, three years before I took the management of De Beers. The shock was felt from Dutoitspan to the farthest limits of the west end of the camps, and terror-stricken people rushed out of their houses to see a vast heaving cloud of smoke rising hundreds of feet into the sky.

The magazines were dashed to pieces, as the Kimberley papers reported, by the terrible power of the explosives. In most instances the galvanized iron was broken into tiny atoms as if by myriad hammers, and cartridges were scattered far and wide through the débris, exploding in volleys or scattering blasts for many minutes after the explosion. One large stone was thrown as far as the Central Company's offices, a distance of two miles, and smaller ones to the West End, three miles from the magazines. In the most distant parts of the camp there was a startling breakage of windows, lamps, and chandeliers, and the hotel bars and canteens were so heavily pelted that "the floors were swimming with what we might call dynamite cocktail, composed of every liquor under heaven from Cape Smoke to Heidseck and Pommery." Witnesses of the explosion thought that hundreds had been killed and injured, but almost miraculously, as it seemed, only two persons were killed, one a white, the other a black, both bodies being horribly mutilated. A third sufferer was taken up and tenderly cared for, a poor native deeply gashed and with broken ribs.

No other accidents in the mines have ever approached in loss to life the terrible disaster from the outbreak of fire in De

Beers mine in July, 1888.   When the Consolidated Mines took over the property of De Beers Mining Company, nearly all the blue ground was hoisted from the 500-foot level, through the first large working shaft constructed, known as No. 1 west end incline.   In July, 1888, another shaft, No. 2 incline, had just been completed to the 700-foot level, and skips in the 7-foot compartment were used in hoisting the ground broken on this level.   In addition to these working shafts a small vertical prospecting winze, called the Friggin's shaft, had been sunk from the 500-foot to the 700-foot level.   When a tunnel connection was opened between No. 1 and No. 2 inclines on the 700-foot level, the prospecting winze was no longer needed, and it stood abandoned except as a ladderway.   There was a small disused engine room on the 500-foot level a short distance from the winze.   With the sinking and connection of both working shafts on the 700-foot level, the output of the mine increased until a total of 104,089 loads was attained during the month of June, 1888.

Mr. Lindsay, Mine Manager, killed in De Beers Mine Fire, July 11, 1888.

On the 9th July following, large skips in No. 2 incline began carrying blue ground from the 700-foot level, and continued hauling until the morning of the 11th, when one of them jumped the rails, either because the hoisting was being done at too rapid a pace, or from some obstruction in the shaft.   Examination showed that both skips were off the rails, and that the shaft timbers had been considerably damaged.   In bringing up one of the small skips in the manway, this was also derailed by the débris in the shaft.   The necessary work of repair was begun at once and continued during the day.   During the changing of the shifts in the evening, the

mine manager, Mr. Lindsay, reported that the work was progressing as fast as practicable and that the shaft would be in running order within a few hours.

At about half-past six o'clock in the evening Lindsay and
six miners went down the shafts in one of the small skips.
A few minutes later an alarm of fire was given just as I was
about to drive to my home from the works. It was reported
to me through the telephone that the Friggin's shaft was on
fire. It is probable that one of the native miners had sneaked

Shaft No. 1. Shaft No. 2.
Fire in De Beers Mine, July 11, 1888.

off to the disused engine room on the 500-foot level, and placed
a lighted candle so carelessly that the flame ignited the timbers,
perhaps while the lazy savage was snoring on the floor. The
precise cause of the fire was, however, never determined, but
from the time of its starting, it spread with such swiftness that
it could not be stifled.

Within a few minutes after the outbreak of the fire both of
the incline shafts were filled with dense smoke, as both shafts
were upcasts, and the passage of any of the men through these
exits from the mine workings was hopelessly shut off. When
the alarm was given, there were 685 men at work in the levels
below the fire, and our anxiety for their safety may be readily
conceived. At the first warning of danger two men were sent

down No. 2 incline to notify Lindsay and his companions of the outbreak of the fire, but the smoke came up through the shafts so heavily that both were driven back gasping for breath, and barely reached the surface before they fell on the floor completely exhausted. For several minutes there was a tension of waiting for some signal to hoist from Lindsay, or one of his party, but none was given. Lindsay and his comrades must have been close to the skip in the shaft when the fire started, and a signal bell wire ran through the shaft close at hand. There was time enough for one of the party who went down the shaft in the skip with Lindsay to climb up the shaft by means of the timbers, a distance of 150 feet, and in view of this, the failure of these men to get into the skip and ring a signal to hoist is inexplicable. Seeing at once that ascent through No. 2 incline was probably hopelessly blocked by the outpouring smoke, I hastened to the mouth of the other shaft (No. 1 incline shaft). The smoke was also streaming out of this shaft in dense volumes.

The signal to hoist men by ringing three bells was repeatedly given, but I hesitated to give the order to hoist the skip, which was at the 600-foot level, as the risk of hoisting a skip-load of men through the stifling smoke was appalling. On the other hand, it was impossible to know at the surface in what desperate straits the men might be on the 600-foot level. So, before giving the signal to hoist, I took measures to revive the men who would be overcome by the smoke in ascending the shaft, and water was provided to dash on them if they came up with their clothes on fire. It was a moment when no balancing of probabilities could determine the decision. There was a desperate chance of safety in the swift pulling up of the skip. I could not let the piteous appeals go on apparently unheeded. I gave the signal to hoist at top speed in response to the last pleading signal. When the skip was about 300 feet from the surface, the wire winding rope parted. The broken end came whizzing up through the shaft, but the skip with its load of four poor victims fell crashing down to the sump at the bottom of the shaft, a little below the 600-foot level. When the rope was examined, it was found that the flames from the burning timbers had made it so hot that the tension of the skip drew out the wires to fine needle points which snapped under the strain. When the first signal to hoist was given, there were ten or twelve men in the skip, but the majority left it when the signal to hoist met with no response. It was impossible for the men at the 600-foot level to know that the shaft through which they wished to be hoisted was on fire a hundred feet above them, nor could we on the surface know what was happening 500 feet below.

The mine was ventilated at the time through an outlet into the old open workings, and through the Gem shaft on the east side of the mine. The Gem shaft was a small, old working shaft that had been sunk from a terrace in the blue ground. Unfortunately it had been partially closed by a recent ground slide in that part of the mine. It was, however, still sufficiently open to be of invaluable ventilating service at this crisis, and it could have been opened for the rescue of the men in the mine if there had been no other means of escape through the outlet into the open workings. During the hours of fearful anxiety that followed the closing of the two main shafts, the outlet from the mine to the open workings was intently watched, and daring parties penetrated far within it in the hope of communicating with miners escaping from the range of the fire. Almost all of the men in the mine were well acquainted with this passage to the surface, and it was confidently hoped that many, at least, would contrive to grope their way upward through this outlet to safety. Fortunately the air draught through this passage was downcast, and the inrush of air cleared the passage from smoke.

To the immeasurable relief of all, so anxiously expectant, one white man and six native miners came climbing through this passage into the open workings at about ten o'clock on the night of the fire. This showed that a practicable way of escape from the mines was open, but many hours of fearful suspense followed throughout that night and the following day, while the miners were groping their way to the surface through the same opening. Forty-two white men and 441 native miners were thus rescued, but 24 whites and 178 natives lost their lives in levels and passageways charged with deadly smoke. The downcast draught through the Gem shaft was the salvation of the greater part of the rescued men, who spent this fearful night on the level close to this shaft, which was free from smoke. During the afternoon of the following day, July 12, a party of heroic men penetrated far into the mine through the entrance

No. 2 Incline Shaft, looking East.

in the open workings, and rescued a number of natives who were cowering stupefied by the smoke, or paralyzed by fear. In this rescuing party were some who had passed the night in this frightful prison, but who were, nevertheless, among the first to volunteer to go down again in the desperately hazardous venture to save their comrades.

No. 1 incline was completely burned out and caved in during the night of the fire. During the night of the 12th No. 2 incline caved in also for a distance of about 40 feet, near the junction of the shale with the hard rock, shutting off all communication with the mine. Before the latter shaft could be

WAITING FOR NEWS FROM THE MINE,
July 12, 1888.

GENERAL VIEW OF PIT-HEAD FRAME AT NO. 2 SHAFT AFTER THE FIRE OF JULY 11, 1888.

reopened, the water in the mine rose to a depth of 20 feet, filling all the tunnels on the 700-foot level.

Several days after the fire I went down the shaft accompanied by Captain Hambley, Assistant Inspector of Mines, and one of the overmen. I arranged to lower the skip gradually down the incline to make the first inspection. As we went down, an insulated signal wire was lowered, and provision was made so that I could keep the bell ringing continually, and instructions were given to haul up the skip at the moment the ringing stopped, for I feared that we might drop into foul air so suddenly that we would not be able to signal in the usual manner. So we went down in the skip slowly to a point about 150 feet above the crushed ground in the shaft. At this point, some 250 feet below the surface, we saw the body of one of the men who went down with Mr. Lindsay just before the breaking out of the fire. We did not stop, for the moment, but kept on signalling until the skip was lowered to the ground which closed the shaft. Our search for any further trace of the lost miners was fruitless, for we could find no more bodies. Mr. Lindsay and his remaining companions were buried beneath the débris when this part of the shaft caved in. Finding that the further descent of the skip was cut off, I then gave the signal to hoist, and on reaching the surface, gave instructions for men to go down and remove the body seen in the shaft. The poor man had climbed up to the point where he died, in a desperate effort to escape. The other men, as well as the skip in which they went down, were buried deeply under the mass of crushed ground.

The work of repairing No. 2 incline could not be begun until July 19th, for the smoke and heat from the mine made work in the crushed portion of the shaft unendurable. Even then it was only practicable to advance very slowly, and the shaft was not opened until the 3d of August, when the large skips were at once employed to bail out the water. Eight days later the mine was drained, and the reopening of the workings could be undertaken.

It was originally intended that the large skips in No. 2 in-

WAITING FOR NEWS FROM THE MINE.

cline should be used in hauling blue ground from the 700-foot level only, as there were ample facilities in No. 1 and in the 7-foot compartment of No. 2 for hoisting all the blue ground taken from the 600-foot level and the levels above. Consequently no stations had been made ready for the larger skips on the latter level. It was necessary, therefore, to open tunnels, sink passes, and put in chutes to connect the 600-foot level with the surface, besides excavating a pump chamber and erecting new pumps, before the regular output of blue ground could be resumed. During the month of August only 8613 loads were hauled, and this was mostly of poor quality from excavations of the west end of the mine. During September the output was increased to 57,408 loads, in October to 87,225 ; but it was not until the following month of November that the output reached 104,285 loads, or approximately the same amount as in the month before the fire.

This brief sketch may serve to show to the general reader something of the terror, the peril, and the disaster which an outbreak of fire in any great mine may cause. As soon as practicable after this fire, the previously designed systematic and thorough opening of the mine was advanced. In addition to No. 2 incline, the rock shaft (elsewhere particularly described) was completed and connected with the mine by a tunnel on the 800-foot level. A vertical escape shaft was sunk from one of the terraces in the open mine to the 700-foot level. It had a ladderway and a single cage compartment, and was connected with seven levels in the mine. The Oriental shaft, situated on the east side of the mine, was connected with it at the 500-foot level, from which all parts of the mines were reached by ladderways. This shaft serves to ventilate the mine, and as an important passage for escape in case of need. Besides these four shafts there was a tunnel into the open mine, which was connected with the lower workings by a double ladderway. The Oriental shaft and No. 2 incline were upcasts. The rock shaft, escape shaft, and the tunnel into the open mine were downcasts.

The first consideration in working a mine is to have a safe exit for the workmen, in case a fire breaks out or the mine becomes flooded by suddenly tapping a large quantity of water, and at the diamond mines this precaution is strictly carried out. In the early days of underground mining, when many of the levels had exits into the open mine, it was necessary, in providing numerous escapes for the workmen, to guard against sly sallies of natives when there was no danger, because they could leave the mine with stolen diamonds, or could go out for the purpose of obtaining intoxicating drink, and bring back bottles of Cape brandy, called "Cape Smoke," into the mines with them. Although the numerous escapes from the mine were guarded by watchmen, the dusky Kafirs would come, at times, in squads, and overpower the guards and make their escape. An ingenious device was invented by our electrician, Mr. Drummond, by placing a small copper rod directly above the iron rungs of the ladders, and connecting both with a battery. Then when a man placed his hand or foot upon the copper rod, it bent down, completing the circuit, and rung an alarm bell in the mine and on the surface. The natives could never quite understand why they were always met by a posse of white guards at the particular place where they were trying to escape.

In later years, since the mines became deeper, all workmen are taken in and out of the mines by means of cages. There are double ladderways in the shafts which may be used in case of emergency, but there is always a sufficient number of white men employed about the tops of these shafts to prevent the escape of natives.

In view of the responsibility resting upon me from my acceptance of the General Managership of De Beers Mines in the year preceding this great disaster, and the common duty of all connected with the mines to do everything practicable to save life, to prevent the outbreak of fire, and to guard against all contingencies, it is proper to note the warmly appreciative recognition accorded by the presiding chairman, Mr. Barnett Isaacs Barnato, at the adjourned first annual meeting of the shareholders of the De Beers Consolidated Mines Limited. Mr. Barnato said in his address to the shareholders : —

MEN ESCAPING THROUGH TUNNEL MARKED + AT THE 380-FOOT LEVEL.

" I suppose you all remember about the sad calamity by which so many poor fellows lost their lives. At this point I feel I must pay a tribute of respect to the brave men who worked and risked their lives on behalf of those poor fellows who perished in the disaster. I remember on that sad occasion, which will never be effaced from my memory, and from the memories of many who lived in Kimberley at the time — I remember seeing our respected and able general manager, Mr. Gardner Williams, a gentleman to whom I believe no person can attach the least blame, working night and day, and doing all he possibly could for the relief of the sufferers. That calamity was an act of God, or at least we must conclude so, for on the very day of the disaster there was an accident in No. 2 shaft, which blocked it up to some extent, and the Gem escape shaft gave way only a week previously. I, therefore, think that calamity was an act of God, and I hope a similar disaster will never again be witnessed in Kimberley or elsewhere. In paying a tribute of respect to Mr. Williams, who worked all through the night when the fire broke out, and to the brave men who went into the mine, to try and save their fellows, we must not forget that those men risked their lives, that they went down into the mine, when millions and millions of loads of reef were hanging over them, to open up the shaft so that the men might escape. And the result of their work, we know, was that out of about seven hundred men in the mine, five hundred escaped. Therefore, in passing this tribute of respect to Mr. Williams and the men, I feel sure that it will be universally indorsed by the shareholders. [Applause.] No more need be said about this matter, except that the state of the mine after the calamity necessitated a considerable expenditure of money. I think it took us three months to get the mine in proper working order, and we lost three months' labor, at a cost of something like £250,000. The balance sheet only shows about £30,000, but by the loss of blue, etc., the loss to the company was, as I have stated, not much short of a quarter of a million of money."

Providentially, and by the exercise of every feasible precaution, there has been no serious spread of fire in the mines since the occurrence of this great disaster.

The chief peril to life and damage to the workings of the mine, for a number of years, has come from the destructive "mud rushes," as the miners call them. There is no water in the blue ground or the mine itself, but the water flowing into the mine from the surrounding reef made a muddy mixture of the disintegrated shales, decomposed basalt, floating reef, and low grade blue ground, which had fallen into the worked out section of the open mines. At times the tremendous pressure of the shifting ground above forces this mud in vast quantities into the working levels of the mine, and the miners do not have time to escape this inrushing mass even by instant flight. On several occasions tunnels in the mine have been filled to the extent of thousands of feet by these rushes in a few minutes. As the work in the mines reached the deeper levels, these rushes became so frequent that the working of the mines was seriously interfered with, and no watchfulness could avert the loss of life.

In June, 1897, one of the worst mud rushes known in the record of the mines occurred in De Beers mine, filling up almost instantly a large number of tunnels on the 1000-foot level. Two native miners were overtaken by the rush, and shut up in a drainage passage that was in progress to tap the water in that section of the mine. For a stretch of 28 hours they were held fast in this narrow prison chamber, momentarily dreading a further rise of the mud that would bury them alive. Meanwhile the most daring efforts were made to rescue them from their stifling prison, and two heroic men, Thomas Brand and John Brown, finally burrowed through 200 feet on the top of the mud, and brought the two natives out safely at an appalling risk to their own lives. The rescue was barely in time, for the next morning another rush followed, filling up the tunnels again still further, and rising to the top of the passage that had given breathing room to the imprisoned men. For this signal heroism medals of the Royal Humane Society were very fitly given to Brand and Brown.

THE SURVIVORS COMING UP THE TERRACES IN THE OPEN MINE, ON THE MORNING OF JULY 12, 1888.

In May, 1898, there was another great mud rush through the 1120-foot level, from which a whole gang of native workers barely escaped alive. On this occasion "Jim," one of the best of the "baas" boys, was almost buried alive with his gang of 15 men. The rush shut this working party up in a narrow passage on this level for more than 64 hours. When the men were rescued at length from their stifling quarters, where they were imprisoned for more than two and a half days, without a morsel of food to eat or a drop of water to drink, all were greatly exhausted, as might be supposed. But in spite of his sufferings, the brave leader, Jim, went back at once into the mine to grope back over the mud in search of one of his gang whom he supposed was missing, and he would not return to the surface until he learned beyond doubt that all had been rescued.

The endurance of the native miners under such circumstances is remarkable. In July, 1898, a Basuto boy, "Joseph," was almost buried in a mud rush, and was completely shut in the "dead end" of a tunnel, on the 960-foot level. The attempt to clear a passage to rescue him was begun at once, and the work was pushed without a respite night and day, but it was late on the third day before the place of his entombment was reached. He was found lying crouched beneath some timbers resting on an overturned truck, around which the mud had risen to the depth of two and a half feet. The rescue party had given up all hope of finding him alive, and were about to blast the enveloping mud in order to pull out the truck, when a faint cough was heard, apparently coming out of the dense mass of mud. The natives at work were badly frightened at this weird sound, and called up the contractor in charge, who finally succeeded in digging out the poor Basuto boy nearly lifeless. One of his legs had been pinned beneath the truck so heavily that the circulation of the blood was stopped, and mortification set in, necessitating its amputation. The boy bore the operation with the characteristic fortitude of his race, and is stumping about to-day with a wooden leg. He had been shut up for more than three days in a little hole in the ground wholly without food and drink, and with only a few cubic feet of compressed stagnant air to breathe.

When a tunnel is being driven there is only one way of escape, and the working face is called a dead end, though not on account of its deadly nature in cases of a mud rush, for it is a common term in miners' parlance. In point of fact these dead ends are the safest places in the vicinity of a mud rush. The mud, which first fills the mouth of the tunnel, forces the air ahead of it, and compresses it to such an extent that it checks the advance of the mud. Hence, if a native is hemmed in, he has sufficient air to breathe until he can be rescued. On more than one occasion when natives have been caught in the rush of mud, their narrow cell would not have held sufficient air to keep them alive had it not been that a large quantity of air was compressed into the small space.

On one occasion two natives were shut up in the dead end of a tunnel for *ninety-five hours*. They had no food, but managed to obtain a small quantity of water as it trickled down from the roof and sides of the tunnel after finding its way through the blue ground from the level above. These men had more air space than is usually the case, and the temperature in the ends of the tunnels ordinarily ranges from 75 to 90 degrees. When rescued they were greatly exhausted, but after a few days of medical treatment they were quite fit again, and resumed their work in the mine. At another time, when natives were shut in for nearly two days, they swallowed small balls of soft mud, and when rescued it took a considerable time to bring their digestive organs back to their normal condition. On several occasions the white miners have been victims to similar experiences, and now and again a white miner has lost his life by being overtaken and enveloped in the mud. The longest period of time that a white man has been confined in the end of a tunnel is about two days, and there were a dozen or more natives with him. By giving the usual miners' signal of tap-tap — tap-tap-tap, on the walls of the tunnel, we knew he was alive, and it may be imagined that no time was lost in extricating him and his men from their perilous position.

Of recent years the measures described in the preceding chapter have proved effective in freeing Kimberley mine from this peril.   The water which finds its way into De Beers mine has not yet been entirely taken up, but by driving tunnels around the mine to tap the water the danger has been minimized.   On the 1st of October, 1899, six natives were overcome by a mud rush and killed.   Wherever there is the least sign of mud, the workmen are withdrawn, and the places fenced off until the mud has come out or the water is drained off, leaving the places safe for the miners to reënter them.

As there have been from ten to twelve thousand men employed in the mines and workshops and on the depositing floors, three-fifths of whom are underground workers, who are to a greater or less extent raw and untrained natives, the percentage of deaths and injuries has not been excessive.

In the painstaking and valuable reports of Dr. C. Le Neve Foster, H. M. Inspector of Mines, he compares the returns of casualties in the South African mines with the like statistics of mines in which trained Englishmen are employed.   This comparison bears hardly in its application to the diamond mines, in view of the fact that the great majority of the native workers in these mines are " raw hands."   There is probably a change of half the workers in the mine every year, and the men coming in to offset the outflow are mostly natives who have not worked in the mines, and are familiarly known as " green hands."   In time these men are trained to a fair measure of proficiency, but it is to be expected that the proportion of accidents to the numbers of such workmen will be greater than the average in English mines.

From the carefully prepared statistics of Sir Frederic Augustus Abel, covering the loss of life in English mines,[1] it appears that the greatest loss occurs from falls of the roof and sides

[1] Supplement to Forty-fifth Annual Report of the Registrar General of Births, Deaths, and Marriages in Great Britain.

of mine workings, amounting to 40.77 of the total. The loss of life from explosions comes next, with a showing of 23.17 per cent. In the records of fires in mines from all causes, it is shown that only a very small percentage of men are actually burned to death, fully 90 per cent of the deaths resulting from suffocation.

Contrary to the popular impression, it has been shown by Dr. C. Le Neve Foster, that the ore miner has nearly as dangerous an occupation as the coal miner; and in Cornwall and some other metalliferous districts the average losses from accidents were higher than in coal mines. Dr. Ogle has pushed this comparison farther by his statistical demonstration that, in spite of accidents, the death rate of coal miners is not high. In comparative mortality these miners ranked only thirtieth in a list of ninety-four occupations; but the mining in Cornwall, at the time of this report, was exceptionally perilous, standing ninety-first on the list. In other words, only three of the ninety-four occupations exceeded the mining in this district in deadliness. This peculiarly high mortality was ascribed to inadequate ventilation and excessive climbing of ladders from deep mines.[1] These conditions, of late years, have been bettered.

[1] Supplement to Forty-fifth Annual Report of the Registrar General of Births, Deaths, and Marriages in Great Britain.

KEY TO PLATE OF GENERAL MANAGER AND HIS STAFF

1. SEYMOUR DALLAS,
   Manager, Kimberley Mine Compounds.

2. W. AUSTIN KNIGHT,
   Manager, Bultfontein Floors.

3. T. R. ENGLISH,
   Chief Buyer.

4. J. H. MURPHY,
   Assistant Buyer.

5. S. TIDD-PRATT,
   Manager, Workshops Compound.

6. J. SWANSON,
   Manager, Premier Mine Compounds.

7. C. E. HOPLEY,
   Sub-Manager, Stables Compound.

8. D. CANTY,
   Acting Manager, De Beers Mine Compounds.

9. A. E. ROGERS,
   Assistant Manager, Premier Mine.

10. C. M. HENROTIN,
    Assistant Manager, Kimberley Mine.

11. W. TUDOR,
    Assistant Manager, De Beers Mine.

12. J. LIDDELL,
    Mechanical Engineer.

13. W. NEWDIGATE,
    Chief Land Surveyor.

14. JAMES STEWART,
    Manager of Pulsator.

15. A. F. BRIGHAM,
    Chief Mine Surveyor.

16. C. L. PORTER,
    Underground Manager, Premier Mine.

17. CAPT. A. J. GARRETT,
    Manager. Bultfontein Mine Compound.

18. R. ARCHIBALD,
    Manager, De Beers Floors.

19. W. McHARDY,
    Manager, Kimberley Floors.

20. A. F. WILLIAMS,
    Assistant General Manager.

21. GARDNER F. WILLIAMS,
    General Manager.

22. P. A. ROBBINS,
    Consulting Mechanical Engineer and Electrician.

23. W. NICHOL,
    Manager, De Beers Mine.

24. T. J. WOODBURNE,
    Manager, Kimberley and Bultfontein Mines.

CHAPTER XIV

NOWHERE else on the face of the earth is there an assemblage of workers of such varied types of race, nationality, and coloring as are to be seen in the South African Diamond Fields. There is hardly a nation of Europe or Colony of the British Empire that has not some representatives. There are adventurers from the United States, Mexico, and South America; and white men from all the Colonies of South Africa mingle with the masses of native Africans of every shade of dusky hue shown by the tribes that range from the Cape to the equator. Even the American Indian is not unknown in the fields, one specimen at least having resided there for many years. Add to this motley throng a sprinkling of dark East Indians, Malays, and Chinese, and the kaleidoscopic shifts and coloring of this babel in the Diamond Fields may be dimly conceived.

Only about a sixth of the workers in the mines are whites, and the larger part of these are employed above ground on the floors, in the workshops, and in the offices of the mining companies. The majority of the white miners are of English descent, largely coming from the hematite mines of Cumberland, and the tin,

J. M. Jones, Manager, Premier Mine.

lead, and copper mines of Cornwall. They come to the fields in search of employment, which is given as occasion arises. Experience in other kinds of mines is soon adapted to the conditions in the Diamond Fields, and the men in the De Beers mines show a high average of efficiency. The nationalities of the mechanics, engine-drivers, and others working about machinery are Scotch, English, and colonial, with a sprinkling of Americans and other nationalities. Those working on the floors and about the washing machines are largely of colonial birth — English and Dutch, — the balance being mostly home-born Englishmen.

The majority of the white workers above and below ground have their homes in Kimberley and the other neighboring min-

The Engineers, Mechanics, and Workmen who built De Beers Crushing Plant.

ing towns. Wages paid to European day laborers on the surface range from 10s. to 15s. a day; mechanics receive higher pay, which ranges from 16s. 8d. to £1 per day, and white miners are paid the same rate. Miners who prove their competence are given contracts for specified work, by which their earnings are usually materially increased. Since 1892 all underground work has been done by the men working eight-hour shifts. The length of the working day above ground varies with the class of work done. Engine-drivers and men employed in general service at the mines work from ten to twelve hours daily.

On the depositing floors work begins in the summer at six
o'clock in the morning; time is given for breakfast, which is
brought to the men, an hour's rest is allowed
at noon, and work generally ends between 5
and 5.30 in the afternoon. All mechanics work
54 hours in the week, stopping at 1 o'clock on
Saturday, at which hour all work on the surface
ends for the day. Sunday is a full holiday
above and below ground for every one except
those in charge of pumping engines, pumps,
boilers, man cages, etc., which must have attention on Sundays as well as week days, and a few hands employed

Frank Mandy, Mana-
ger, De Beers Mine
Compound.

underground on necessary repair work to the shafts and mines,
which cannot be done during the week while the mines are in
full work. Extra time is allowed mechanics, miners, and others
working under exceptional conditions. The pay of the men
enables them to live comfortably in the mining towns, and as
they are little given to dissipation, the thrifty are enabled to add
to their savings yearly, as the work, except for the interruption
by the war, is continuous and regular.

Employés' houses in Kimberley are scattered through the
city, and many of them own their own homes. Some of the
miners' houses cost £500 or over. They are commonly made
of brick, or with corrugated iron sides and roofs — the division
walls being of unburnt brick and the outside walls being of the

R. G. Scott, Superin-
tendent, De Beers
Convict Station.

same material. The rental of a house in town
ranges from £4 to £8 per month. The price
of board at the boarding houses is about 25s.
per week. The price of meat has commonly
been about 6d. or 7d. per pound, although since
the war, and owing to the devastation caused
by rinderpest, the price of beef has nearly
doubled. To supply the urgent demand for
cheaper meat, the De Beers Company has
erected large cold-storage plants at Cape Town and Kimberley,
and is now importing meat for sale to butchers at Kimberley.

Beef and mutton make up the bulk of the meat sold.   From March to August the markets are well supplied with game, chiefly springbok, stembok, guinea fowl, partridges, bustards, korhaan, and sand-grouse.   Vegetables of all kinds are fairly plentiful and to be had at reasonable prices.   For potatoes the current charge is from 15s. to 30s. per sack of somewhat less than 200 pounds.   Cabbages, cauliflower, beets, beans, parsnips, carrots, onions, sweet corn, and celery are among the vegetables chiefly sold.   Melons and fruits of all kinds are also plentiful in season.   All vegetables and fruit brought from the neighboring

Rinderpest.

farms to Kimberley for sale are taken to the market square and sold under the supervision of the market master to green-grocers, East Indian hawkers, and the public generally.   Flour has nearly a fixed value, being cheaper when the production in Basutoland and other grain-producing districts is plentiful, but never exceeds a certain price, fixed by the competition for imported flour upon which the government levies a duty.   The flour chiefly used by the natives and by many of the white people as well is what is called Boer meal, which makes a brown bread, for only the bran has been removed.   There are a number of roller mills in the country that produce flour which compares favorably with

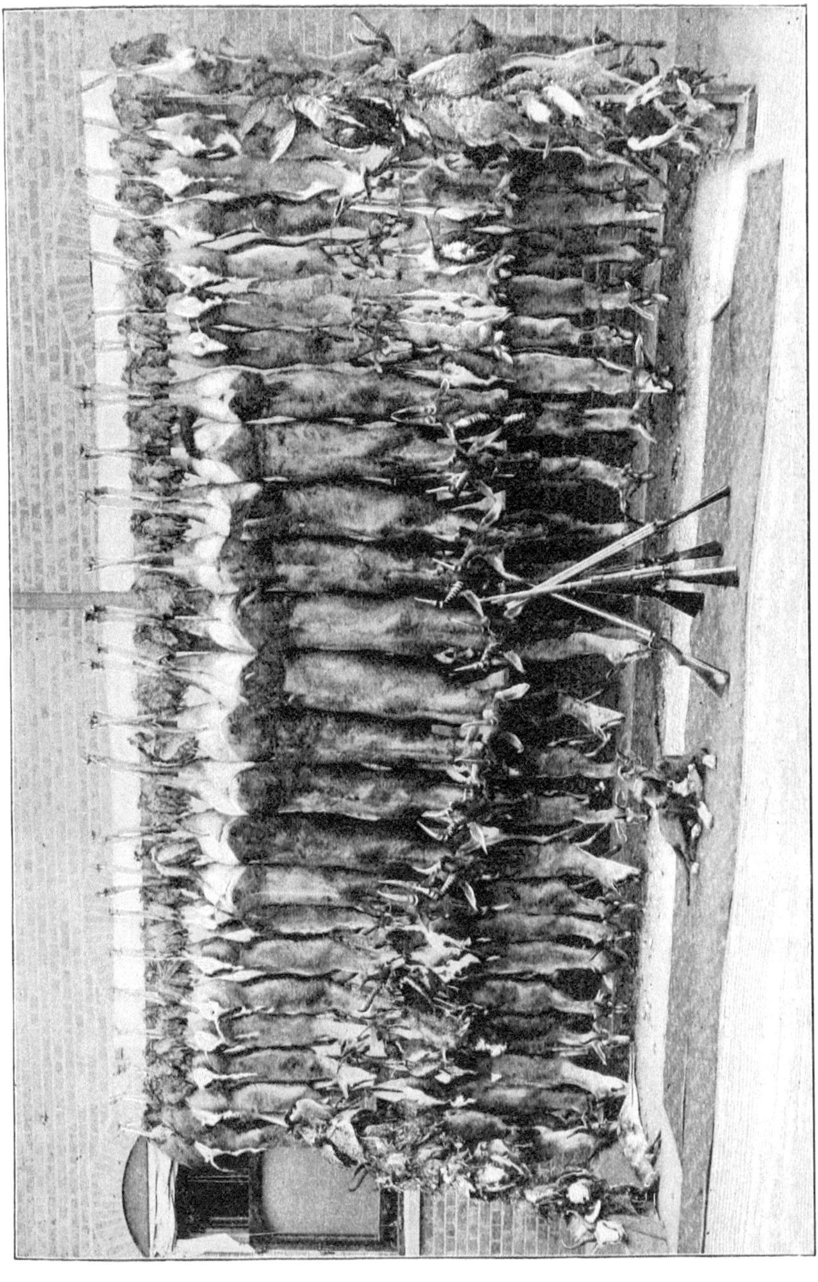

THE GAME OF THE COUNTRY.

imported flour. There is an understanding between the Government, the local dealers, and De Beers, that De Beers Company shall only sell the necessaries of life to the natives in the compounds, and that the price shall range about the same as local prices in town. Any profits derived from these sales is to be distributed among public institutions and charities.

Chiefs of the Batlapin Tribe.

In the mines operated by the De Beers Company alone, more than eleven thousand African natives are employed below and above ground, coming from the Transvaal, Basutoland, and Bechuanaland, from districts far north of the Limpopo and the Zambesi, and from the Cape Colony on the east and the south to meet the swarms flocking from Delagoa Bay and countries along the coast of the Indian Ocean, while a few cross the continent from Damaraland and Namaqualand, and the coast washed by the Atlantic. The larger number are roughly classed as Basutos, Shanganes, M'umbanes, and Zulus, but there are many Batlapins from Bechuanaland, Amafengu, and a sprinkling of nearly every other tribe in South Africa. Many travel hundreds of miles, and some more than a thousand miles, in order to reach the Diamond Fields, and many of these arrive half starved, and so weak and emaciated that they are almost worthless as laborers for weeks afterward.   The natives, as a rule, are generally muscular, sinewy men, but not fleshy. Their feet are broad and flat, but their legs and arms are commonly well rounded, and their thigh and shoulder muscles are large.   The living skeletons who come in from the far interior districts of Africa gain flesh, as rapidly as lean cattle do in green pastures, when they reach a field flowing with meat and porridge.   In the early years of the mines, the raw recruits were hooted at and sometimes pelted with stones by their kinsmen at the mines, as before noted, but of late years this rough greeting and hazing has very largely passed away.

For the lodging and feeding of this great force of native Africans, special provision is made by the erection of large walled enclosures, called compounds, at the mines and on depositing floors.   There are seventeen of these compounds on the Diamond Fields, twelve of which are owned by the De Beers Company.   The largest of all is the one at De Beers mine, and the description of this will serve for all, as they are essentially alike, except in size.

Fully four acres are enclosed by the walls of De Beers Compound, giving ample space for the housing of its three thousand inmates, with an open central ground for exercise and sports. The fences are of corrugated iron, rising ten feet above the ground, and there is an open space of ten feet between the fence and the buildings.   At the northern end of the compound there is an entrance gate.   Iron cabins fringe the inner sides of the enclosure, divided into rooms 25 feet by 30 feet, which are lighted by electricity.   In each room twenty to twenty-five natives are lodged.   The beds supplied are ordinary wooden bunks, and the bed clothing is usually composed of blankets which the natives bring with them, or buy at the stores in the compound, where there is a supply of articles to meet the sim-

DE BEERS COMPOUND, SHOWING SWIMMING BATH IN CENTRE.

DE BEERS COMPOUND.

ple needs of the natives.   Besides these stores there is a hospital and dispensary, where any needed medical attention is promptly given, and a church for religious services, conducted by missionaries delegated by the various church denominations.   During week days this church is also used as a school for the instruction of the natives.   Compartments, with entrances opening through the walls, are set apart for latrines, and cared for with strict attention to sanitation.   In the centre of the enclosure there is a large concrete swimming bath, in which most of the natives are at times found diving and swimming, as is vividly shown in the accompanying illustrations (see also page 440). If any fail to show the necessary regard to cleanliness, they are compelled to keep themselves clean.

A competent manager is in charge of the compound, and his assistants are intrusted with the charge of preserving order and enforcing the compound regulations.   The natives look upon the manager as their great white chief.   He settles any disputes which may arise among them, and in conjunction with the mine manager investigates any complaints in reference to the amount of pay which has been allowed them, or any punishment or ill treatment by their white " baases," which, needless to say, is contrary to the regulations.

The compound is lighted by electricity, arc lights being hung within and without the enclosure.   When a newcomer or a number of natives, for they usually come in little troops, apply at the gate of the compound for employment, the applicants are admitted into the compound only by the immediate direction of the manager or his assistants.   As soon as they enter, their clothes are searched to prevent the smuggling in of liquor, playing cards, or other forbidden articles ; then the officer in charge of the dispensary examines each separately and carefully.   No diseased man is given work, and any suffering from contagious diseases are sent at once to a quarantine building outside the compound, where a temporary provision for such cases has been made.   Within twenty-four hours, a second examination of every one admitted who shows any symptoms of disease is made by a physician in the employ of the company, who daily visits the compound.

To enter the service of the company, each applicant must sign a written contract, binding himself to live in the compound and work continuously and faithfully for a period of at least

De Beers Compound.

three months, or longer if he so desires. At the expiration of a contract, the applicant may leave if he chooses, or his contract may be renewed indefinitely. Some of the natives in De Beers Compound have been employed continuously for ten years or more in the service of the company, for the more industrious prefer the certainty of wholesome food and steady pay to the

2 E

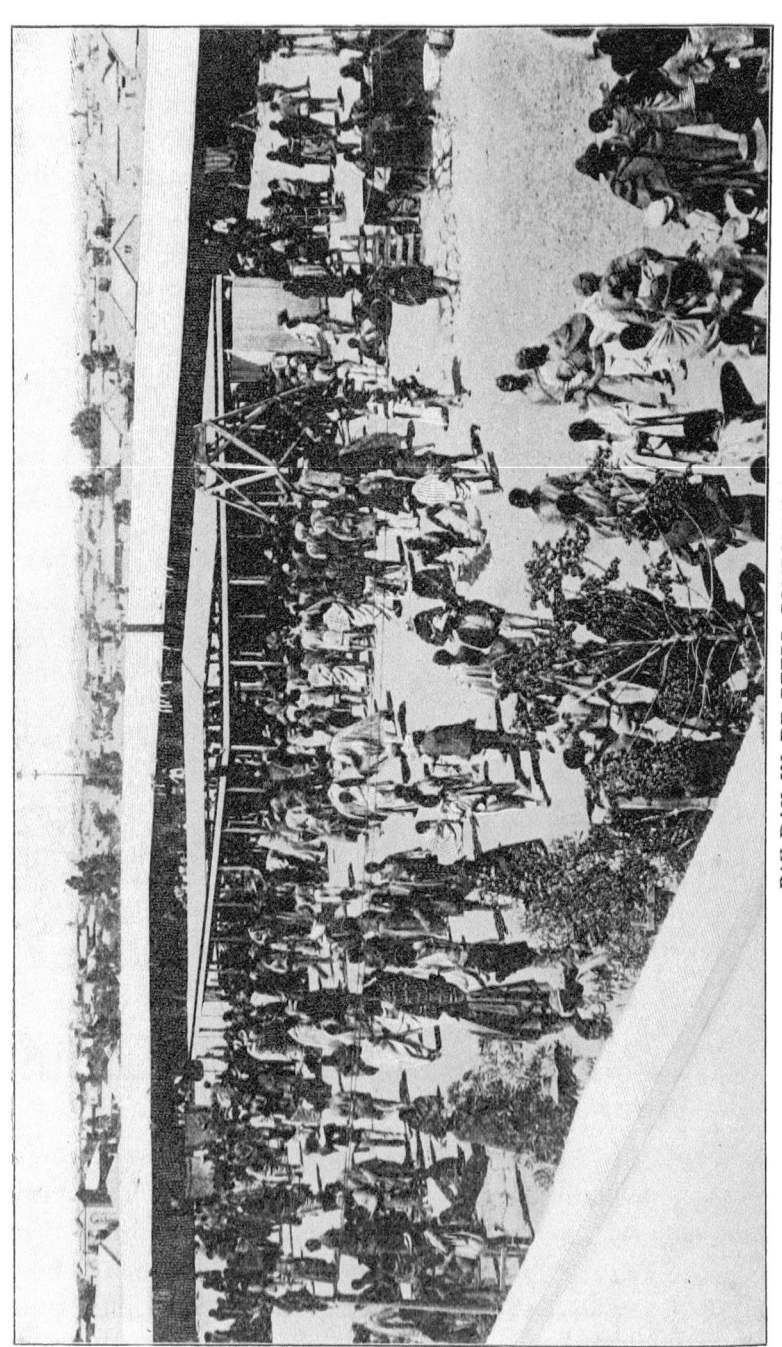

PAY-DAY IN DE BEERS COMPOUND.

DE BEERS COMPOUND, SHOWING THE COVERING OF WIRE NETTING TO PREVENT DIAMONDS BEING THROWN OUTSIDE THE COMPOUND.

shifting to any other occupation that is open to them, or to return to their old savage life. All contracts are filled out in behalf of the natives by an officer delegated for this purpose by the Registrar of Natives, a Government Official, in order to keep a record of all additions to the inmates of the compound, and provide assurance that the contract is signed with a full understanding of its provisions. In consideration of this service the native pays a registration fee of a shilling, and a shilling per month during the term of his employment. All receipts from

De Beers Compound Musical Band of Natives.

this source, except the registration fee, go to the Kimberley Hospital Fund for the care of sick and wounded natives. As the company provides for the natives in its own hospitals, where free medical attendance and nurses, as well as free food, are furnished, the Kimberley hospital receives a very large monthly contribution without being at any expense for the care of sick natives in the compounds. After his signature or mark has been affixed to this agreement, a native cannot leave the compound until the specified term has expired, except by the permission of the compound manager, which is rarely given because of the opportunities that would be opened for taking out diamonds.

A Fireside Gathering, Kimberley Compound.

Underground work in the mine is carried on both day and night by three shifts, under the supervision of the mine manager and overman and three assistant overmen, one of whom is detailed to take charge of each shift. The shaft is reached through an underground passage leading from the compound, and a partition in this passage gives separate entry and exit ways to and from the mine. All laborers are taken up and down the shafts in cages. Each "boy" wears a number on his wristband for easy identification, and when he passes into the mine his number is taken by a guard, and a tally machine records each native as he leaves the compound to go to work; on his return, daily, he brings a ticket noting in what working gang he was employed and what pay he had earned for the day. The natives commonly work for the contractors, who mine and tram the diamond-bearing ground at a price per load which is arranged by tender, and the natives are paid a fixed wage per diem; but a worker must drill a certain number of feet of holes for blasting, which in soft ground is about twelve feet, or he must load a fixed number of trucks, in order to earn his daily pay. The natives usually work in the mines in gangs numbering from ten to thirty men and boys. The limit of age

for the employés in the mines is fixed by government regulation, which provides that no boy under twelve shall be
employed. Another regulation prohibits the employment of
females in mining work. It is further provided that no native
shall be employed underground, or in any of the compounds,
except under the responsible charge of a white employé of the

Natives making Coffee, Kimberley Compound.

company. The handling of the dynamite cartridges used in blasting is intrusted solely to white employés, and all work done by
the native gangs is laid out and directed by white overseers.

The drilling in the blue ground is done for the most part
with long hand drills, — jumpers, — which are sharpened at both
ends, and which the natives readily learn to use effectively; where
the blue rock is hard, the natives use single hand hammers.
Their sinewy frames and powers of endurance enable them to

labor day in and out without any apparent injury to their health. As a matter of fact, nearly all gain strength and flesh in the mines. All the "drill boys" in De Beers mines are now natives, and are scattered through the mines on various levels

Open Mine Workers, Kimberley Compound.

while working, the number at any one point depending upon the size of the working face or stope of blue ground. At points half a dozen boys may be working together with drills, industriously pecking away at the diamond-bearing ground. Natives are also employed in clearing away the excavated ground, and loading the trucks, which run on tramways to the hoisting shafts

when working on a main level, or to chutes on the intermediate
levels. If the roofs of the levels were transparent and a view
were possible of the workers, — whites and blacks, — toiling day
and night in these underground passages and stopes, gleaming
with the white rays of electric lamps, or plunged in darkness,
only relieved by the flickering yellow flame points of straggling
candles — this vast underground hive of workers would be a
greatly stirring and impressive sight. As it is, some conception

Kimberley Mine Compound.

of the great mine may be built up piecemeal in the mind's eye
by combining the illustrations of the men at work which artists
in the mines have been able to make, some of which are given
in the pages of this work.

There is a certain racial resemblance in the temperament,
character, and often in the speech of all these native miners, but
there are also marked tribal distinctions. The natives are clan-
nish, and it is rare to see members of two different tribes lodging together. " Boys " of the one tribe always prefer working
together, and this natural liking is humored to some extent in

selecting gangs to work, although the mixing of the tribes in the mines is inevitable, and often desirable. The Zulu, sprung from the warlike tribes moulded by Chaka, is one of the best of the native workmen, tall, straight, and erect in bearing, proud of the tribal traditions of the Amazulu, " the people of the sky," and, but for an exceptional fit of passion, a good-tempered, cheery,

Natives drilling in the Open Mine.

and ever willing and capable worker. The Amashangaans, coming chiefly from Portuguese East Africa, are closely akin to the

The Last of Open Mining in Dutoitspan Mine.

Zulus, and resemble them in form, temperament, and working efficiency. The Transvaal Basutos rank with the other two as workmen, for most are industrious and capable, and form the most obedient class of native laborers, and nearly all become skilled in drilling. The men of most of the native tribes range over 5 feet 8 inches in height. Many are fully 6 feet tall, and several

Native drilling Underground.

of the old hands are from 6 feet 4 inches to 6 feet 6 inches in height. To this high range the Batlapins from Bechuanaland are the most notable exception, for they often are not much larger than the dwarfed Bushmen of the Kalahari desert. They are not favorites at the mines with the other tribes, or with the whites, for they are often impudent and meanly selfish, and difficult to instruct in underground work.

The ordinary dress of the natives in the compound is a woollen shirt, trousers, and shoes. They rarely wear any underclothing, and when at work in the mines, a pair of ragged

Drilling Underground.

trousers, a blanket, or old breech cloth will often be their only covering. Occasional visitors to the mine are startled by the native disregard for cover ; but the natives are commonly alert to pass the word " umfas " (woman) from one to another when a lady visitor is seen in the mines, and then the native workers on the level ahead scramble for cover or hiding.

The Midday Meal.

When any injuries happen to the men from accidents in the mines, the suffering natives show remarkable fortitude in bearing pain and enduring the necessary surgical operations. Their blood is warm and pure, and cuts in their flesh, or bruises, heal very rapidly. They suffer most from diseases of the lungs, especially phthisis and pneumonia, which are common maladies of the native tribes outside of the mines, as well as within the compounds. They can readily obtain fresh vegetables and fruit, but the common choice of food, such as mealie meal and meat, exposes them to attacks of scurvy. In spite of the careful and repeated medical examinations before men are admitted to the compound, cases of leprosy are occasionally found. In such cases provision is made at once for the isolation of the sufferers. The Government officials are notified, and the diseased men are transferred to Robin Island, where the Government has a permanent leper station. Outbreaks of other contagious or infectious diseases are met by the isolation of the patients in a special lazaretto outside of the town, which is under the supervision of the board of health. Natives suffering from any disease that is not infectious are cared for in the hospital of the compound, which has several wards, — one for cases of fever, one for convalescents, and one for surgical treatment. A qualified dispenser is in charge of the hospital and dispensary, and physicians engaged by the Company are in daily attendance.

At the shops in the compound any articles of food and clothing which the inmates commonly want are supplied. The staff of life is corn, or mealie meal in some form, sometimes baked in hoe cakes, but generally made into porridge. A considerable quantity of brown bread made from Boer meal is also eaten, with meat, vegetables, and fruit in season. Meat is commonly cooked by boiling or by roasting over wood fires. The prices are never permitted to be in excess of the common market prices in Kimberley. If a " boy " does not want the trouble of cooking for himself, he can buy ready cooked food, which is supplied by the company or at any one of a number of coffee shops in the compound. One of the favorite resorts belongs to a Zulu, popularly known as " Roast Beef," who had the misfortune to lose his leg in an accident in the mines. He does his cooking over an open wood fire with the aid of a few kettles and pans ; and a bare wooden table, usually made from dynamite cases, serves for his dishes ; but he is a chef in his line, in the eyes of the compound, and is making more money than he earned before he was crippled.

There are a number of native tailors on the ground, who can fit and make a suit to order, or repair one, with no little dexterity. Native mining suits are usually made of the English cloth known as moleskin, and the tailors, in accordance with South African custom, put large patches on the seat and around the foot of the trousers. Sewing machines are commonly used, which the natives buy in Kimberley through the compound manager. Some work in the mines during the week, but like to earn additional shillings by cloth cutting and sewing during their leisure hours, when their machines may be heard clicking from morning till night.

ZULU WORKMEN, DUTOITSPAN MINE.

There are native barbers and hair-dressers, also, of whom the chief is " Sandy," a Cape boy, who struts about on Sunday in a khaki jacket with the airs of a tonsorial artist on the crest of fashion, and is reputed to make more on his holiday with his clippers than he can earn in a week with the drill below ground. He has not as much range for his art as a French barber, for most of his patrons want their hair cut off close to the scalp ; but

Native making Bangles.

he is justifiably vain of the speed with which he lops off one bushy head of hair, and makes room for the next to fall.

Pedlers of all sorts, dealing in cakes, tobacco, and ginger beer, have their stalls in the moving throng, especially on Sundays and other holidays, and here and there are to be seen workers in Kafir adornments, principally in armlets or bangles, and bands for the legs. These are usually made of fine copper and brass wire rolled upon rings of horse hair. The rings are about one eighth of an inch in cross section and from four to five inches

MR. ROULIOT, MANAGER OF THE "COMPAGNIE GÉNÉRALE," DUTOITSPAN MINE, AND HIS NATIVE WORKMEN.

in diameter, varying with the size of the hands over which they must be slipped. The wire is wound round the hair very skilfully. European visitors occasionally supply gold wire to these workers, which the natives wind around the hair centres into fanciful bangles, some of which are very pretty.

All the workers in the compounds are supplied with Bibles, printed in various tribal languages, which the natives are taught to read by missionaries. At any and all times De Beers Compounds are open to these teachers, who are specially delegated by English and German missionary societies.

When a "boy" is once moved to apply his mind to any study, he will commonly plod on persistently, and there is among the natives generally an unfeigned respect for teachers, and pride in the attainment of any advance in learning. There is only the crudest notion of religion in the minds of these negroes, and the missionary must have unwearied patience who seeks to impress them with the idea of an invisible, omnipotent, omnipresent God and Father of all. It is very difficult for the missionaries to prove by the Bible that these savages should have only one wife, and this has been a great stumbling-block in teaching them Christianity. The native argues that, if he has only one wife, she is continually wrangling with him, but if there are two or more, they occupy themselves by wrangling with one another. And again, he says, the more wives he has, the more crops he can raise. The women do all the work at the kraals, and the men idle their time away in peace and plenty.

The preachers at the compound chapel or elsewhere in the compound often call together their flocks with stirring notes of drum and trumpet, and at gatherings of natives lime-lights and lantern slides are also effectively used in vivid and telling illustrations. Sometimes an interpreter stands at the preacher's elbow, to make his meaning clear to native listeners, for the tribal dialects in the compound are like the confusion of tongues in Babel. The missionaries are somewhat vexed by the Kafir "doctors," who keep before the natives the vision of old superstitions, as they squat on the ground in the compounds, solemnly laying out their "bones" and muttering incantations. They are so tricky with their impostures that it is difficult to bring any of them patently into contempt.

Almost all of the natives are fond of sport.   They have plays of various kinds which may be seen every day in the compound, but the chief show is naturally on Sunday, the holiday for all.   Then a number of the tribes put on their native dresses, and there are vivid spectacles of native dances, chants, and games.   The Zulus often arm themselves with clubs or wooden assagais, or any long canes which they can brandish and strike upon their ox-hide shields, while they circle about in a ring, marking time with a stamp of the foot that makes the earth quake.   It is the traditional report that no one is admitted to this war dance who has not killed a man ; but the chances are that, in recent days, unquestionable evidence of this qualification is not strictly required.   Nevertheless, the pretence of bloodthirstiness is very exciting, as warriors

'Mshangaan in War Attire.

spring forward, one after another, swinging their assagais or knob kerries, and advancing their shields, while they show a pantomime of attack upon an imaginary enemy almost as vivid and thrilling as actual battle.   When this dance begins, a circle of native spectators gathers about, shouting and crying with the passion of the scene, till the noise at times is deafening.   Other natives, less particular than the Zulus, dance about in rings and crescents, waving any kind of stick in their hands, from a miner's candle-

2 F

NATIVE WAR DANCE, IN THE PRESENCE OF HIS EXCELLENCY THE GOVERNOR, SIR ALFRED MILNER, AND SUITE.

stick to a twig or old hatchet. Among these figure the fantastic
Machopis, dancing to the music of native imbilas, or Basutos
blowing their little reed or bone whistles and swaying about with
strange contortions, accompanied by monotonous tapping on a
crude drum made by stretching a raw ox-hide over the end of a
barrel. The 'Mshangaans chant while dancing, but the Basutos

are not gifted with
musical voices and
have no evident
ear for music, al-
though they are
so fond of their
own harsh and
discordant blow-
ing that they will
pipe away on their
hollow bones and
dance for hours at
a time on Sunday
to their own pip-
ing.

Among the
other native tribes
there are many
boys with fine
voices, sweet
toned or robustly
sonorous, ranging

'Mshangaans in War Paint.

from the highest tenor or falsetto to the deepest bass; and some
are readily trained to part singing. In De Beers and other of
the larger compounds there are native choral societies under the
charge of white instructors. The most popular songs are the
familiar American negro minstrel and concert hall melodies.
These are freshly ludicrous to one who pictures the black singers
"climbing the golden ladder" and "wearing the golden slipper"
on their big flat feet. The climax is reached when the high

voices sing, "What are you goin' to wear?" and the reply comes from the deep bass voices, "I'se goin' to wear a standin' collar." Native African chants are rarely heard in the compound, except sometimes as an accompaniment of native dances.

At all hours of the day, until the stir and buzz throughout the big compound are hushed in the sleep of its thousands of inmates, the rattling and humming and squeaking of imbilas and gubos, and various other crude instruments of native fashioning, are to be heard, more or less widespread. The "imbila" is the same as the maninba noted by Dr. Livingstone in his travels in Africa. In the native villages it is made by fixing strips of board across dry calabashes. By grading the size of these gourds, different notes are produced when the overlaid strips are struck by a drumstick with an elastic gum knob. In the compounds empty dynamite boxes with tin cans fastened underneath the strips of wood supply the lack of calabashes, and the striking knob is imitated by twisting a piece of rag tightly round the end of a stick. The native "gubo," as the Zulus call it, is an instrument also common throughout South Africa. This is a bow of bamboo with a tightly stretched string. The player holds the end of the bow against his parted lips with one hand and strikes the tight string with a slip of split bamboo. A peculiar effect is obtained in playing on this bow in the compound by attaching a calabash to the back of the bow, and holding this improvised sounding-board against the breast. These are the favorite instruments, but there are others, like the bone whistles of the Basutos, which are much cruder, and grate far more harshly on the ear of listening white men.

That the native African has an inborn fondness for music is signally shown by its persistent pursuit in the compounds, even through refuse boxes and bones. It may advance in time, with education, to high artistic appreciation and accomplishment. Even at its present barbaric stage the Kafir may be greatly moved by the art of a great singer, as was evident when Madame Albani came to the diamond mines, for she never saw an audience so passionately enraptured as the black men massed about her within the walls of De Beers Compound.

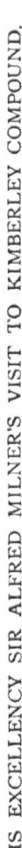

HIS EXCELLENCY SIR ALFRED MILNER'S VISIT TO KIMBERLEY COMPOUND.

A Quiet Game of Cards.

There would probably be a common resort to gambling as well as to music, if the practice were not sharply restricted by the compound regulations and oversight. It was necessary to prohibit the playing of cards, because native sharpers were fleecing the tyros too unmercifully. There is still, probably, some covert card playing, for many of the natives understand a few of the games familiar to white men. Faro was played with the top of an empty dynamite box as a table, upon which cards were tacked. The game was probably introduced by natives from the Portuguese possessions. The native African has only a few games of his own devising. The most popular of these in the compounds, and Africa at large, is "umtshuba" or "chuba," the Syrian "mancala," or, as the Nubians call it, "Mungala." The widespread knowledge of this game is noted by Schweinfurth as one of the links of evidence of "the essential unity that underlies all African nations"; and it has been shown by the investigations of Mr. Stewart Culin for the Smithsonian Institution of the United States,[1] and by other reports, that the same

[1] "Mancala, the National Game of Africa," by Stewart Culin, 1896.

game with essential variations is played throughout Africa and extends along southern Asia as far as the Philippine Islands. For this game a long strip of board is provided, edged with two parallel rows of holes scooped in the wood. When a board can-

not be procured, the rows of holes are made in the ground. The number of holes in a row varies widely, the Nubian "Mungala" having sixteen, while the board common on the Diamond Fields has from thirteen to seventeen holes and four rows. Each player has

Natives playing Chuba.

about two dozen pebbles in hand, and the play is in shifting the pebbles from one hole to another. Stanley calls the game an African "back-gammon," and speaks of the board as a "backgammon" tray. The word "mungala" is of Arabic origin, derived from "nagal," "to carry from one place to another." There is no apparent interest in the

Natives playing Mancala.

game to the ordinary white man's eye, but native players in the compounds and African negroes generally will keep on moving the little stones for hours at a time with evident satisfaction, taking up their opponents' pebbles, as certain combinations occur, until one or the other has won all.   The spectators usually offer advice to the players.

There is some running, jumping, and wrestling in halting imitation of English athletic sports; and on special holidays, like Christmas, they have obstacle races, sack races, walking the greased pole, which lies horizontally over the swimming bath, and other comical features for the general amusement of the native and white spectators.   But the workers in the mines are rarely nimble enough to figure with any distinction in these sports, and the only English games that can be called popular in the compounds are the counterfeit of cricket and football.   The native wickets are made of empty paraffin tins, and the fine points of the game are not in evidence; but there is plenty of hard swiping and sharp bowling, to the delight of the native players and the spectators.   Christmas is the great holiday of the year for all, for everybody in the compound then receives for his Christmas box a loaf of bread, a bottle of ginger beer, and a piece of meat, and sports of various kinds are specially provided for

Swimming Bath, De Beers Compound.

Natives smoking Indian Hemp.

their amusement. Grabbing and diving for money thrown into the swimming bath by the directors and managers form a lucrative sport for the natives, and amusement for the lookers-on.

At every gathering for dances, sports, or games of any kind there are more lookers-on than participants, for the African dearly loves a spectacle of any kind, and is commonly well pleased to stand or loll on the ground where he can get a view of the contributors to his entertainment. Some of these indolent ones will be smoking cheap cigars, and more rarely pipes. A native usually puts the lighted end of a cigar in his mouth, inhaling and blowing out the smoke, and shifting the hold of his teeth as the tobacco burns. Sometimes sets of boys are seen squatting on the ground and passing from mouth to mouth a lighted pipe filled with dry dagga, a native herb similar to the Indian hemp, that burns with pungent and stupefying fumes. The natives inhale the smoke, and, after a few puffs, a fit of violent coughing comes on which brings tears to their eyes. The use of this herb is not so extended as to cause any serious ill effect, but the native becomes stupefied for a time, though he soon recovers.

There is some tribal jealousy and vanity, but the inmates of the compounds live together on good terms, as a rule.   In their occasional fights they use bottles or stones or clubs, or anything they can lay their hands on quickly; but, as soon as the guards come up, they hurry off to their rooms, where they are put under strict oversight for a time.   Even these short encounters often leave many with sore heads and bruised bodies.   Only once has there been the threat of a serious insurrection in the compound.   This was at Christmas time, when the compound manager was absent for a few days.   After the usual Sunday holiday several hundred natives, chiefly from Kaffraria, refused to go to work on Monday morning, as the following day was Christmas, demanding the grant of Monday also as a holiday.   I went at 5 A.M. to the compound and urged the leaders of the strike to take their followers into the mines.   The Basutos were willing to support me, and offered to drive the reluctant Fingos, or Amafengu, underground.   After some protracted but ineffective appeals, I sent word to Mr. Rhodes, who happened to be at Kimberley, that the Fingos refused to go to work, and suggested that he might come over and try his persuasive power on them.   So he did, but after an hour of fruitless parleying we determined to try the demonstration of force, for the Fingos not only refused to work themselves, but barred the other natives from entering the mine.   As they numbered from five to six hundred, they were rather a formidable barrier at the underground entrance.

We then decided to call in the assistance of the police and our own guards, Mr. Rhodes riding to the police station, while I rode to a station where a number of extra guards were posted. When we came back into the compound with a force of fifteen men armed with carbines, the Fingos instantly began to pelt us with bottles and stones, and anything else which would serve as a missile.   At this outbreak I asked the officer in charge to fire a few blank shots at the crowd of rioters, and in less than a minute there was not a native to be seen in the open area of the compound, for all scurried off like frightened sheep to their rooms. We then went around the compound, picking out the ringleaders, thirty-three in all, ranged them in line, and sent them to jail. They were soon brought up before the magistrate and each was fined £3, which they obtained by a little begging from their brothers in the compound. Meanwhile, it was difficult for us to restrain every native left in the compound from going to work that day on the first shift.

After the ringleaders came back to the compound, they wanted a meal, but they were forced to go underground and work eight hours before any food was provided. Then they were singled out and led around the compound, one by one, as an exhibition or warning to others, before they were finally discharged from the employ of the Company and sent away from the works. One of our interpreters had been taken along with the rioters by mistake. He was so vociferous that some one put him in with the other noisy boys. A few days later, when I wanted an interpreter, the unlucky one said, " All right, Baas, I don't mind interpreting for you, but I don't want to be run in for it."

No corporal punishment of the natives by white employers is allowed. If a boy is unruly, he may be placed in a room by himself until he can be taken to jail, and charged with whatever offence he has committed. The most common offence is petty thieving. There can be no doubt that the covert purloining of diamonds would be a frequent practice, and cause heavy losses to the diamond mining companies, if it were not for the compound system, which makes it impossible for natives to take any diamonds out of the compounds with them.

A fine wire netting is stretched over the top of the compound to prevent the sly tossing of precious crystals over the walls, to be picked up by confederates outside the mining areas. Precautions are also taken to prevent the smuggling away of diamonds from the compounds, and all communication by the natives with persons outside the walls is carefully restricted. Until the expiration of his contract, no native can go through the compound gate, except by special permission, or when he is taken under guard before a magistrate for some offence.   If convicted, when his term of imprisonment expires, or after he has paid his fine, he must return to the compound and complete his contract.   Before leaving the compound his clothes and person are thoroughly searched to prevent the disappearance of diamonds with them.   Gems were sometimes found secreted in clothing, or shoe heels, or canes, or cans with false bottoms, in fact, in anything that the natives were allowed to take out with them.   Even this close inspection did not bar

the practice of stealing, and there was an inexplicable trickle of fine diamonds from unlooked-for quarters, until it became known that natives on the point of leaving the compound were swallowing diamonds and conveying them away.

In 1895 one native had the nerve and capacity to swallow a lot of diamonds worth £750, and did not appear

Diamonds which a Native had swallowed, and which were recovered by the Guards in the Compound.

to suffer by this strain upon his digestion.   There has been only one authentic instance where a native has embedded diamonds in his flesh — this was done by a native in De Beers Convict Station, who made an incision under the shin bone and concealed several small diamonds wrapped in a rag.   This native had symptoms of tetanus, and the visiting physician (Dr. Otto) searched the man's body, and, finding an ugly-looking wound on his leg, cut it open, and to his great surprise found a rag full of diamonds.   The native soon recovered, a wiser, if poorer, man.   The largest yield

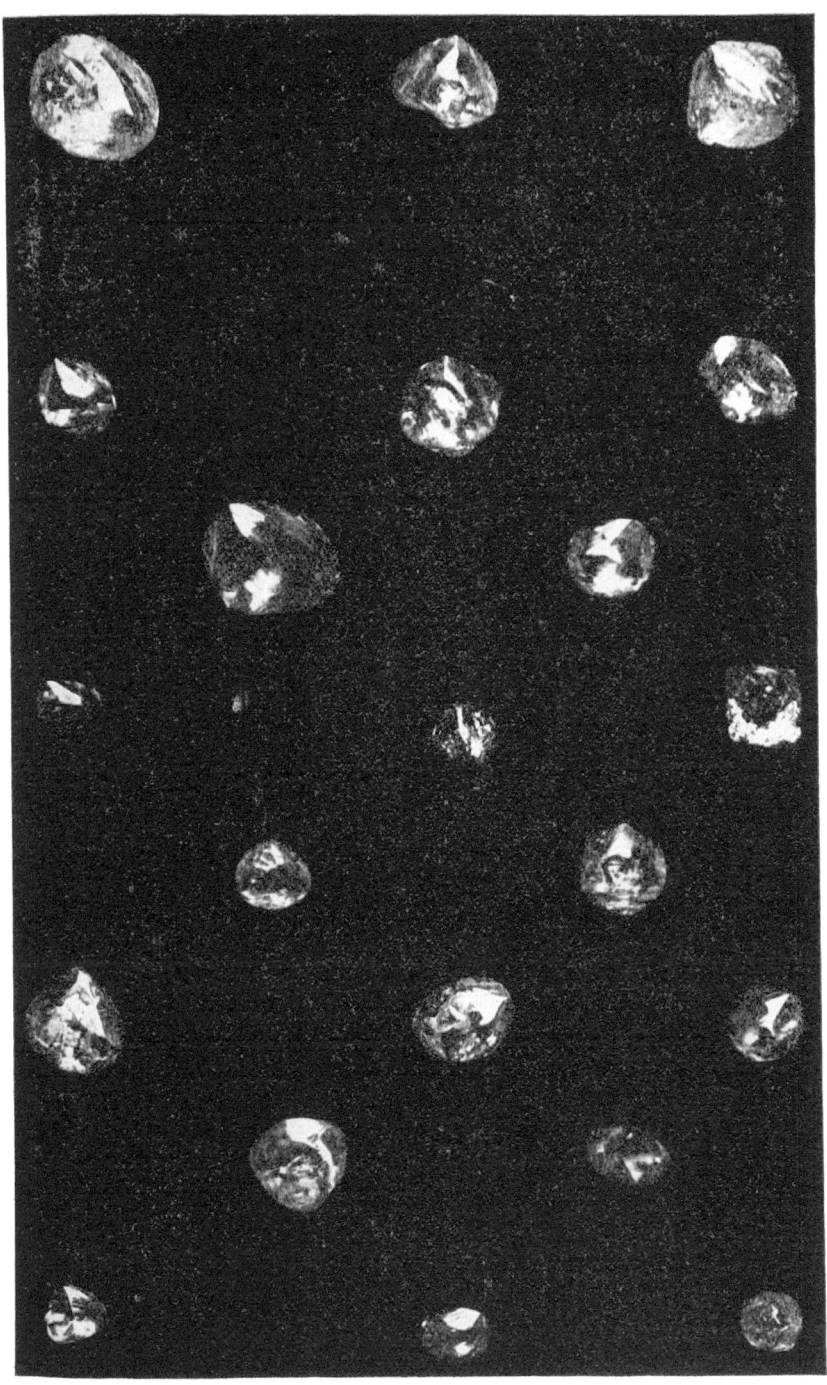

DIAMONDS SWALLOWED BY A NATIVE AT ONE TIME, AND ALL RECOVERED
AFTER FOUR DAYS. TOTAL WEIGHT, 348 CARATS ; VALUE, £1067 : 4 : 6.

of diamonds which a native had swallowed is represented by the illustration on page 445, each diamond being drawn the

Diamond Thief.

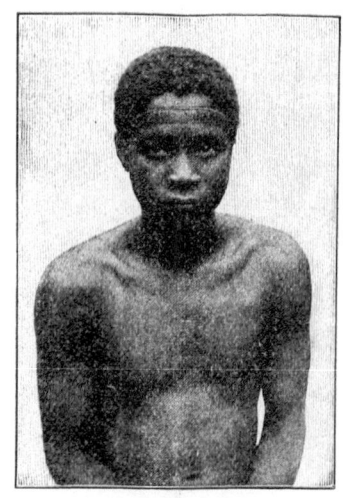

Diamond Thief.

exact size of the original. There is no apparent fear of swallowing any stone which can be forced through the throat,

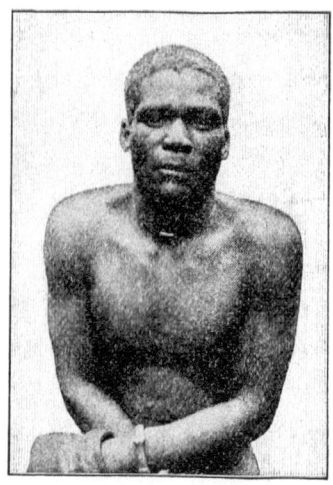

Diamond Thief.

Diamond Thief.

and in one instance a diamond as big as a large chestnut and weighing 152 carats was hidden for over seven days by this means.

The swallowing of a rough diamond is evidently so easy, but so difficult to detect, that it was necessary to put an end to the practice by providing a longer period of detention and search. At the close of their contracts, natives whose terms of service have nearly expired are placed together in a commodious room capable of holding two hundred men or more. They enter this room entirely naked. Their clothes and baggage are deposited in sacks marked in accordance with the number on the arm band. Blankets are supplied for clothing, and as wraps when sleeping. They are fed, and generally well cared for, free of

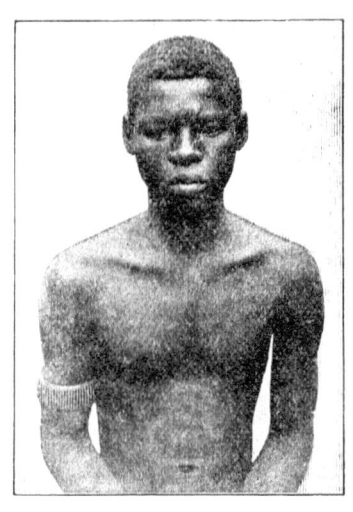

Diamond Thief.

cost to themselves. While in the detention room they are under strict supervision of white guards, so that any diamonds they may have swallowed must be left behind before they leave. Natives have been known to keep diamonds in their bodies for over seven days. At the end of five days of detention, generally on Saturday morning, they are released. Meanwhile, the clothes placed in the sacks have been thoroughly searched; and departing natives are not allowed to take away with them anything but soft goods. In fact, they are even required to leave their boots behind, for cunning smugglers used to insert diamonds in their boot heels so neatly that the trick could not be detected without

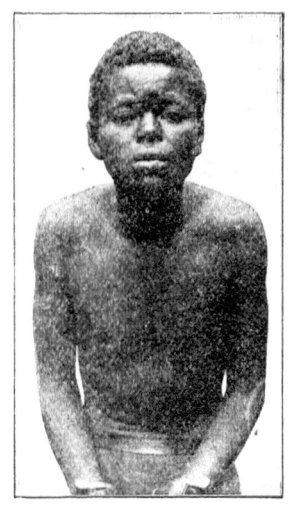

Diamond Thief.

cutting away the greater part of the sole of the boot. Boots and shoes, and other articles which are not allowed to be taken from the compound, are sold or given away to customers or friends before their owners leave.

It may be that De Beers Compound is a "Monastery of Labour," as was wittily said by a lady visiting the fields as a

 correspondent of the London *Times;* but the testimony of all careful observers on the ground affirms the beneficial effect of the restrictions from dissipation, and the general good cheer of the workers. Mr. Thomas H. Leggett, an entirely independent and competent American witness, wrote of his inspection of the men in the compounds, in *Cassier's Magazine*, September, 1898 : "These chaps are well cared for, contented, and happy, as proven by the fact that many have been there for years ; and the secret of it

Diamond Thief.

lies in their not being able to get drink."

Occasionally a visitor at the fields is less observant and candid. One such was a member of the Legislative Assembly of Cape Colony, who came to Kimberley to investigate the conditions of life and treatment of the natives in the compound. On arriving at De Beers Compound, in company with his wife, he first impressed upon the natives whom he met that he was a member of the Cape Colony Legislative Council. He had come to the fields in their behalf, and he wanted them to tell him freely everything of which they had to complain. With the aid of an interpreter he interviewed a number of natives in the compound, asking searching questions about their treatment. One native told him that he had been working for eight years in the mines and had been outside the compound only three or four times in all that period. When asked if he was well treated in the compound his answer was, "If I didn't like it, Baas, I wouldn't be here." The visitor's wife meanwhile kept tugging at his coat continually, saying in Dutch, "They treat the Kafirs altogether too well here; they will be spoiled by such good treatment as this." Before leaving, the legislator said that he was glad to have the opportunity to inspect fully the operations of the compound. From what he had heard he had been much opposed to compounds, but he now saw with his own eyes that he was wrongly informed, and henceforth he should be a strong advocate of the system. Yet a year or two later, when questions affecting De Beers Company and the compound system arose in the Upper House, this gratified member was one of the first to denounce the system in an intemperate speech.

De Beers Machine Shops.

CHAPTER XV

Kimberley

KIMBERLEY, the largest of the cluster of diamond towns on the Fields, is, like the rest, the
natural efflorescence of the mines near which it
is situated, and from which it derives its birth
and being. Its mushroom growth must have
withered like so many other pretentious upstarts
from the mining fields, had it not been for the fact of its rising
on ground of such sustained richness and promise. While the
diamond-studded blue ground continues to show a persistent
extension in depth and in richness, and while man's energy and
art avail to pierce and extract it, the Kimberley of the surface
will surely continue to flourish.

It might indeed be said, without any stretch of imagery, that
the modern Kimberley is literally as well as essentially built up
on the yield of the mines. This has been brightly noted by the
late Rev. James Thompson in his pleasing sketch of the modern
Kimberley. "Kimberley, as we know it," he says, "with its
streets and warehouses, and shops and schools and churches, is
largely built upon that strange mixture known as débris, every
atom of which has a story to tell if it could only speak. As in
any English town you can go down foot after foot through the
different strata representing the pavements or pathways upon
which successive generations of ancestors pressed their feet; so
in Kimberley we have beneath the present surface of our roadways the red soil on which our fathers pitched their tents, and
which their labor soon covered up by spreading out all around
them the heaps thrown out of that great hole which now looks

so desolate, but which was once the centre of activity and throb-
bing life which made Kimberley famous throughout the world." [1]

Dr. Thompson marks the middle age of Kimberley as the
period when decent buildings of iron and wood, with here and
there more pretentious brick, had replaced the age of canvas;
but when there were no softening or beautifying surroundings,
when every tree and bush had been cut down, and when the
veld once dotted with thorn trees had become a vast expanse

Snow in Kimberley, 1876.

of wind-swept dust as gray as the iron dwelling places which
alone seemed to convert the desert into a town. This was the
period preceding the introduction of an abundant and pure
water-supply that wrought such a transformation in the appearance of the city. Now the upspringing of flowers of varied hue,
and green thickets and vines and trees in the gardens that now
surround nearly every house in town outside the business
quarter, has made during many months of the year a beautiful
country town of the old and barren Kimberley.

In spite of the visible yield of the mines and the consequent
prosperity of the town there was, for many years, a prevailing

[1] Christmas number, *D. F. Advertiser*, 1898.

Theatre Royal.

distrust of the permanency of the diamond-bearing deposits and the consequent stability and future of the city that was founded upon them.   But later, as systematic development gave substantial assurance of the endurance of the mines, the advance in the architectural beauty of the residences and public buildings in Kimberley has been marked.   Now many of the residences of the more wealthy townspeople are not only substantial, but distinctly ornate in character, with widespreading verandas rising in the midst of green lawns and lovely gardens.   Some of the public buildings already erected or in process of erection need not fear comparison with any like structures in any city of its size in the world.

Among these structures is a handsome and well-appointed theatre, built of burnt brick with stone facings, excellently situated for the accommodation of theatre goers. This building, the Theatre Royal, was designed to introduce all the latest improvements in theatrical construction, and its acoustic properties are particularly fine. The commodious stage has a face of 54 feet and a depth of 38 feet, and is so arranged that the whole stage is

Theatre Royal Interior.

in full view of the audience in the box stalls, dress circle, family circle, and gallery. The theatre is lighted by electricity, and its fire exits are so complete and well placed that in case of need the whole audience could leave in a very few minutes.

The Town Hall is another building that deserves special mention. It was erected by resolution of the borough council on the Market Square after the destruction by fire of the old town hall. This building is designed in the Roman-Corinthian style and its appearance is notably pleasing. Its site is in the

centre of the Market Square, a particularly convenient position.
There are three entrances to the main hall, which is finely pro-

The Town Hall.

portioned, — 105 feet in
length, 50 feet in width,
and 35 feet in height. At
one end there is a stage
25 feet wide, and a hand-
some proscenium and
space for the orchestra is
also provided.

There are emergency
exits opening into large
yards that afford abundant protection in the event of the outbreak of fire. Passages along the building lead to suitable administration offices for the borough engineer, market master,

Dutoitspan Road, Kimberley.

sanitary inspector, and native officials. At the back of the main
hall extends the market house, over 83 feet wide and running
the full width of the building. In the east wing of the building
is a council chamber, 50 feet long by 26 feet wide, and, opening

DUTOITSPAN ROAD. KIMBERLEY.

Kimberley Hospital.

out of the chamber, rooms for the mayor and councillors.   In
the other wing of the building, accommodation is provided for
the town clerk and his assistants.   The building is substantially
constructed of the best burnt brick covered with cement and
enriched with cornices.

On the site of the old Kimberley Hospital, established in
1871, a new and spacious building has been erected, with sev-

The Kimberley Club.

eral outlying wards.   The main building, about three hundred
feet long, contains the operating rooms, convalescent room,
and the Merriam, Victoria, and Lanyon wards for the reception

Hall of Kimberley Club.

of European patients only.   The detached buildings comprise
native medical and surgical wards, each containing fifty beds;
the Southey ward for colored women and children; and isolation

wards for infectious cases; male and female contagious disease
wards, and mortuaries. The offices of the resident officials, a
dispensary and doctors' quarters, nurses' home and chapel, with
a further provision of European and native kitchens, make the
hospital complete and comfortable. This hospital has accommodations for 250 patients, European and colored, and from
the day of its
erection it has
been of indispen-
sable service.
During the single
year of 1897,
2683 patients
were admitted,
798 of whom were
Europeans, and
the remainder na-
tives and persons
of color. Six hun-
dred and sixty-
three patients
were admitted
free, or on sub-
scribers' letters.
Besides this ser-
vice it should be
noted that the
number of day

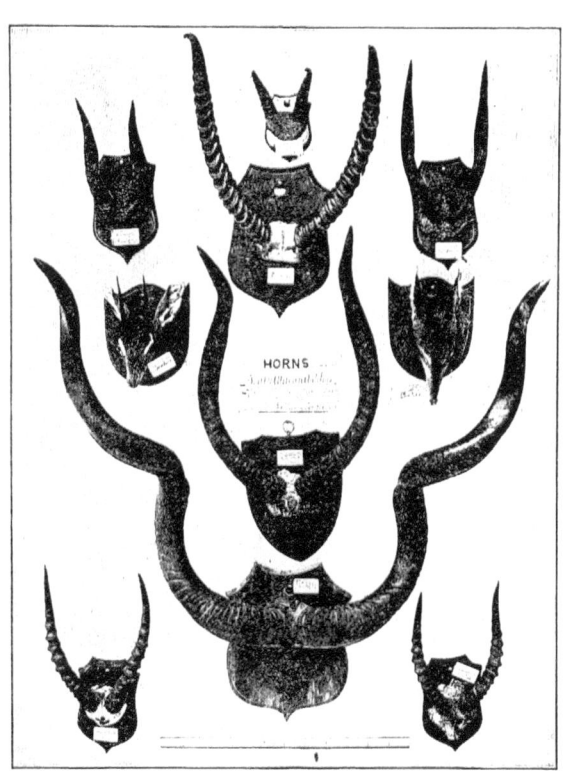

Horns of South African Antelope.

patients treated during the same year was 1220; one of the hospital doctors is in attendance in the day-patients' room for an
hour every morning to give advice without charge to the poor.
To all who cannot afford to pay for treatment, medicines are furnished free. Every subscriber is entitled to give a letter of ad-
mission to one patient for every £2 2s. subscribed, upon the sole
stipulation that the person receiving the letter must be too poor
to pay for his or her own treatment. The staff of the hospital

consists of two resident house surgeons and a visiting body of seven local practitioners. The matron and forty-two nurses constitute the nursing staff. A recent addition has been made to the original hospital, in which will be the maternity ward, for the sake of providing the needed accommodation and the training of experienced midwives. The cost of this hospital with its enlargements has been upwards of £30,000.

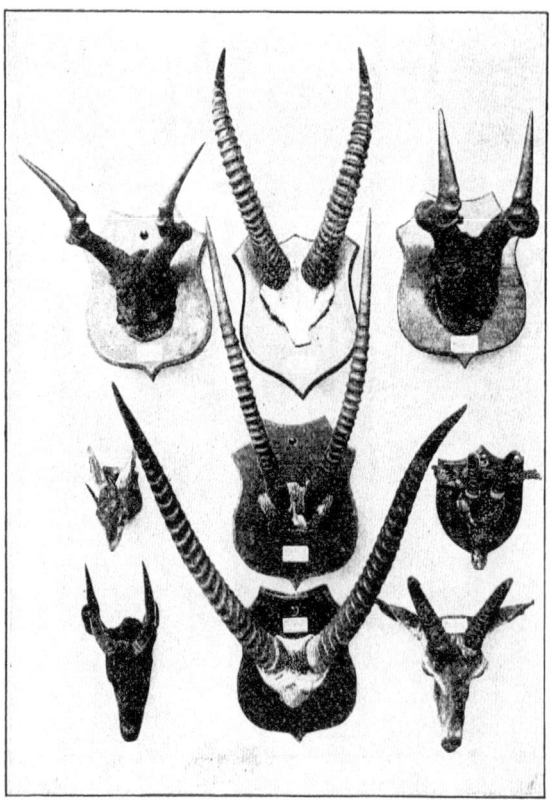

Horns of South African Antelope.

The Kimberley Club has a commodious and finely furnished house on Dutoitspan Road. This building was erected in 1896 on the ashes of two predecessors which had been unfortunately destroyed by fire. It possesses a unique collection of trophies of the chase, and its list of visitors bears the name of many of the most notable men in the British Empire.

Besides these structures a government building of massive stone and brick on the north side of the Market Square deserves mention as one of the conspicuous edifices in the city. Here the High Court of Griqualand is held. The magistrates' courts are arranged on either side of the entrance, and rooms are provided for the Civil Commissioner, Judges, and Magistrates.

The Kimberley Public Library is a well-built building, containing three large rooms, of which one is free to the public, and the others reserved for subscribers. Smaller rooms are provided for the librarian and committee. It is especially notable for its remarkable store of reference works, which is esteemed to be the best in South Africa. It contains in all twenty-two thousand books, many of which would be irreplaceable if destroyed. The building up of this library is justly credited to the fostering care of Mr. Justice Lawrence, the Judge President.

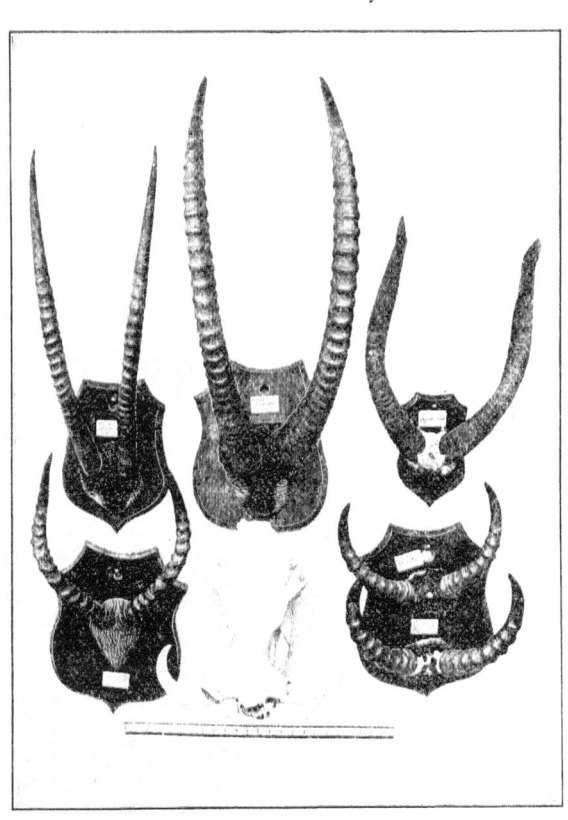

Horns of South African Antelope.

Midway between Kimberley and Beaconsfield stands the Kimberley Sanatorium, a superb structure erected by the liberal contributions of De Beers Consolidated Mines Limited at a cost, with its furnishings, of £26,000. Its fine enclosing grounds, the gift of the London and South African Exploration Company, were artistically laid out under the direction of Mr. Fenner of the De Beers Forestry Department. The larger part of the building is designed for the accommodation of guests, and the smaller block contains the billiard room, smoking room, kitchen, servants' and store rooms. The buildings are of burnt brick, two stories in height, with ample verandas and balconies; all the rooms are large, lofty, and handsomely furnished, and in the construction the best sanitary knowledge has been applied. The building is lighted throughout by electricity, and abundantly supplied with pure water.

The Masonic Temple was erected in 1889 by the combined lodges of the city. Its main hall is spacious and admirably lighted by electricity, and the interior throughout is very handsomely decorated and furnished. At the top of the staircase there is one of the finest stained glass windows in South Africa, which was presented to the lodges by Mr. Rhodes.

Horns of South African Antelope.

The post-office, police barracks, and railway station have no special pretension to architectural beauty, but they are serviceable structures for the uses to which they are applied. The offices of the De Beers Company are in the centre of the business section of the town, and are, as might be expected, excellently designed buildings, and stand out notably among the business edifices that surround them.

The South African School of Mines was established at Kimberley in 1896. The first two years' studies are taken at the

South African College, Cape Town, or at similar colleges at Grahamstown and Stellenbosch, the third year at Kimberley, and the fourth at Johannesburg. The object of the school is to train young men in South Africa as mining engineers. Suitable buildings were erected at Kimberley at a cost of £9000, De Beers contributing on the pound for pound principle with the Educational Department of the Colony. There were twenty students in attendance during the year 1901. De Beers mines and workshops are open to the students, where they are given practical instruction in mining and mechanical engineering. Their theoretical training is under the supervision of Professor J. G. Lawn, assisted by Professor Orr. The management of the school is entrusted to a local committee, consisting of the four members of Parliament representing the

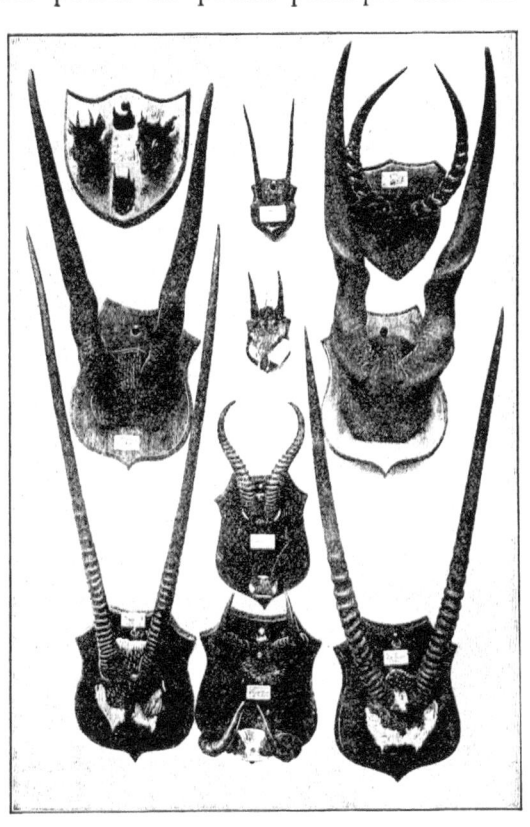

Horns of South African Antelope.

Kimberley district, the member of the Legislative Council for Griqualand West, the Inspector of Mines, the Mayors of Kimberley and Beaconsfield, the Chairman of the Public Schools Committee, and myself. I have the honor of being chairman of this committee.

There are six distinct church establishments in Kimberley, —the Anglican, Presbyterian, Baptist, Roman Catholic, Wes-

Professor J. G. Lawn, Kimberley School of Mines.

leyan Methodist, and Griqualand West Hebrew Congregation. The Anglican denomination has three churches in Kimberley, St. Cyprian's, St. Augustine's, and De Beers, besides churches at Beaconsfield and at St. Matthew's, Barkly Road. The largest church provides accommodation for 650 attendants. The first edifice of the Church of England was built at Dutoitspan, the pioneer town on the fields, and subsequently transferred to Beaconsfield. When Kimberley became the principal city of population, St. Cyprian's Church was erected on Church Street and removed to its present position in 1878. Kimberley became part of the diocese of Bloem-

fontein, and gave its name to an Archdeaconry comprising Griqualand West and Bechuanaland. The Archdeacon of Kimberley is the head of the church organization in this part of the diocese.

The Kimberley Presby-

Professor Orr, Kimberley School of Mines.

terian Church was founded in September, 1877, and has over four hundred enrolled communicants and a still larger number of adherents.

Sir Alfred Milner passing the Offices of De Beers Consolidated Mines Limited on his First Visit to Kimberley.

De Beers Offices decorated in Honor of the Governor's Visit.

Kimberley Public Library.

In 1889 the Rev. James Hughes, of Port Elizabeth, at the invitation of the Baptist Union of South Africa, came to the Diamond Fields and held the first denominational meetings in the Good Templars Hall in Kimberley. Through his efforts a

The Sanatorium.

church was formed, and in 1892 the foundation stone of the present commodious Baptist Church in Dutoitspan Road was laid.

The foundation stone of St. Mary's Roman Catholic Church was laid on the feast of All Saints, 1879, by the Vicar Apostolic of Natal and Griqualand West. For many years previously a Catholic Church had been maintained on the fields, but its building was too small for the growing congregation. The foundations of the new building had just been completed, in August, 1879, when the old building was levelled to the ground by a terrific hail-storm. This was looked upon as a significant

Masonic Temple.

warning to replace the iron sides of the new church with brick, and the present edifice was accordingly erected, which will accommodate about five hundred people. It is adorned with stained glass windows, paintings, and oak altar and reredos, the gifts of its parishioners.

Wesleyan Methodist missionaries were probably the first to visit the River Diggings in 1869, and representatives of this energetic denomination were among the first also that flocked to the Dry Diggings at Dutoitspan and Kimberley The first regularly appointed minister of this church came in 1871, and the succession since that year has been unbroken. The Methodists erected their first church at the West End, but as the town moved eastward, a new church, Trinity, was built to meet

The Post-office, Kimberley.

the call from that quarter. The original Trinity Church was blown down by one of the fierce gales sweeping over the Fields, but a second Trinity has now taken its place. There are now in Kimberley three Wesleyan churches for whites, two for natives, and one for other people of color, and a missionary

Nazareth House, Kimberley.

is in daily attendance at the compounds. It is estimated that there are probably not less than three thousand persons under the charge of these seven ministers. At a very early date in the history of the Fields the foundation stone of the Hebrew Synagogue was laid on the Dutoitspan Road, occupying a site donated by the London and South Africa Exploration Company.

The Gardens of the Intermediate Pumping Station of the Kimberley Water Works Company.

The Author's House at Kimberley.   Wistaria in Bloom.

Another View of the Author's House and Garden.

"Smell my Flowers." (The author's daughter, Dorothy, and Jim, a good specimen of Basuto boy.)

Experience has shown that Kimberley has special attractions as a health resort in spite of the occasional intense heat of its summer days and the blasts of its high winds laden with dust. It has the pure atmosphere of the high karoo plateau, and even in the hottest days the bright starlight evenings are usually cool and refreshing, inviting the people to live for the greater part of the year in the open air on verandas and balconies. During the winter months the nights are often extremely cold, and well protected dwelling rooms are essential for comfort and health; but during the day the atmosphere is commonly clear, and so still that the severity of the cold is not felt, and all kinds of active outdoor exercise are agreeable in the bright sunlight of the unclouded skies. It is noted by the medical officer of health in Kimberley that the number of days of unbroken

Kimberley Race Course.

The Grand Stand, Kimberley Race Course.

sunshine are particularly enjoyable to newcomers. They will find that the air they breathe is never heavy, damp, or oppressive, but always dry and light, and, outside of the centre of the town, pure and invigorating. The heavy thunderstorms that occasionally occur bring deluges of rain, but the water rapidly flows off the surface, and as vegetation is scanty, the soil remains exceptionally dry.

It is this marked climatic attraction which, in connection with the pleasure resorts of the city, suggested the establishment of the Kimberley Sanatorium.

Beaconsfield

Dutoitspan, as before noted, was the original town on the Diamond Fields. When crowds flocked to the Fields and a demand for greater accommodation arose, the London and South Africa Exploration Company laid out the town of Beaconsfield, which adjoins Dutoitspan on the north. It was laid out as a business town, and has grown to be a place of considerable size containing several thousand inhabitants.   The town limits extend to the farm Dorstfontein, but the business and residence quarters are all within the farm Bultfontein.   The main street in Beaconsfield leads direct to Kimberley.   Many of the houses are of brick and iron, but the larger number are of unburned adobe brick, made of clay dug directly from the soil on which the house stands.   With few exceptions all are unpretentious, one-story buildings.

The town originally belonged to the London and South Africa Exploration Company, the organization which laid out the town, but together with all that company's property passed into the hands of the De Beers Company in 1898.   According to the common practice houses are put up by the tenants on lots leased from the Company.   Beaconsfield is laid out in wards, and has a distinctive Municipal Government of its own, consisting of a Mayor and Town Council and the usual town officers. The Mayor is a member of the Council and elected annually. Although Beaconsfield has thus a distinctive individuality, the business firms are very largely branches of corresponding firms in Kimberley.   The town transacts considerable business, chiefly in stocks which are carried for the use of the mines ; but there is also a large number of shops which carry supplies of all kinds for the consumption of the white residents as well as for the native population which lives in locations near the town.

Wesselton

Close adjoining to Beaconsfield lies the little village of Wesselton.   This was laid out by the owner of the Wessels estate on Benaauwdheidsfontein farm.   Its buildings resemble those of Beaconsfield, but are commonly of a poorer order of adobe brick structures, built like the Beaconsfield houses on leased lots. Wesselton has now only a few hundred inhabitants, mostly natives and East Indians.   The natives are chiefly workers for debris washers about Dutoitspan and Bultfontein mines, while the East Indians are commonly kitchen gardeners and small shopkeepers and pedlers. The various vegetables that are raised are sold in the little greengrocer stores, or hawked about by the pedlers in handcarts. Some of the East Indians also peddle clothing and knickknacks more or less industriously.

Kenilworth

On Kenilworth farm, about two and a half miles from Kimberley, the so-called model village of Kenilworth is built. This

Bachelors' Quarters, Kenilworth Village.

village was planned in the latter part of 1888 by Mr. Rhodes, and laid out under his general direction by the late Mr. Sydney Stent, an architect then residing in Kimberley. It covers a space about half a mile long and a quarter of a mile wide, upon land owned by De Beers Consolidated Mines.

The land was divided up into lots of about 80 by 100 feet, and upon these lots semi-detached houses were built, of brick with corrugated iron roofs, by De Beers Company. Nearly all of the houses are built of red burnt brick made at the brickfields in the neighborhood of Kimberley. The cottages rent from £2 10s. to £5 a month. The houses occupied by the

Passenger Train for Workmen and their Families.

unmarried men contain six rooms, and the other houses occupied by families contain from four to seven rooms. In the unmarried men's quarters each occupant has a room to himself. Nearly

The Road to Kenilworth, and Kenilworth Reservoir.

all of the houses are built with verandas, and all the lots are planted with fruit trees, vines, and flowers, supplied by the Company. Most of the residents take a keen interest in their gardens and have added largely to their beauty by purchasing plants on their own account. In laying out the town, the village was originally planned with four main avenues, bounded on the north by a bordering avenue, on the south by the main road to Kimberley, and intersected by a central avenue. Only two of the avenues are at present completed. They are broad, well-made roads lined with blue and red eucalyptus, beefwood and pepper trees, and provided with wide sidewalks fronting the semi-detached villa-like residences. These avenues are finely macadamized and the streets watered by distributing carts.

Kenilworth Club-house.

Supply pipes are laid out along the streets and every garden is supplied with free water for irrigation from the Premier mine or Kenilworth reservoir. Separate pipes are laid to carry water for drinking purposes, and for this water a light charge is made, averaging about 10s. for 1500 gallons.

A circle at the junction of No. 1 and Central Avenues divides the residences of the married people from the quarters of the single men, who occupy a row of houses on the south side of the circle in the heart of the village. One of the main houses on this circle is occupied by the Cape Government for a post-office, telegraph-office, and post-office savings bank. On the other side of the circle bordering on the central avenue is a club-house, a large brick building containing a reading room, dining room, kitchen, and manager's rooms. This building, like the residences, has a veranda in front, and is surrounded by trees. It is open to any white employé of the Company, but it is, of

Kenilworth Village, the School in the Foreground.

course, principally used by those living in Kenilworth. Citizens of Kimberley may visit it, and join in the social gatherings arranged by the residents in the village. On the north side of Central Avenue, opposite the club-house, is a schoolhouse containing three rooms, in which the library of the town is placed, and this is open after school hours for the distribution of books. The library has its own store of good books, but in addition to this stock, the Kimberley library contributes books by special arrangement, and it is practically operated as a branch of the Kimberley library. The school of Kenilworth is a primary school connected with the Kimberley public schools, and the children of the village are taught the usual elementary studies ranging up to the common English grammar school. When this grade is attained, arrangement is made for the attendance of the children at the higher schools in Kimberley. To assist them, the Company provides free monthly tickets to and from Kenilworth via the Kimberley-Kenilworth tram line.

The village is wholly given up to residences ; there are no stores or shops of any kind. All supplies come from Kimberley, and by special arrangement the schoolhouse is used on Sundays for worship and mission work, and on evenings during the week by the various philanthropic and social organizations. The village is lighted by a few large arc lamps, and the houses by paraffin candles and kerosene.

Arrangements are made by which the unmarried men take their meals at the club-house at a cost averaging about 25*s*. a

Kenilworth Village, with Meteorological Station on the Left.

week. The men come in from their work to dinner, which they take in the dining rooms of the club ; their breakfast and lunches are sent out to the depositing floors or other places of work. The breakfast at the club is like that served at the better class of miners' boarding houses at Kimberley, consisting of bacon and eggs, chops, or steaks, or other substantial dish, bread and butter, coffee or tea. Lunch consists usually of a meat dish with bread, vegetables, fruit, and tea or coffee. Dinner is the main meal, at which roast beef, roast mutton, and vegetables of all kinds are served.

The shade and fruit trees of Kenilworth and adjoining plantations are the special pride of the village and of the De Beers Company, which has been indefatigable in introducing, acclimatizing, and maintaining every variety that will thrive. Just

View of Kenilworth Village.

adjoining Kenilworth on the north is the orchard of the Company, containing about 8000 trees, — oranges, lemons, apricots, peaches, plums, pears, apples, quinces, and other fruits, as well as shade trees and grapevines. Most of the grapevines are trained on trellises. The first one built by the Company was 975, and the second 1800, feet long. On these trellises all the best varieties of grapes are grown. The ripening season is from the end of December until the end of February, or during the summer months of a season stretching from October to May. Grapes and fruit from these orchards are largely distributed to employes, and sent to hospitals and charitable institutions. Some fruit is sold in the compounds to natives at a price hardly reaching the cost of production. At times apricots have been sold at a shilling a hundred from the trees, and for sixpence when they were picked off the ground. In favorable seasons trees and vines are very prolific.

The difficulties met with in raising fruit are frost in the early part of the season, when the trees are blossoming, and hail-storms

Preparing Trenches for planting Vines and Trees.

in the beginning of the year, when the fruit is young. Locusts come in millions and at times devastate the whole orchard, leaving the fruit exposed to the sun and at times badly eaten. There are two kinds of these locusts: one comes and stays for a day or so, doing what damage it can for the time being; the other one alights on the trees for permanent occupation. They first appear in the early spring as small insects. The little dark-brown, wingless creatures are commonly known as voetgangers (walkers), and come out of the ground when they are hatched, hopping along in countless myriads. The locusts plant their eggs in the sands to hatch during the months of September and October. Sometimes all Kenilworth and the adjoining fields are swarming with these insects. In order to protect some of the gardens from young locusts, sheets of corrugated iron twenty-six inches wide are placed along, and leaning against, the fences. The locusts cannot climb up the smooth surface of the iron. In that way many residences are also protected. Sometimes servants are employed continually from morning till night in driving away the insects. They destroy all the vegetation over which they pass. The natives are very fond of eating them. They go out into the veld in large parties, and drive the voetgangers from all directions upon blankets, and then empty them into sacks which they carry to their huts. Flying locusts develop in about six weeks from the dark-brown little insects. The other variety that scourges the fields is a species of locusts with red wings, and their damage is the greater from the fact that they stay in one place until every green plant upon which they alight is destroyed. Swarms of these locusts occasionally appear, at times darkening the horizon, and following the wind. For the past seven years these swarms have been very troublesome. During one season, after consuming all the leaves, the leaf and fruit buds on the trees were entirely eaten off by these pests, destroying the fruit not only for that year, but for the following season. In spite of these drawbacks to fruit raising, the efforts of the Company have been unflagging.

CHAPTER XVI

The Diamond-bearing Deposits

VER the basin now extending as an arid karoo for hundreds of miles to the south of the Kimberley Diamond Fields the waters of a great lake once spread. It is apparent that the diamond mines are on the northerly rim of this basin, for the beds of shale that everywhere underlie the basaltic trap surface or country rock are notably thinner in the northern mine openings than they are farther south at Bultfontein and Dutoitspan,[1] and shortly after passing Kimberley fields the shale terminates at the edge of the "bed rock" of the Vaal River diggings, an amygdaloidal trap which Dr. Stelzner[2] determined to be olivine diabase.

By the great open excavations and the extension of the underground workings, the rock formations of the karoo basin are very clearly revealed. The red soil that covers the surface of the country to the depth of from one to five feet is evidently the result of the decomposition of the friable face of the underlying basalt, which is scattered in fragments over the country in jutting boulders and rounded stones. This rock at De Beers and Kimberley mines is from twenty to ninety feet in thickness, but very much decomposed throughout. Below the layer is a bed of black shale, ranging in thickness from two hundred to three hundred feet. In this bed there is a considerable amount of carbon and a large quantity of iron pyrites.

[1] "Diamonds and Gold in South Africa," p. 19, Theodore Reunert, M.E.
[2] Dr. A. W. Stelzner, Professor of Geology at the Freiberg Mining Academy.

Geological Sections of
DE BEERS AND KIMBERLEY ROCK SHAFTS.

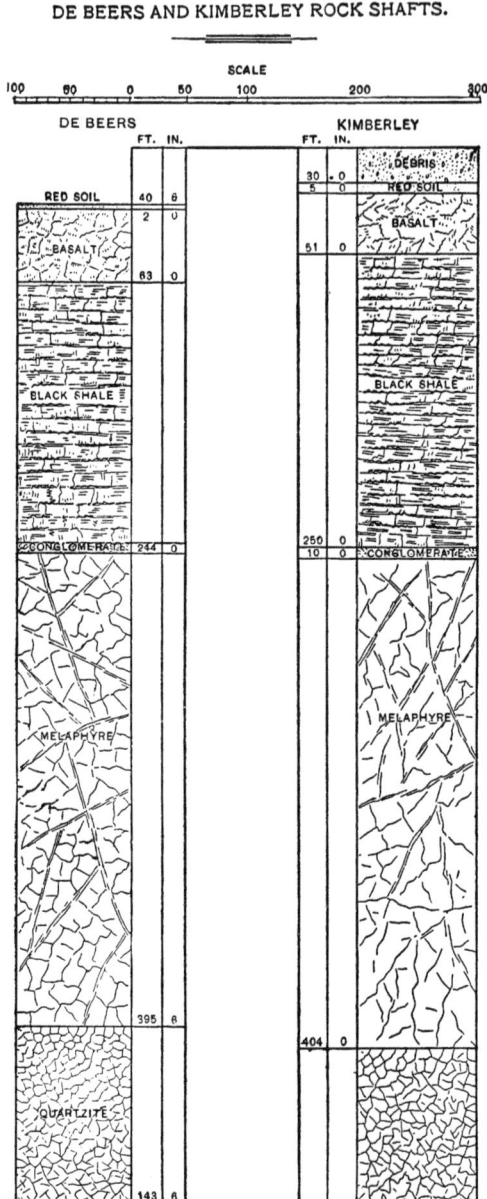

Underlying the shale is a thin bed of conglomerate, composed of small stones, some well rounded and others angular, and firmly cemented together. Its thickness, measured in the rock shaft in the Kimberley mine, did not exceed ten feet. This band has been styled by Professor A. H. Green the basement conglomerate of the Kimberley shales,[1] and it is assumed by Mr. E. J. Dunn to be of the same origin as the Dwyka conglomerate belt on the northern base of the Zwarte Berg and Witte Berg mountains, forming the southern boundary of the old lake basin.[2] He holds that this conglomerate is a glacial deposit marking the shore line of the ancient lake.

Below the conglom-

---

[1] "A Contribution to the Geology and Physical Geography of the Cape Colony." Quar. Jour. Geol. Soc., Vol. 44 (1888), p. 245.

[2] "On the Mode of Occurrence of Diamonds in South Africa." Quar. Jour. Geol. Soc., Vol. 30 (1874), pp. 54-59.

erate is a very hard amygdaloidal rock, called melaphyre by M. A. Moulle,[1] which was finally determined by Dr. Stelzner[2] to be olivine diabase. Its mineral composition is the same as melaphyre, — plagioclase, augite, and olivine, but one is granular and the other porphyritic. It is about four hundred feet in thickness and is very hard. Underlying the amygdaloidal rock is quartzite, the thickness of which is not yet determined. The Kimberley rock shaft has passed through fourteen hundred feet of it, and the bottom of the shaft is still in the same formation. All these strata lie nearly horizontal, but dip slightly to the north. They are graphically presented in the sectional views of the rock shafts of De Beers and Kimberley mines shown on page 480.

Through these layers of rock extend from an unknown depth the huge pipes containing the diamond-bearing deposits, or blue ground, which is a breccia filled with fragments of shale and other minerals. These immense funnels are obviously extinct craters filled with volcanic mud from below. All evidence to hand points to an aqueous formation, and the upheaval is shown by the upturning of the enclosing shales at various places in contact with the blue ground.[3] Many boulders are found in the blue ground of the same composition as the surrounding rock, but others have undoubtedly come up from greater depths than have yet been reached by the sinking of shafts. It is, however, highly remarkable that there was almost no apparent overflow in the filling of these craters, for the diamond-bearing ground is either level with the surrounding surface, or rises, usually, only a few feet above it in kopjes or hillocks. Outside of the mouths of the craters no diamonds have been found except at Dutoitspan, where the upheaval formed quite a hill, and some diamonds have been taken from the surrounding ground within a few yards from the margin of

---

[1] " Mémoire sur la géologie générale et sur les mines de diamants de l'Afrique du Sud." Annales des Mines, 8th Series, Vol. VII (1885), p. 193.

[2] Dr. A. W. Stelzner, Professor of Geology at the Freiberg Mining Academy.

[3] Still to be seen at De Beers Mine.

the mine.   It is also evident that the mines were not all fi led with the same material at one and the same time.   Each mine

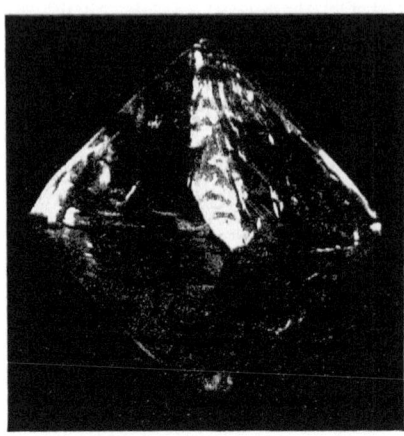

has its distinctive character-istics, and even in the same mine all the blue ground does not seem to have been forced up at one time.   This is par-ticularly demonstrated by the striking fact that, in both De Beers and Kimberley mines, the west side blue ground is wholly unlike the other por-tions of the mines, and carries fewer diamonds, and these are unlike the diamonds that are found in other parts.   The

De Beers Diamond.   Found March 28, 1888. Weight, 428½ carats.

blue ground which filled the west ends of these mines must have come up first, filling the whole crater.   Afterward there was a second upheaval which filled the eastern parts of the craters with a richer deposit.   The reason why the west end was not mixed with the better blue ground was because the west end parts of the mines formed benches, and were not ver-tically above the second boiling mass.   Mr. Rhodes suggested this solution, and I quite agree with him.   This peculiarity is noticeable in the other mines.

Another View of the De Beers Diamond.   After cutting, its weight was 228½ carats and it was sold for £13,600.

The composition of the blue ground, which is the principal filling of the volcanic pipes, has been carefully determined by

Dr. Stelzner. This ground, he says, must be designated as a breccia. Most of the small or large angular-edged or rounded fragments of this breccia are composed of a green-black or blue-black serpentine-like mass. Fragments of rock which are found in the karoo formation, such as sandstone, shale, and diabase, are to be found in the blue ground. There are also other rocks in the shape of boulders of greater or less size, which are not known in the karoo formation, and have doubtless come from a much greater depth than the karoo beds, possibly from rocks upon which these beds lie. The mass of blue ground consists of olivine more or less changed by oxidation, with the following minerals: chromic diallage, bronzite, pyrope containing chromium, flesh-colored zircons (locally called Dutch boart), cyanite, biotite, chrome, titanium, and magnetic iron, and also small crystals of perofskite.

The Largest Diamond ever found in the Kimberley mine. It weighed 503 carats, but was full of spots.

In the Jagersfontein blue ground corundum is said to have been found. This was for a time held to be cordierite. The existence of small crystals of tourmaline and rutile is also reported. Professor J. G. Lawn, Kimberley School of Mines, reports that he discovered rubies and sapphires of inferior quality in the Frank Smith mine near Kimberley. Iron pyrites and barytes are found in the deposit resulting from washing the blue ground. The pyrites come from the country rocks, and become mixed with the diamond-bearing ground during the process of mining. The barytes is a secondary formation of small veins in the blue ground, or at its junction with the country rock. Beautiful crystals of doubly refracting or Iceland spar are occasionally found also near the junction of the blue ground and the rock.

In Professor Lewis's discussion of the genesis of the diamond in 1886, he designated the blue ground variously as "dunite porphyry," "Saxonite porphyry," and "diamantiferous peridotite." His application of the term "Kimberlite," now generally accepted by geologists, first appears in his paper of the following year, 1887, at the British Association meeting at Manchester.[1] Dr. Stelzner thought this name should be adopted as concisely covering "a porphyritic volcanic peridotite of basaltic structure."

In the mass of diamond-bearing blue ground in De Beers mine there is a curious dyke of igneous rock which extends from the southeast part of the mine around the east and north sides, and is lost in the unexplored poor blue ground of the west. Owing to its taking a serpentine course across the mine, it has received the local name of "snake." The upper end of this snake is at or near the surface, and the body extends down to the lowest workings. It does not adhere to the blue ground, and is very easily separated from it. It stands like a vein, nearly vertical, varying in thickness from two to seven feet. No diamonds have been found in it, yet Dr. Stelzner's investigations show that its composition is substantially the same as the surrounding breccia. It was difficult to obtain slides of the blue ground for microscopical observations and comparison, but after many trials Dr. Stelzner succeeded in getting a few sections which revealed these interesting facts : —

"The main body of the blue ground is entirely analogous to the snake rock, naturally more decomposed, but in essential points the microscopic features of blue ground and snake (not taking into consideration the numerous little slate fragments in the blue) are in an extraordinary degree alike. It therefore impresses upon one's mind that the "snake" is a younger eruptive formation coming from the same volcanic source as the blue ground."[2]

[1] "The Matrix of the Diamond," Henry Carvill Lewis, M.A., F.G.S., Professor of Mineralogy in the Academy of Natural Sciences, Philadelphia, U.S.A., at meeting of British Association at Manchester, August and September, 1887.

[2] Letter of Dr. Stelzner addressed to Gardner F. Williams.

PIECE OF BLUE GROUND.
Showing Diamond embedded in it.

On the 1000-foot level of Kimberley mine a tunnel driven in the quartzite outside the margin of the mine shows several dykes of similar rock. Wherever these dykes exist there is a considerable quantity of water at the junction of the dykes and quartzite.

There was a large mass of country rock in De Beers mine, which in the upper levels covered several claims, or approximately an area of 3000 square feet. It continued down to a depth of about 750 feet. It was an olivine diabase, and was the same as the amygdaloidal rock, except that it was filled with numerous veins of zeolites. The " Island," as it is called, was a gigantic horse of country rock embedded in blue ground, and has disappeared in depth. Islands of the same rock appeared in the Kimberley mine near the surface and at a depth of 1200 feet, and near the surface in Dutoitspan, Bultfontein, and Premier mines, where they have been left standing as the blue ground which surrounded them has been removed, and form huge islands in a sea of blue ground, which are locally known as Mount Ararats.

Floating shale appeared at or near the surface of the mines and covered many claims. This was originally volcanic mud, and it contained no diamonds. It gradually became smaller in depth, and has disappeared in the lower levels.

In the early descriptions of the mines fossil wood and plants are reported to have been found in the blue ground. I am of the opinion that these came either from the shale surrounding the craters, which was constantly falling into the open mines, or from the pieces of shale which became embedded in the blue ground at the time the craters were filled. The only fossils which have been found in the mines since they have been under my management are the fish which are shown in the illustration on page 486. They are embedded in sandstone which was found on the 185-foot level of Premier mine.

The surface shales and basalt surrounding the pipes are called reef, and the masses of shale and igneous rocks, scattered through the blue ground in the upper levels of the mines, are commonly spoken of as floating reef.

After careful microscopical observations, Dr. Stelzner and others have reached the conclusion that the blue ground is of volcanic origin, and was forced up from below. This conclusion accords with the opinion which I formed of the origin of the

Fossil Fish from Premier Mine.

diamond-bearing deposit, during my visits to the Diamond Fields in 1884 and 1885. I then thought that the filling of the pipes was due to aqueous rather than igneous agencies, possibly to something in the nature of mud volcanoes.

The Genesis of the Diamond

The chemical composition of the diamond has long been determined, at least approximately. Sir Isaac Newton conjectured it to be of vegetable origin and combustible, but it was not until 1694 that Newton's assumption of its combustibility was actually proved by the famous burning glass experiment of the academicians of Cimento, at the prompting of the Grand Duke Cosmo III.

Lavoisier, Guyton de Morveau, and others practically determined, later, that the burning of a diamond with a free supply of oxygen converted it into carbon dioxide; and, finally, the experiments of Sir Humphry Davy, in 1816, showed that the diamond was almost entirely pure carbon. Davy's conclusions

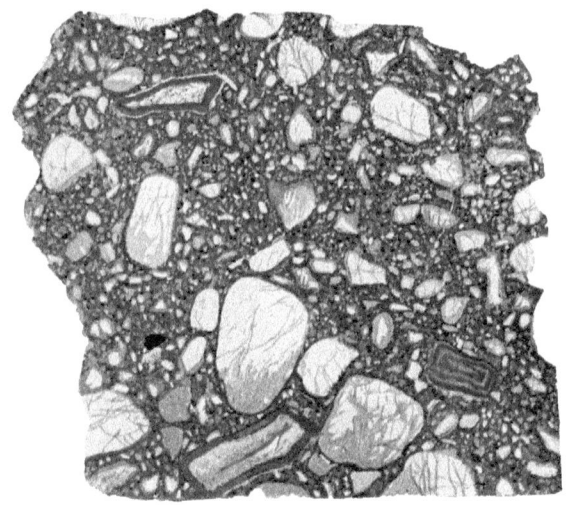

A Thin Section of Diamond-bearing Rock (enlarged 3½ Times) from the
1320-Foot Level of the De Beers Mine.

By kind permission of Sir William Crookes.

have been confirmed by Dumas, Stas, Friedel, Roscoe, and other eminent chemists, apparently fixing with extreme precision the chemical composition of the diamond. It is, however, noteworthy that the diamond is a non-conductor of electricity, while graphite and charcoal, substances so closely similar in chemical composition, are excellent electrical conductors. By the application of friction the diamond can be positively electrified, but Streeter says that it loses its electricity completely in the course of half an hour.[1]

So much, it may be claimed, we know; but the process of the formation or crystallization of the diamond carbon is still uncertain. The proofs are most conclusive that the diamonds in the South African mines were not formed *in situ*, but have come up from below with blue ground. The frequent occurrence of broken crystals embedded in the blue is sufficient evidence that the diamonds are not in their original place of crystallization, for it is impossible for nature to produce a fragment of a diamond.

The late Dr. W. Guybon Atherstone, F.G.S., whose identification made known the first diamond of the South African Fields, presented his theory at a meeting of the South African Geological Society, as follows : —

" The succession of the strata in the Kimberley mine is precisely the same as that of the lacustrine sedimentary beds, beginning from the quartzite base of the carboniferous rocks and shales, through the ecca and karoo formation, the coal-bearing shales of the Stormberg, to the dolerite, capping and protecting the surface, as proved by the rock shaft recently sunk out of the influence of the Kimberley mine to a depth of one thousand feet, where a thickness of four hundred feet of amygdaloidal lava with the trappean ecca conglomerate above it represents the prevailing rocks of the Vaal, Riet, and Orange rivers for a great distance below Hopetown. Incredible as it was deemed at the time, my story of the small rounded river stone which fell out of the unsealed letter placed in my hands by the post-boy, has since

[1] " Precious Stones and Gems," p. 58.

proved to have been the key that has unlocked the vast underground wealth of South Africa.[1]

" The story I have now to tell of its birthplace and subsequent history will, I know, appear still more incredible, as fabulous indeed as was that of Sindbad, the Arabian voyager, who,
with the talisman and magic lamp of Aladdin the Seer, unlocked
the caverns of Africa's fairy land, and viewed in prophetic vision

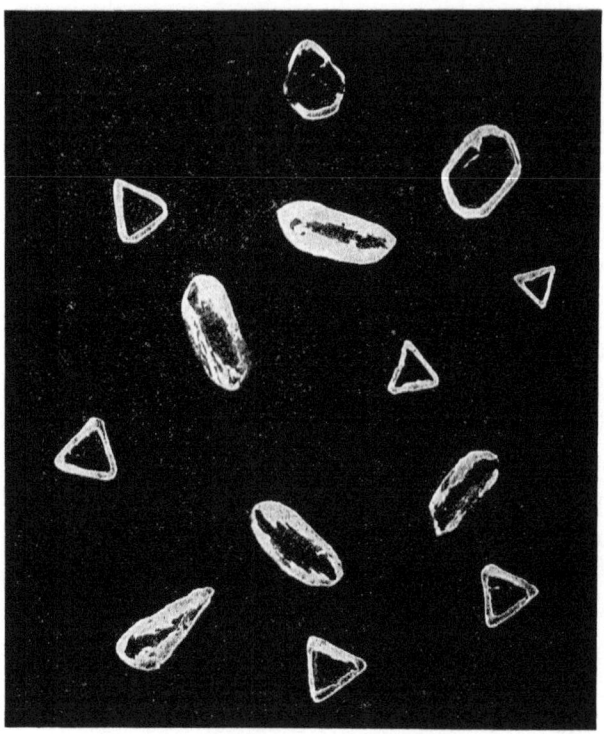

Irregular Crystallization of Diamonds.

the vast stores of buried treasures, — gold, diamonds, and other
gems, — just as we see them now with our magic electric lamp a
thousand feet down in the dark recesses of the extinct volcano,
yielding millions of the purest gems upon earth.

" How came the diamond there in its hard blue matrix of
ashes and lava, with its accompanying gems, — garnets, rubies,
sapphires, agates, and other gems, — the products of solution and

[1] *Geological Magazine*, Vol. VI, p. 208, May, 1889.

heat? For a substance to crystallize, its molecules must be free
to move under polarizing and other metamorphic forces influenc-
ing crystallization; but the diamond we know is neither soluble

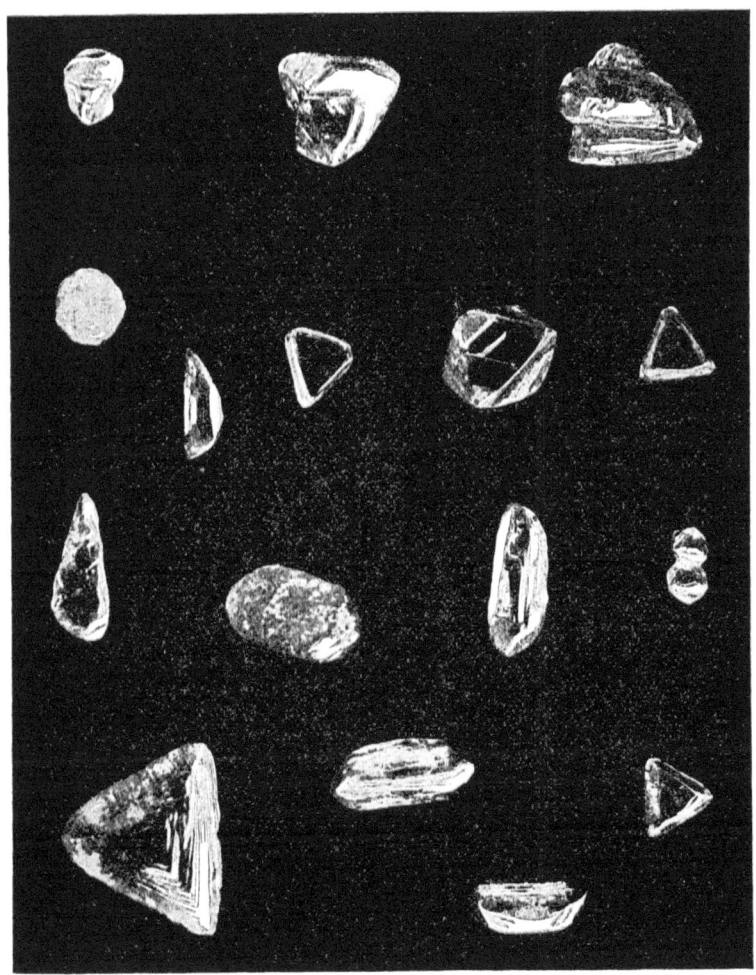

Irregular Crystallization of Diamonds.

nor fusible. It is the element carbon crystallized, and is consumed by heat. How, then, could it survive as a crystal in the
crater of a volcano?

"The key to solve this mystery was placed in my hand over
half a century ago, by one of the greatest philosophers of the

age, whose lectures I had the privilege of attending. But it was not until I had examined a diamond mine in South Africa and speculated upon the apparently irreconcilable phenomena attendant upon the origin of the diamond in its matrix, that the practical application of Faraday's discovery began to dawn upon me. 'Hold out your hand,' said he, at the close of the lecture that fairly electrified the world of science, as with a loud hiss a snowy substance, burning like a coal but in reality intensely cold, escaped into the palm of my hand from the strong iron vessel in which, with a pressure of fifty atmospheres, he had liquefied carbonic acid gas — the very gas resulting from the combustion

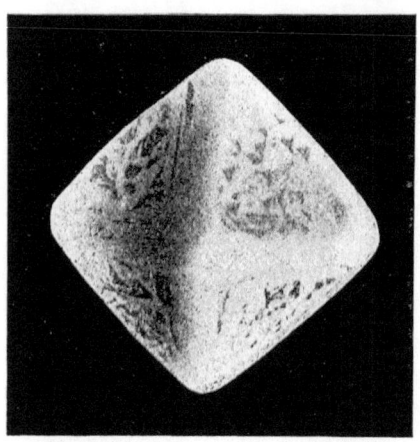

Crystal of Diamond, showing Rounded Edges.

of the diamond, consisting of one atom of carbon and two of oxygen.

"I have shown that the sedimentary beds deposited from this vast freshwater lake attained a thickness of about eight thousand feet. The lake itself, therefore, probably equalled that depth. (?) Now the experiments of Wyville Thomson and Carpenter, made during the voyage of the *Lightning* and the *Porcupine*, proved that at a depth of three to four hundred fathoms, the pressure is equal to half a ton on the square inch; at a mile to one hundred and fifty-nine atmospheres, and at seven thousand feet it amounts to two hundred atmospheres, or four times the pressure under which Faraday liquefied carbonic acid gas, the temperature at such great depths being very few degrees above freezing point. In the carbonic acid gas generated from the carbonaceous shales by heat, and interspersed as gas bubbles in the cavities of the viscid, ferruginous amygdaloid, and in the admixture of steam, lava, and ashes known as the 'Kimberley Blue' — reduced to the liquid state by the enormous pressure in the subaqueous volcano — we have the constituents of the diamond in a form admitting of crystallization, and the subsequent absorption of its oxygen by the iron always present in its containing walls during long intermittent periods of volcanic inactivity. There are proofs in the Kimberley mine that such alternating periods of activity and repose have occurred at long intervals, as shown by the four or

Twin Crystal Formations.

five distinct and separate layers of diamonds lining its walls, of varying size and quality, known and recognizable by diamond buyers."

In this presentation, which Dr. Atherstone seemingly regarded as conclusive, there is a lack of the clear, logical reasoning which in other discussions has distinguished his views. He dogmatically puts the carbonic acid gas evolved from the carbonaceous shales into cavities of the amygdaloidal rock which lies outside of the volcanic pipes.   Then he reduces this gas by enormous pressure to a liquid state, and, having gotten it into a form, as he thought, admitting of crystallization, he absorbs the oxygen of the carbonic acid by the iron in the containing walls of the craters.   Now, as a matter of fact, there are no cavities in the amygdaloidal rock underlying the shales, for all interstices are filled with silica in the form of agates, or with calcite.   Furthermore, if carbonic acid had been left in the olivine diabase to crystallize, then the resultant diamonds would have been enclosed in this formation, which is also contrary to fact, for no diamonds have ever been found in the amygdaloidal rock.   His main contention, too, is the derival from the shales of the carbon necessary for the formation of diamonds.   It will be made clear, subsequently in this discussion, that this assumption is not justified.

The late Henry Carvill Lewis, M.A., F.G.S., Professor of Mineralogy in the Academy of Natural Sciences, Philadelphia, U.S.A., advanced the proposition that the diamond is the result of the intrusion of igneous rocks into and through the carbonaceous shales, and the crystallization of the carbon throughout the rocks, as it cools, from hydrocarbon, distilled from the shales that had been broken through.[1]

In support of such a theory, it is claimed that the diamonds in the various mines or pipes have different characteristics.   It is quite true that large parcels of diamonds from the various mines have distinctive characteristics, and it can be easily told from which mine a parcel of diamonds comes ; but it is very difficult to tell in which mine a single stone may have been found, though each mine has stones in a great measure peculiar to itself.   Some observers claim that the broken diamonds which are extracted are broken during the process of winning them.   It is admitted that diamonds may be broken in the process of mining and the subsequent operations of winning, but these cases are exceptional.   Fragments of diamonds are very frequently found embedded in the blue ground, and there

[1] "Gems and Precious Stones of North America," George F. Kunz.

is no doubt in the mind of any one who has had practical experience in finding these fragments that they were not crystallized where they are found.   The fact that no diamond has ever been found embedded in the shale itself strikes one as conclusive proof that Professor Lewis's theory is wrong.

Diamonds of Regular Forms.

Again, would not the intrusion of an igneous rock through carbonaceous shales have altered these shales in the vicinity of the igneous rock?   There is, however, no difference that can be detected between the shales at the junction of the pipe and at a distance of one thousand feet.   Moreover, would not the fragments of shale enclosed in the blue ground have changed, and have lost the carbon which they contain, if diamonds were formed from them?   One sees no difference between the shale which forms the country rock, and the fragments embedded in the blue ground.   If such a theory as is attributed to Professor Lewis by Mr. Kunz had a shadow of foundation, it is dispelled by the occurrence of diamonds in the Jagersfontein mine in the Orange Free State, some eighty miles from Kimberley. In this mine there are no carbonaceous shales surrounding the diamond-bearing deposits.   The pipe, as far as developed, is in quartzite, and it is apparent that the shales never existed here, or were denuded before the formation of the diamond-bearing pipe. If such denudation had taken place after the filling of the pipe with a diamond-bearing matrix, the alluvial deposit of the country surrounding this mine must contain diamonds, but no such discovery of diamonds has been made.

The Jagersfontein mine is not the only diamond-bearing pipe that has produced diamonds without having shale as a country rock. Other pipes or veins have been found both in the Free State and the Transvaal, which are, however, of little commercial value, owing to the small quantities of diamonds found in them, but they are most useful in refuting existing theories, if not in the determination of the genesis of the diamond.

An important contribution to this discussion was made by Professor Molengraaf, state geologist of the South African Republic, in a monograph on the diamonds at Rietfontein in the Transvaal. He stated that " the diamond-bearing breccia on the farm was of the same nature as the well-known blue ground of the Kimberley mines. The geological position of the volcanic chimney at Rietfontein was very different from that of the other diamond pipes in South Africa. The latter, of course, all occurred in a higher or lower horizon of the karoo formation, whereas the chimney at Rietfontein seemed to occur in the upper parts of the Pretoria beds in a system of strata overlying the Magaliesberg quartzite. If that position, which was almost certain to his mind, was proved to be correctly determined by a later and more careful geological survey of the surrounding country, this fact would be of high importance in the discussion of the genesis of diamonds. Of the different theories regarding this genesis he would only mention three principal ones.

" He would take up first the theory agitated by Messrs. Stanislas Meunier,[1] M. Chaper,[2] and in a somewhat modified form

[1] " Composition et origine du sable diamantifère du Du Toits Pan, Afrique australe." Comptes rendus de l'Académie des Sciences de Paris. Vol. LXXXIV, No. VI, p. 250. " Examen minéralogique des roches qui accompagnent le diamant dans les mines du Cap de Bonne Esperance." Bulletins de l'Académie Royale de Belgique, 3d series, Vol. III, No. 4.

[2] " Note sur la région diamantifère de l'Afrique australe." Paris, 1880.

lately by Professor Garnier. They denied the igneous origin of
the blue ground and the diamonds in it, and considered the blue
ground to be a kind of mud, or peculiar alluvial deposit, which
had been forced up by a hydrostatic process. That theory, to
his mind, had already been proved untenable by several eminent
geologists.   The two remaining theories agreed as far as the
igneous origin of the blue ground. According to one of these,
the diamonds belong to the primary constituents of the eruptive
rock itself, and had crystallized at a great depth under very high
pressure and high temperature, before an eruption of an explosive character brought the igneous rock to the earth's surface.

"According to the second theory, which was discussed by
Mr. Harger at a meeting of the Geological Society of South
Africa, the diamonds were formed in the blue ground, during
its ascension, from carbon borrowed from the carbonaceous
shales through which the eruptive rock forced its way. Now,
that theory, although rather weak in his opinion, had been maintained, hitherto, mainly because the geological position of the
known diamond pipes was such that it could be proved, or, at
least, be accepted as very probable, that the blue ground had
forced its way through carbonaceous strata. The discovery at
Rietfontein deprived that theory of its strength. As already
pointed out, the chimney at Rietfontein was found in the upper
Pretoria beds. But in the Pretoria beds, as well as in the formations underlying these, strata containing any notable quantities of
carbon were nowhere to be found in the Transvaal ; so that the
conclusion might safely be drawn that the igneous blue ground, in
forcing its way from great depths toward the place where it was
found, could not borrow any carbon from the surrounding strata
in order to convert it into diamonds. The discovery at Rietfontein might afford a valuable argument in favor of the formation
of diamonds as a primary constituent in breccia, or ultrabasic
magma at great depth, and geologists were entitled to derive
from it an argument in favor of the following more general
thesis : 'The elements of carbon, under the conditions of heat
and pressure ruling at great depth in the interior of the earth,

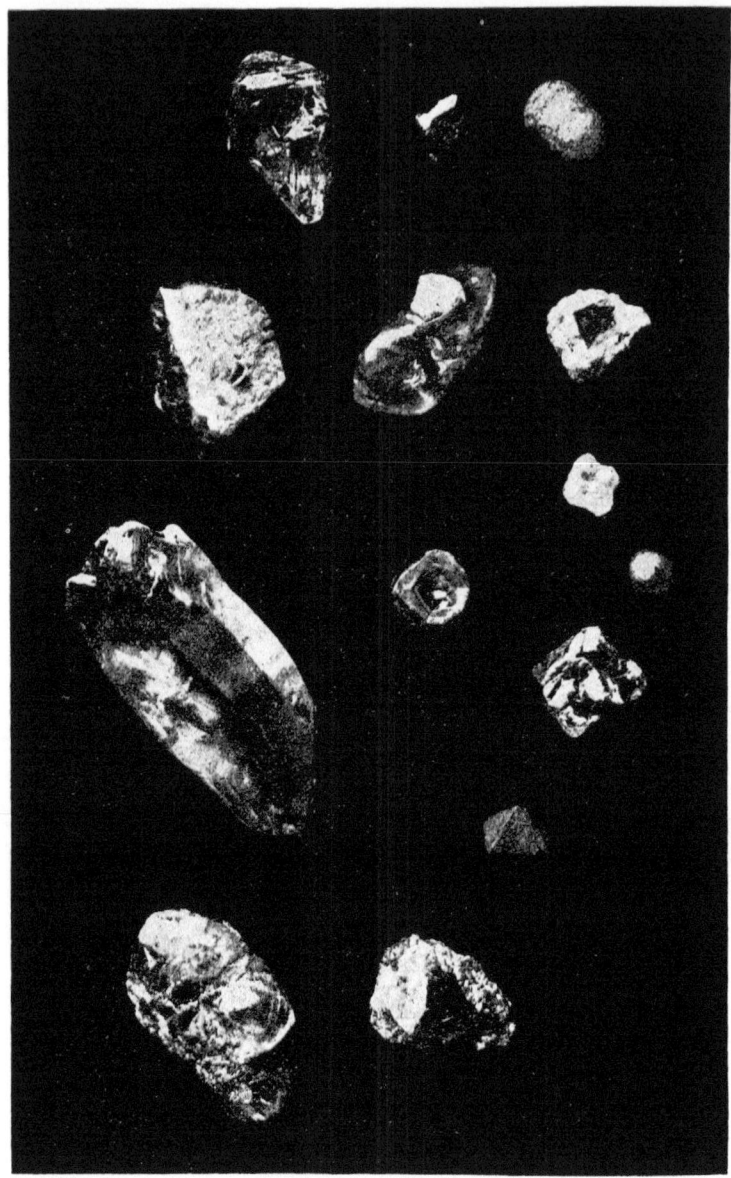

Diamonds of Irregular Forms.

can only exist and crystallize in the modification called diamonds.'
This thesis was, of course, in perfect harmony with the latest
scientific discoveries, especially with the famous experiments of
Moissan."

It was the opinion of the late Dr. Stelzner that the diamond was crystallized at great depths and came up with the magma or matrix.   The following liberal translation from a lecture delivered by Dr. Stelzner before the Isis Society in Dresden on April 20, 1893, gives the views of this celebrated geologist: —

" Before I give my own opinion, may I be allowed to recall three well-known geological facts : first, that various minerals which compose many of the eruptive rocks, for instance the olivine of certain basalts, contain liquid carbonic acid, and we must come to the conclusion that the molten magma under some circumstances must have been impregnated with carbonic acid ; second, that the blue ground of Kimberley, as already mentioned by Lewis, has a known resemblance to many meteorites ; and, third, that a modified form of carbon, besides graphite, similar to the diamond, has been met with recently in meteorites.

" If we take these three facts into consideration, and also remember that in most of the localities in which diamond-bearing alluvial deposits appear (Ural, India, Borneo, New South Wales, and in the United States), serpentine (especially peridotite) is to be found, we come to the conclusion that the carbon of the diamond itself crystallized when this molten mass, rich in magnesium silicate, became cool.   In support of this opinion we find that in some instances diamonds and garnets (pyrope) are found together, showing that they have the same origin."

For the illumination of the problem of the formation of diamonds the experiments of Mr. J. B. Hannay of Glasgow, Professor Dewar, and M. Moissan, and later of Sir William Crookes, are of the greatest interest to the scientific world.

The conversion of a diamond into graphite was effected by Professor Dewar, publicly, in London, as far back as 1880.   Sir William Crookes repeated the same experiment in a lecture at the Royal Institution of Great Britain, on June 11, 1897, by placing a diamond in the electric arc where the temperature was 3600° C., when it was converted into graphite.

Among the first attempts to make artificial diamonds may be mentioned that of Mr. J. B. Hannay of Glasgow, who com-

2 K

menced his experiments in 1879, and after many trials, some of which resulted in violent explosions, he is said to have succeeded. The method adopted by Mr. Hannay is described as follows:—

"A tube twenty inches long by four inches in diameter was bored so as to have an internal diameter of half an inch.   In the tube was placed a mixture of ninety per cent of rectified bone oil, and ten per cent of paraffin spirit, together with four grammes (about sixty-two grains) of the metal lithium.   The open end of the tube was welded air-tight, and the whole mass was heated to redness for fourteen hours ; on opening it a great volume of gas rushed from the tube, and within was a hard, smooth mass adhering to the sides of the tube.   It was quite black, and appeared to be composed of iron and lithium, but on a closer inspection small transparent pieces were found embedded in it. The mass was dissolved, and the small transparent pieces proved to be 'crystalline carbon,' exactly like diamonds but almost microscopical.

"Out of eighty complex and extensive experiments only three succeeded.   Violent explosions were frequent, steel tubes burst, scattering their fragments around, and furnaces were blown up.   'The continued strain on the nerves,' writes Mr. Hannay, 'watching the temperature of the furnace, and in a state of tension in case of an explosion, induces a nervous state which is extremely weakening, and when the explosion occurs it sometimes shakes one so severely that sickness supervenes.'"[1]

Sir William Crookes attributes the possibility of making artificial diamonds to the facilities afforded by the enormously high temperatures which are obtainable in recent years by the introduction of electricity.   While electricity has, no doubt, played an important part in the scientific researches during the last decades of the nineteenth century, Mr. Hannay's experiments would indicate that it is not absolutely essential to have enormous temperatures or pressures to produce artificial diamonds.   Still, Sir William Crookes shows that by means of these high temperatures substances such as carbon obey the

[1] *Glasgow News.*

common laws which govern other substances, and can be made volatile and fusible under certain conditions. He has demonstrated that the temperature necessary to volatilize pure carbon is about 3600° C., and that it passes into the gaseous state without liquefying, and he infers that, if sufficient pressure were applied with the high temperature, liquid carbon would be produced which upon cooling would crystallize in diamonds. For this product the absence of oxygen is absolutely necessary, as the carbon would readily unite with it in the form of carbonic acid. It is a well-known fact that iron when melted dissolves carbon, and while Moissan discovered that other metals effect this dissolution, he found that iron was the best solvent.

Sir William Crookes went through the process of producing diamonds before the eyes of his audience, but was only able to show them the result of this experiment by reproducing a lantern slide of microscopical diamonds which he had made in the same way previously, for it takes a fortnight to separate them from the iron and other substances in which they are embedded. The scientific principle upon which this experiment rests, according to Sir William Crookes, is that molten iron absorbs carbon, and as iron increases in volume as it passes from the liquid to the solid state, if the outer crust of the iron is suddenly cooled and the centre remains in a liquid state, the enormous pressure caused by its expanding while cooling affords the two factors necessary for the crystallization of the diamond — heat and pressure.

Authorities differ somewhat as to the exact moment when molten iron expands on cooling, but it is the generally accepted theory that expansion takes place at the moment of solidification. It is also a well-known fact that shrinkage or contraction takes place as the solidified metal cools. It is therefore possible to obtain enormous pressure in the molten centre of a casting by the contraction of the outer shell which has been rapidly cooled and the expansion of the inner mass just as it begins to solidify.[1]

[1] American Society Mechanical Engineers, Vol. XVIII, pp. 419 and 431. American Institute of Mining Engineers, Vol. XVII, 126 and 1015.

Sir William Crookes says further, that it has been "conclusively proved that the diamond's genesis must have taken place at great depths under enormous pressure. The explosion of large diamonds on coming to the surface shows extreme tension." According to my own experience, a diamond never explodes. Light brown, smoky diamonds often crack on exposure to the dry air, but they will remain intact if kept in a moist place. The cracking is, therefore, more probably the result of heat or drying than of tension or inward pressure. It is possible, however, that the greater heat to which the diamond is exposed when brought to the surface may expand contained gases sufficiently to crack the stone.

Sir William holds the same view of the formation of the diamond-bearing pipes which I suggested at the time of my visit to the Diamond Fields in 1885,[1]—that these pipes were volcanoes which were filled with the mixture which they now contain while it was in the form of mud. My reasons for this theory are fully set forth upon another page. Continuing in his lecture, Sir William says: "The ash left after burning a diamond invariably contains iron as its chief constituent, and the most common colors of diamonds when not perfectly pellucid show various shades of brown and yellow from the palest 'off color' to almost black. These variations accord with the theory that the diamond has separated from molten iron."

I have a collection of diamonds of all colors, see colored Plate opposite, and recently made exhaustive tests in order to ascertain whether they contained any iron either in the metallic or oxidized state. These experiments were made upon a magnetic separating machine, the field magnets of which attracted any mineral which contained iron in a metallic or oxidized state. Although some of these diamonds had the appearance of being coated with iron in some form, and others were colored dark brown and deep yellow, they were in no way attracted by the magnet even when excited by a strong electrical current. These

[1] Transactions of the American Institute of Mining Engineers (October meeting, 1886), Vol. XV, pp. 392–417.

experiments do not, perhaps, disprove the existence of iron in the diamond, but they do establish the fact that, if iron does exist in an oxidized state, the quantity is infinitesimally small.

One more theory of the deposit of diamonds in the South African fields is deserving of special mention, more for the purpose of showing to what heights of imagination the human mind may soar, than for any scientific value it may have. This is an assumption that the diamond deposits came from a fall of meteors, "a direct gift from heaven," and was first advanced to notice, it is said, by Meydenbauer. Such a theory seems highly fantastic and is the most improbable of all. The occasional inclusion of black diamonds in meteorites is well attested, but these occurrences are very far from accounting for the formation of the South African diamond-bearing deposits. "Bizarre as such a theory may appear," says Sir William Crookes, "I am bound to say there are many circumstances which show that the notion of the heavens raining diamonds is not impossible." The

"Ava" meteorite which fell in Hungary in 1846 contained graphite in cubic crystalline form which G. Rose thought was produced by the transformation of diamonds. Later Weinschenk found transparent crystals (diamonds) in the Ava meteorite.

Since it became known that diamonds (infinitesimally small, it is true, but nevertheless diamonds) occur in meteorites, a general

A Microscopical Diamond (magnified 200 times) from a Meteorite from Canon Diablo.

search has been made for the minute crystals in meteorites from Australia and Russia, and from Cañon Diablo, Arizona, and diamonds and graphite have been found.[1]

From the above facts and from observations which Sir William Crookes made at Kimberley, he concludes that the genesis of the diamonds found in the South African mines was by crys-

[1] Sir William Crookes's lecture.

tallization of pure carbon in molten masses of iron which form a part of the internal regions of the earth.

The theory that the diamonds must have crystallized in a matrix of iron is not new. That small diamonds have been produced in this way there is no doubt, and in the absence of further proof to the contrary one might assume that such was the origin of the diamond. Iron in the form of magnetite and other similar minerals forms a considerable part of the concentrates from the washing machines; but all proof that these minerals, which may have been derived from metallic iron by oxidation, were the matrix in which the diamonds originally crystallized is wanting. As a matter of fact, I am positive that neither the iron nor, as others have asserted, the olivine found with the diamonds is the original matrix of the diamond; and my assurance rests upon the fact that no diamonds, however small, have ever been found in the iron combination, or in the other minerals which accompany them, although these concentrates have passed daily under the eyes of hundreds of keen-eyed sorters for more than thirty years, and thousands upon thousands of tons have been looked over, not once, but at least four times. The pieces of the iron minerals and especially of the olivine are often very large, quite large enough to contain diamonds weighing several carats, which in many cases would have been exposed to view had these minerals been the original matrix. We must, therefore, look to other sources for the genesis of the diamond. I have been of the opinion that diamonds crystallized in very much the same way as quartz or other minerals, but under peculiar circumstances possibly of pressure and heat. Professor Crookes states that diamond crystals are almost invariably perfect on all sides. As a rule this is the case. Quartz crystals have been found which have been formed without any attachment to other substances, that is, with both ends showing pyramidal facets. The same formation may be seen in a great many other minerals, and this is usually a characteristic of the diamond, but diamonds are found which have been crystallized with some portion of the surface resting upon or adhering to some other substance. The several reproductions of the various forms and sizes of diamonds will give the reader some idea of the eccentricities of these stones.

The experiments of Herr W. Luzi[1] of Leipsic in the production of artificial figures of corrosion on rough diamonds are of exceeding interest in the light which they throw on the crystallization and the probable matrix and genesis of the diamond.

Until lately the only appearance of chemical corrosion upon the surface of rough diamonds was the regular, triangular, negative pyramids, which were produced through heating the diamonds in the open air, or under oxygen flame. Herr Luzi has succeeded in producing different and peculiar kinds of figures. He discovered that the breccia from the South African diamond mines (that is, the matrix or blue ground), when in a molten condition, possesses the property of absorbing the diamond or of changing its shape.

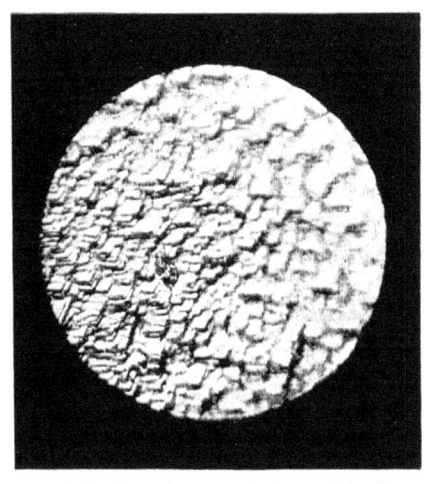

Smooth Surface of Diamond dissected by Combustion. (Magnified 100 times.) From a photograph by Sir William Crookes.

He describes his experiment as follows : A small quantity of blue ground was melted in a crucible placed in a Fourquinon-Leclerq furnace at a temperature of 1770° R., which was the highest temperature attainable. A diamond with perfectly smooth natural faces was submerged in this molten mass. A further quantity of blue ground was then added to the contents of the crucible until it was completely filled. A tightly fitting cover was placed on the crucible, which was placed in the furnace and again exposed for thirty minutes to the greatest heat attainable. When the crucible was cooled the diamond was removed and found to be

[1] "Artificial Figures of Corrosion on Rough Diamonds," *Berichte der Deutschen Chemischen Gesellschaft*, 1892.

covered with irregular oval and half-round grooves of various depths. In one experiment, the diamond was found to be deeply eaten away on one side, so that the depression nearly penetrated through the stone.

Diamonds thus magmatically corroded have a similarity, as regards the appearance of the corrosion, to hornblende and kindred materials. A small spot or scar was, at times, found at the bottom of a large indent. The diamonds were usually found, after the experiments, to be blackened, or covered with a red coating, which proved to be oxide of iron.

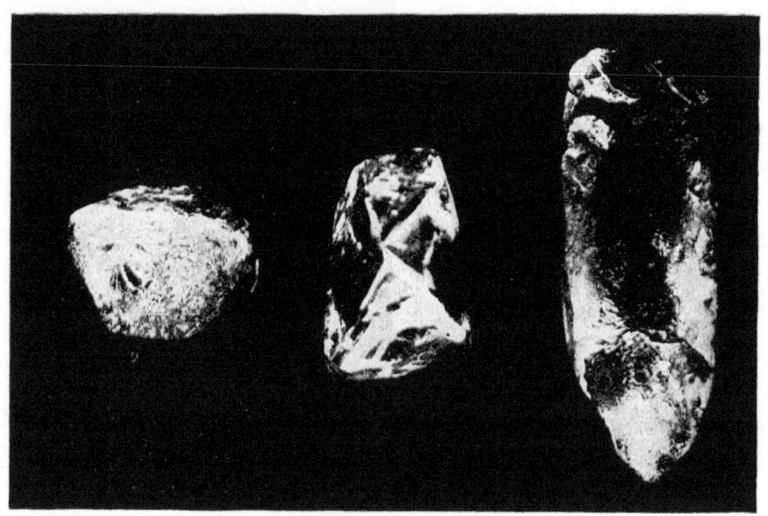

Diamonds of Irregular Forms.

Some of the diamonds showed little black or greenish black balls located exactly in the centre of the holes. The formation of the balls is doubtless connected with the creation of the grooves. These little balls are magnetic, and when treated with hydrochloric acid, in which they are only partly soluble, they evolve a gas.

The quantity of these balls was too limited to permit of any very exact investigation of their nature. Herr Luzi presumes that they are transformed diamond-carbon, *i.e.* a different modification of carbon, which contains either oxide of iron or metallic iron reduced out of the oxide. He was, however (owing to the cost of the material to be experimented upon), unable to determine positively what chemical action took place during the time the diamonds were heated in the complicated silica flux. Some of these partly absorbed diamonds, upon which Herr Luzi experimented, are deposited in the mineralogical museum of the Leipsic University.

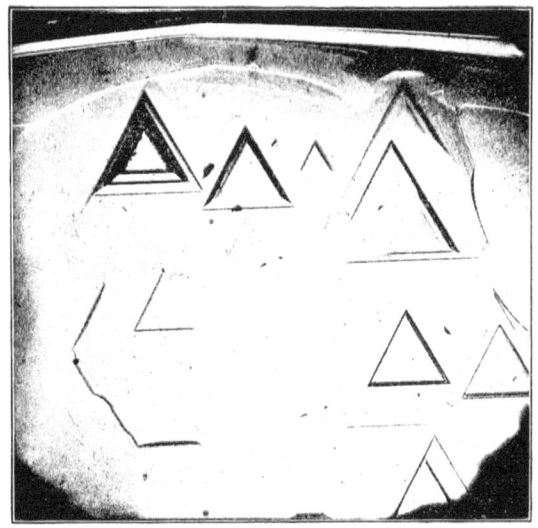

Herr Luzi further remarks that perhaps other molten silica combinations, or those of a similar nature to the blue ground, may have the same power of attacking the diamond.

The knowledge that diamonds can be absorbed by a silicate magma makes one inclined to investigate further the genesis of the diamond, which many claim was formed under great heat and pressure. If such

Two Views of the Face of a Rough Diamond, as seen through the Microscope. (Magnified 100 times.)

was the genesis of the diamond, Herr Luzi's experiments would indicate that the original matrix was not a silica combination such as the present blue ground. They tend to prove, rather, the theory, which I advanced more than sixteen years ago, that the blue ground which contains the diamond owes its formation, as it at present exists, more to aqueous than igneous agencies. If the diamond is unable to withstand the corroding influence of the silica magma at the comparatively low temperature given above, — how could it possibly have retained its forms of crystallization and perfect faces at the far higher temperature and pressure which must have existed under the volcanic or igneous theory?

It seems a pity that Herr Luzi did not state the exact weight of the diamonds upon which he experimented both before and after his experiments. The burning or absorption of the diamond in its matrix would be a strong argument against the diamond having been crystallized *in situ*, or that it came up in its present matrix when such matrix was in a molten state. If a diamond, subjected in its own matrix or magma in an ordinary graphite crucible to a temperature of 1770° R., changes its shape and appearance as described by Herr Luzi, — could it be expected that many diamonds in our mines should be found perfect in shape, without a flaw or spot, and with clear, transparent sides, so smooth that they have the appearance of having been polished? Nevertheless, such is the appearance of nearly all South African diamonds. It would seem from the evidence brought forward that only one conclusion is possible, namely, that the blue ground in its present state is not the magma of the diamond. What the original magma or matrix was is unfortunately far less certain. Some years ago a diamond, weighing 28½ carats, was found at Kimberley. The external surface of the diamond was smooth and crystallized, showing no other mineral except the diamond itself. The interior of the diamond was white, but not transparent, and, owing to its peculiar appearance, the valuator broke the stone in order to satisfy his curiosity. The result of the breaking is shown in the full-size illustration on page 507. A small perfect octahedral diamond was enclosed in the centre of the larger diamond. Nor was this all. There were flakes of a white mineral, not diamond, attached to the fragments of the broken diamond. A few grains of these were collected and analyzed by Professor Lawn, of the Kimberley School of Mines. In appearance the flakes were white, translucent, and crystalline, and about as hard as the steel blade of a knife. When heated in a closed tube, moisture was given off. The mineral was very slightly effervescent, probably due to a trace of carbonate of lime. It fused readily on platinum wire to a white bead.

Diamond bearing a Smaller Crystal in its Centre.

The mineral was determined to be apophyllite, a silicate of lime and potash with 16 per cent of water. If a mineral, which is fusible at the ordinary temperature obtained with a blowpipe, and which contains 16 per cent of water, was formed at the same time the diamond crystallized, it is certain that this did not take place under the condition mentioned above, *i.e.* under enormously high temperature. How, then, one may ask, did the apophyllite become a part of this stone?

Von Tschudi describes a beautiful crystallized Brazilian diamond, in the centre of which is a little gold leaf.   He had the information from Dr. Mills Franco, who maintained that there was no deception in its being gold.[1]   Occurrences of this nature tend to veil the genesis of the diamond in still further mystery.

Professor T. G. Bonney lately obtained specimens from the Newlands mines, some forty miles northwest of Kimberley, of a coarsely crystalline rock studded with garnets, technically "holocrystalline allied to eclogites," which were embedded, as he says, in typical blue ground.   In this eclogitic rock he found a number of small but perfectly formed diamonds.   At a meeting of the Royal Society in July, 1899, he presented his conclusions: "The blue ground is not the birthplace, either of the diamond or of the garnets, pyroxenes, olivine, and other minerals, more or less fragmental, which it incorporates.   The diamond is a constituent of the eclogite, just as much as a zircon may be a constituent of a granite or a syenite.

" Though the occurrence of diamonds in rocks with a high percentage of silica (itacolumite, granite, etc.) has been asserted, the statement needs corroboration.   This form of crystallized carbon hitherto has been found only in meteoric iron (Cañon Diablo), and has been produced artificially by Moissan and others with the same metal as matrix.   But in eclogite the silica percentage is at least as high as in dolerite ; hence it is difficult to understand how so small an amount of carbon escaped oxidation.

" I had always expected that a peridotite (as supposed by Professor Lewis), if not a material yet more basic, would prove to be the birthplace of the diamond.   Can it possibly be a derivative mineral, even in the eclogite ?   Had it already crystallized out of a more basic magma, which, however, was still molten when one more acid was injected and the mixture became such as to form eclogite ?   But I content myself with indicating a difficulty and suggesting a possibility ; the fact itself is indisputable : that the diamond occurs, though rather sporadically, as a constituent of an eclogite, which rock, according to the ordinary rules of inference, would be regarded as its birthplace.

[1] "Travels in South America," by J. J. von Tschudi.

" This discovery closes another controversy, viz., that concerning the nature of the ' hard blue ' of the mines (kimberlite of Professor Lewis) in which the diamond is usually found. The boulders described in this paper are truly water-worn. The idea that they have been rounded by a sort of ' cup and ball ' game played by a volcano may be dismissed as practically impossible. Any such process would take a long time, but the absence of true scoria implies that the explosive phase was a brief one. They resemble stones which have travelled for several miles down a mountain torrent, and must have been derived from a coarse conglomerate, manufactured by either a strong stream or the waves of the sea from fragments obtained from more ancient crystalline rocks. . . .

" The presence of water-worn fragments, large and small, in considerable abundance, shows the blue ground to be a true breccia, produced by the destruction of various rocks (some of them crystalline, others sedimentary, but occasionally including water-worn boulders of the former), *i.e.* a result of shattering explosions followed by solfataric action. Hence the name kimberlite must disappear from the list of peridotites, and even from petrological literature, unless it be retained for this remarkable type of breccia.

" Boulders, such as we have described, might be expected to occur at the base of the sedimentary series, in proximity to a crystalline floor. The karoo beds in South Africa . . . are underlain in many places by a coarse conglomerate of considerable thickness and great extent, called the Dwyka conglomerate, which is supposed to be Permian or Permo-carboniferous in age. It crops out from beneath the karoo beds at no great distance from the diamond-bearing district and very probably extends beneath it. If this deposit has supplied the boulders, the date of the genesis of the diamond is carried back, at the very least, to Palæozoic ages, and possibly to a still earlier era in the earth's history." [1]

[1] Proceedings of the Royal Society, Vol. LXV, pp. 235, 236, July 27, 1899. "The Parent Rock of the Diamond in South Africa," Prof. T. G. Bonney.

I cannot accept the contention that the boulders came from any strata through which the pipes have been formed, unless these strata lie very deep and below the quartzite.

The conglomerate which lies between the shale and melaphyre is only a few feet thick, ten to fifteen at most, and does not contain large boulders such as are found in the blue ground; besides, the quantity of boulders or conglomerate which could have been contained in the area of the mine would not have supplied the amount of stones already found in the blue ground. These must, therefore, have come up from below with the diamond-bearing ground. If the boulders came from the Dwyka conglomerate, it must lie very deep beneath the surface, for nothing of the kind has been found at a depth of over twenty-one hundred feet.

Professor Bonney says above that the statement of the occurrence of diamonds in itacolumite[1] needs corroboration. There is no doubt in my own mind that diamonds in Brazil have been found in itacolumite, and the consensus of opinion is that it is not the original matrix, but that the diamonds were washed from their volcanic origin and became bedded in this sandstone when it was being formed.

I have been frequently asked, " What is your theory of the original crystallization of the diamond?" and the answer has always been, " I have none; for after seventeen years of thoughtful study coupled with practical research I find that it is easier to 'drive a coach and four' through most theories which have been propounded than to suggest one which would be based upon any more unassailable data." All that can be said is that in some unknown manner carbon, which existed down deep in the internal regions of the earth, was changed from its black and uninviting appearance to the most beautiful gem which ever saw the light of day.

---

[1] Brittle quartz sandstone of slaty (schistose) character. — Heusser.

CHAPTER XVII

N preceding chapters the extraction of the blue ground and the winning of the precious stones have been fully described. It remains to trace the handling of the diamonds from this point until they reach the hands of the jewellers and are spread broadcast in glittering array over the face of the world, or applied to uses less showy than adornment.

After the diamonds are separated and collected at the Pulsator, they are cleaned and sent under guard to the diamond office, which is in the general offices of the Company. Here the crystals are boiled in a mixture of nitric and sulphuric acid to remove any particles of earth which may adhere to them. They are then thoroughly rinsed with clear water to get rid of the acids, and finally washed in alcohol and spread out on tables to dry. The alcohol seems to clean the diamonds and leaves them brighter than when water alone is used.

The daily productions of diamonds are put away in parcels until there is an accumulation of about 50,000 carats of De Beers and Kimberley diamonds. The diamonds from these two mines are mixed and are known locally as "pool goods." When the requisite quantity is at hand, the mixed stones are screened to grade the sizes, after first taking out the larger diamonds by hand. They are then ready to pass to the hands of the sorters, who separate and classify them for accurate valuation. The chief classifications in use are —

1. Close goods. 2. Spotted stones. 3. Rejection cleavage. 4. Fine cleavage. 5. Light-brown cleavage. 6. Ordinary and rejection cleavage. 7. Flats. 8. Maacles. 9. Rubbish. 10. Boart.

"Close goods" are pure, well-shaped stones; "spotted stones" are crystals slightly spotted; and "rejection" stones seriously depreciated by spots.   Broken stones are grouped under the head of "cleavage."   Flat crystals formed by the distortion of octahedra are classed as flats, and flat triangular crystals, which are in reality twin stones, are marked as maacles. "Rubbish" is the refuse, ranking a little better than the lowest grade of all, ordinary "boart" the material used for polishing purposes.   Round or shot boart is found in the mines at

Diamond Sorters and Valuators, through whose hands £4,000,000 worth of Diamonds
pass annually.

Kimberley and is very valuable for use in diamond drills since the Brazilian carbonado has become so scarce.   Well-formed shot boart, averaging about the size of peas, sells readily for £6 a carat.

After this separation has been made, the first eight classes are each further subdivided according to their shades of color.   The scale is given below in descending order of purity—

Blue White, First Cape, Second Cape, First Bye, Second Bye, Off Color, Light Yellow, Yellow.

Only the first grade, or close goods, are carefully distinguished by separation of all eight shades.   For other classes a smaller number of shade divisions is noted. It may be perceived that the minute distinctions of this separation can only be made by the trained eyes of experts. No magnifying glasses are used by the sorters, all being able to make the distinctions with the naked eye. Ten sorters are employed, all Europeans, two women and eight men. To replace any who leave, apprentices are trained to the work at Kimberley. The sorters determine the quality of diamonds with notable accuracy and speed.

De Beers mine is noted for yielding an exceptionally large percentage of ordinary "yellows," a very small percentage of very "dark yellows," a limited number of brilliant "silver Capes," and considerable "light-brown cleavage" of a delicate shade. The very "dark yellows" are ranked as "fancies" and highly valued, and the "silver Capes" are also rated highly, as they have great lustre when cut as brilliants, but absolutely white or colorless stones are rarely found in this mine.

Kimberley mine yields a fair proportion of "white crystals," a good percentage of "white cleavage," and quite a remarkable percentage of large "maacles." It also produces a fairly large proportion of "yellows," generally somewhat lighter in color than those from De Beers.

Dutoitspan mine yields some very fine blue-white stones, "silver Capes" and ordinary "white" stones and "cleavage" of comparatively fine quality, together with large "yellows," showing an exceptional proportion of large stones, and a comparatively small percentage of very minute crystals.

Bultfontein's product is very largely composed of white stones, but many of these are spotted more or less; its diamonds are also comparatively small, usually ranging from two to three carats downwards.

The diamonds from Premier mine are mostly octahedron crystals, or fragments of these, with a large percentage of rubbish and boart. Beautiful, deep-orange colored diamonds are frequently found, and blue-white stones are not uncommon.

When the sorting has been completed, the diamonds are

2 L.

placed in little heaps on a long table covered with white paper. In all cases, except in small sizes and boart, where the weight and value only are recorded, the number of diamonds in each heap and their average weights and values are carefully recorded in a book kept for that purpose. This exhibit was previously made also for the benefit of buyers calling at the diamond office, who could thus readily value the stones; but of late years the entire product has been sold to a syndicate composed of the leading diamond merchants of Holborn Viaduct and Hatton Garden, London. The careful sorting and arrangement are nevertheless continued in order to determine precisely what the relative quality and value of the diamonds are in passing from level to level as the mine grows deeper. The buyers know the exact value of every shipment they make, and the De Beers Company must also be informed of any changes for better or for worse in the value of its production, so as to take advantage of them in the former case, or make allowances to the syndicate upon the renewal of the contract, in case the quality should become poorer. These are perhaps remote conditions, for, up to the present time, the average monthly or annual production of diamonds has been remarkably regular in quality.

For the safe-keeping of the gems in the Company's office there is a strong room or vault, built of very thick concrete walls, which are fire and burglar proof. The door of the vault is secured by several bank locks of the latest and best design. The keys fitting these locks are kept by several officers in the secretary's department of the Company, who must all be present at the opening and closing of the strong room. Inside the strong room are burglar-proof safes, with doors also secured by several locks, which can only be opened by two or more persons having separate keys. In addition to these safeguards, the strong room is protected by the application of an electric alarm system. Two armed guards are on duty at the offices at night, and connections are made by which they can signal for help should an attempt be made to break into the building. Even

if both men should be overpowered before they could give
a signal, no robbery could be effected; for, as soon as they
should cease to send test signal reports at regular intervals, an
armed force would soon arrive on the ground and frustrate any
attempted burglary. Under existing conditions for the sale of
diamonds only a small quantity of precious stones are kept at
the diamond office; but, in former years, the quantity, at times,
has been very large and the most stringent precautions were

A De Beers Group.

necessary. It may be noted further that adequate measures
have been taken also to protect the office from assault in the
daytime.

Of late years, with improved methods of working, a larger
percentage of small diamonds has been recovered from the blue
ground. In order to have an average quantity of these in
each parcel made up for the buyers, a fixed percentage of small
stones is included in the parcel. If there is any surplus, it
is valued in the ordinary way and sold to the buyers at a valuation agreed upon between the seller and buyer. After the

diamonds are sorted, they are put into square tin boxes, fitting into tin cases like despatch boxes, which have tightly fitting, locked covers.   A despatch box will contain about forty tin boxes.

All De Beers diamonds are delivered to the buyers at the diamond office of the Company and paid for at once in cash or in bills on London, as the Company may prefer.   After delivery to the buyers the diamonds are sorted over again for the London market, which desires a classification of the stones for different purposes than valuation simply.

They are reassorted according to quality into from 350 to 400 different parcels.   Each parcel is put into specially made papers bearing on their face a description of their contents.   Then these parcels are packed in tin boxes which are securely wrapped in cloth-lined packing paper, carefully sealed and delivered to the post-office, which forwards them to Europe as registered mail.   All diamonds so forwarded are insured with insurance companies in Europe.

Classification is made into —

Pure goods
Brown goods
Spotted goods
Flat-shaped goods } Completely formed or crystallized stones.

Pure cleavage
Spotted cleavage
Brown cleavage } Broken crystals or split stones.

Naats or Maacles { Flat, triangular crystals, in reality twin stones.

Rejections or Boart { Uncuttable diamonds used mostly for splitting and polishing more perfect crystals.

Most of the above classifications, except rejections and boart, are subdivided into six or seven colors, and each color is again subdivided into eight, ten, or twelve sizes.

THE PERCENTAGES OF DIAMONDS IN THE VARIOUS CLASSES

		DE BEERS AND KIMBERLEY MINES POOL GOODS.	PREMIER MINE.	JAGERSFONTEIN MINE.
Pure Stones	Close goods, stones over one carat	4.1%	8.7% all sizes	11.7% { all sizes, and including slightly colored and brown stones
	Irregular shapes of all sizes.	1.6%	4. %	—
	Melee of all sizes under one carat	3.1%	—	—
	Brown stones of all sizes.	1.6%	4.2%	—
Spotted Stones of all sizes . . .		8.1%	6.6%	20.4% { all sizes and descriptions
Cleavages, pure and spotted, over one carat		38.8%	28.2 { including maacles	39. %
Chips, pure pieces of all sizes under one carat		1. %	1.2%	10.6%
Chips, spotted pieces of all sizes under one carat		12.6%	5.5%	—
Maacles (Naats) and Flat Stones, pure and spotted, all sizes . .		4.3%	—	4.3%
Rejections, lowest quality of above descriptions		17.5%	16.2%	3.5%
Boart, diamonds not suitable for cutting		7.3%	25.4%	10.5%
		100. %	100. %	100. %

When the diamonds arrive in London, they are again re-assorted for sale, *i.e.* in the manner that will best suit the customs and requirements of the trade. The London importers sell (*a*) to merchants of rough diamonds, who again resell the goods in their rough state, (*b*) to merchants of brilliants who get their purchases cut and polished for sale, (*c*) to actual manufacturers who, buying for their own account, cut and polish the goods and then resell with profit as compared to the manufacturer who works for a fixed cutting charge.

It is of interest to compare the present elaborate method of assorting and valuing with that obtaining in the eighteenth century and previously in the European market. It was the custom then to forward diamonds from India in "bulces" or parcels neatly wrapped in muslin and sealed by the sellers. The largest stones were never offered for sale, but reserved by the native owners, as David Jeffries observes, to aggrandize their families. He states further that "the head of the family has a small shallow hole drilled in the surface of the stone, and when he dies the next chief does the same, and so from one to another, and the more of these holes a stone has the higher it is in esteem, although such holes may prejudice it if it were to be manufactured; but as that is never intended, they do not regard such prejudice; and these stones are never parted with, let what will happen, and if they foresee any ruin to the family . . . in such cases they bury these stones, so that they never appear again." The other stones, comprising the small and middle size and some of the large ones, were put in the parcels for sale unassorted and

The Officials who manage the Benefit Society, De Beers Consolidated Mines Limited.

"valued by the lump, as they weigh one with another, by the rule." In the European markets such parcels were generally bought, he states, "by the invoice, that is before they are opened, it being always supposed they contain the value which they were sold for in India; and the buyer here gives the merchant such a profit as contents him. The diamonds being thus bought, the buyer opens the parcel, separates them, and then values them separately as his judgment directs; making to himself likewise such a profit upon the whole parcel as he thinks proper."[1]

This expert jeweller notes with regret that at the time of his

[1] "A Treatise on Diamonds and Pearls," David Jeffries, London, 1751, pp. 118, 119.

A GROUP OF OFFICIALS OF DE BEERS CONSOLIDATED MINES LIMITED.

writing, in the middle of the eighteenth century, there was no uniform standard of valuation and that the purchase of large stones in particular was essentially a gambling speculation. In the East Indian market there was a persistent effort to maintain fixed prices, and there was comparatively little fluctuation in the market rates of the East Indian stones, but the diamonds of Brazil were thrown irregularly on the market, so that the supply ranged from a dearth to a glut, and the prices were so greatly fluctuating that any investment in these stones was extra hazardous.   Mr. Jeffries marked clearly the disastrous consequences of greatly varying products and prices in the marketing of precious stones.   He reached the conclusion " that to maintain as invariable a price of these jewels [diamonds] as is possible must be of the greatest utility to the public," and gave high praise to the owners of East Indian diamond fields and diamond merchants because they did not flood the market regardless of the diamond, like the Brazilian producers.   He notes a shift of fully 33 per cent in the market rate of diamonds in a single year. In 1733 the value of Brazil diamonds fell to a point below 20 shillings per carat for rough diamonds, and within 20 years ran up to more than treble this price.

One of the simplest and oldest divisions in grading and in the measure of values of diamonds and other precious stones is in accordance with their weight.   The transmitted measure of weight is the carat, derived from the Greek κεράτιον, the fruit of a variety of acacia, whose remarkably uniform seeds served as convenient measures of value of diamonds and other precious stones, and is equivalent to 4 grains avoirdupois or 3.174 grains troy weight.   In market quotations from year to year and in contracts for sorted diamonds the valuation is expressed in a stated price per carat.   Von Tschudi states that the word carat is derived from kaura, an African creeping plant, whose red seeds specked with black were used for weighing gold in Africa and diamonds in India.   On the supposition that £2 may be reckoned the general or average price of a rough diamond of one carat weight, Mr. Jeffries gives two methods or formulas for computing the values of "wrought," or as we would say "cut," diamonds. First, the weight of the cut stone should be doubled, to offset the loss of one-half in working; then this figure or figures should be squared, and the product multiplied by the price per carat. Thus a cut stone weighing one carat would be valued by multiplying 2 by 2 by 2, or at £8, and a stone weighing 5 carats by multiplying 10 by 10 by 2, or at £200. By the second method, the calculation is made on the basis of the valuation of a cut stone weighing one carat, at £8, as before determined. Then to find the value of a stone of any given number of carats, multiply the number by 8, and the multiplicand will be the estimated value of every carat in the stone. The total value may then be reached by multiplying the number of carats by this multiplicand. For example, if a given stone weighs $5\frac{1}{8}$ carats, the value of every carat in the stone will be found by multiplying by 8 to be £41. Then multiply £41 by $5\frac{1}{8}$ and the result will be £210 2s. 6d.; the estimated value of a cut stone weighing $5\frac{1}{8}$ carats. It was the expectation of Mr. Jeffries that the general adoption of his method of valuation would go far to fix the price of diamonds, and it did prevail for more than a century before falling into disuse.

Production of Diamonds

De Beers Consolidated Mines. During the eleven years ending June 30, 1899, the yield of De Beers Consolidated Mines has been 24,476,000 carats of diamonds in round figures, which would measure about 72 cubic feet, showing an average of something more than 2,200,000 carats annually. Compared with this product, the production of the other diamond fields of the world, with the exception of Jagersfontein, is comparatively unimportant, not exceeding 5 per cent of the total.

The Orange River Colony. The principal diamond-producing mine in this colony is Jagersfontein, which has averaged about 250,000 carats annually for several years past. The Jagersfontein mine is controlled by the syndicate which has for many years purchased the total production of De Beers Company.

There are a few other diamond mines in the Orange River Colony, but the yield of diamonds from all of them combined is small in comparison with Jagersfontein. The total output of diamonds from Jagersfontein up to March, 1901, was 2,168,399¼ carats, valued at £3,923,940.

Photograph of a Book with the Leaves cut out in the Centre, used by an Illicit Diamond Dealer to send Diamonds through the Mails to England.

Transvaal. There are alluvial diggings along the banks of the Vaal River a few miles above the river diggings in Griqualand West, Cape Colony. The town of Christiana is situated near these diggings, lying just outside the jurisdiction of the Cape Colony and the late Orange Free State (now Orange River Colony), in which Colonies the Diamond Trade Act, which forbids dealing in rough diamonds except by licensed dealers, is in force. A large illicit trade has been carried on at Christiana for many years in diamonds stolen in Kimberley and the river diggings in the Cape Colony. A few years ago the Government of the late South African Republic passed certain laws in reference to the registration of diamonds, but these laws were not stringent enough to stop the illicit traffic. Diamonds have also been found at Rietfontein, near Pretoria, but up to the present time the total yield has been very small. A few years ago there was a remarkable occurrence of diamonds in the conglomerate gold ores from the mines at Klerksdorp, when several green diamonds were found in the battery box. As the conglomerate is a sedimentary formation, the diamonds must have been washed into it from some crater in a similar manner to the depositing of diamonds in the itacolumite of Brazil.

Outside of South Africa the diamond fields of any determined value are in Brazil, India, New South Wales, and Borneo.

Brazil. There was a revival of the diamond-mining industry to some extent in the Brazilian fields, owing to the diminution of the South African product by the Transvaal War. The State places a duty of 16 per cent on the valuation of all diamonds produced, and there is in addition a tax of 1 per cent demanded by the municipalities. Owing to the tax evasion, it is difficult to determine the total annual product. The value of exports from Minas Geraes during the first half of 1900 was reported at 250,000 milreis, $140,000.

Mr. A. de Jaeger has estimated the total production of Brazilian stones from the time of the discovery of the diamond fields at 12,000,000 carats, valued roundly at $100,000,000. It is stated, however, in " The Mineral Industry," presenting probably the best extant record, that the best available statistics show that the total output of Brazil, up to and including 1898, was 13,105,000 carats.[1]

Dr. Le Neve Foster, one of his Majesty's inspectors of mines, in his Annual Report on Mines for 1899, says : " Compared with the output of Kimberley, the total production of

[1] " The Mineral Industry," 1899, p. 222.

diamonds in Brazil for the year, estimated at 40,000 carats, is at present insignificant. . . . The most important diamond districts in Brazil are Diamantina, Grao Mogul, Chapada Diamantina, Bagagem, Goyaz, and Matto Grosso."

India. In the same report the quantity of diamonds produced in India for 1898 is given at 170 carats, valued at 10,873 rupees, and for 1899, 124 carats, valued at 8011 rupees.

New South Wales. The existence of diamonds in New South Wales was made known as early as 1859, by Rev. B. W. Clarke, who received in that year several specimens from the Macquarie River, Burrendong, and Pyramul and Calabash Creeks. It was not, however, until the rush for the gold diggings, seven or eight years later, that any considerable number of diamonds was found, when the gold digging along the Cudgegong River, about nineteen miles northwest of Mudgee, brought to light diamonds in an old river drift, generally covered with a layer of basalt.

The diamonds were sparsely distributed through the gravel, and were usually small, the largest of the stones, a colorless octahedron, weighing only $5\frac{5}{8}$ carats. Later, other diamond fields were opened near Bingera, on the river Hoclon, and in the tin-mining districts near Inverell. The diamonds occur in alluvial gravel wash in the beds of ancient rivers. This gravel carries tin ore or gold in places, and usually one or both of these are won with the diamonds. These ancient river channels resemble those in California, in which diamonds were occasionally found with the gold. Many of these rivers lie buried beneath lava hundreds of feet thick, and the diamonds are won by driving long tunnels and drifting out the gravel lying on the bed rock.

Dr. C. Le Neve Foster gives the production in New South Wales for 1898 as 16,493 carats, valued at £6060, and for 1899, 25,874 carats, valued at £10,350. These figures give an average value per carat of seven shillings and four pence and eight shillings respectively, as compared with forty shillings per carat for De Beers and Kimberley mines.

Borneo. The estimated production of diamonds in Western Borneo was 1190 carats for 1897, and 1950 carats for 1898.

British Guiana. Some attention has been drawn of late to the reported diamantiferous deposits in British Guiana. It is stated that there was a shipment of 282 specimens from this field to London early in 1900, and, later in the year, 400 small stones were brought to Georgetown. The location of the deposits is reported to be on the Mazaruni River, about 250 miles from its mouth. The diamonds have been found in an alluvial formation, consisting of sandy clay mixed with pebbles and fragments of ironstone, quartz, and felsite.[1]

Importation of Diamonds

In the importation of diamonds the United States leads, and England, Germany, France, and Italy follow in the order named. The increase in the demand of the United States has been extraordinary, showing an advance of fully 2000 per cent in the last fifty years. In 1899 the valuation of the total import of precious stones was $17,208,531. In 1900 there was a falling off of about $3,850,000 owing to the interruption in the supply, but the records of the year 1901 indicate a probable importation exceeding $20,000,000, the total for the first two months of the year reaching $3,870,359.31, an increase of $2,674,787.88 over the import of the corresponding months in 1900. The importation is a close measure of the total sale, as the production of precious stones in the United States only reached a valuation of $185,770 in 1899, and this was larger than in any previous year. Nearly five-sixths of this native product is made up of sapphires and turquoises.

Rubies, emeralds, sapphires, and pearls are the gems most commonly used in settings in combination with diamonds. It is estimated by Mr. George F. Kunz, of New York, an expert of international reputation, that the value of the diamonds imported into the United States is approximately 75 per cent of the valuation of all precious stones and pearls imported, and it is judged that this consumption fairly represented the percentage in other countries. The changes in settings from year to year and even from decade to decade are not very pronounced. The resetting of

[1] "The Mineral Industry," 1900.

stones is an appreciable fraction of the jeweller's business, but inconsiderable in comparison with the setting of newly cut stones.

The World's Stock

Diamonds are so highly prized and so imperishable that the amount of these gems in existence to-day may almost be reckoned as the total of the world's production, ranging in value through hundreds of millions of dollars. Mr. Kunz does not estimate a loss of 5 per cent in a hundred years, and the South African Diamond Fields alone have contributed over $400,000,-000, or £80,000,000, in value to the world's stock. Yet the demand advances apace with the world's growth in wealth, and no diversion of the world's fancy is apparent. The plunder of Delhi by Nadir Shah in 1739 has been estimated at $300,000,-000,[1] and a great share of this was precious stones. There may never again be such a collection in the hands of any monarch or nabob as the store amassed by the Great Moguls, but the crown jewels and private treasures of the leading courts of Europe to-day are of immense value and are growing greater.

The crown jewels of France were estimated at $6,000,000 (£1,200,000) more than a hundred years ago, and even this great amount is far exceeded by the value of the Russian crown jewels. The crown of Ivan Alexiowitch contained 881 brilliants, the Empress Catherine had 2536 brilliants in her crown, and the purchases of succeeding Czars have been enormous. At the London Industrial Exposition in 1851 a firm of Russian jewellers exhibited a superb diadem on which were mounted 11 beautiful opals, 67 rubies, 1811 brilliants, and 1712 rose-cut diamonds.[2]

The British crown jewels do not equal the Russian in number or value, though there are other magnificent gems among them besides the Koh-i-nûr, whose romantic story is told in a former chapter. The crown specially made for the coronation of the late Queen Victoria, in 1838, was regarded as a superb showing of the art of the leading jewellers of London as well as

[1] "Great Diamonds of the World," Streeter.
[2] "A Popular Treatise on Gems," Feuchtwanger, New York, 1867.

of the gems displayed. It is fashioned of hoops of silver enclosing a cap of deep blue velvet. Precious stones completely encase the hoops, which are surmounted by a ball covered with diamonds and bearing a Maltese cross of brilliants, with a splendid sapphire as the central jewel. The rim of the crown is clustered with brilliants and Maltese crosses. On the cross at the front of the crown is set the magnificent heart-shaped ruby, which was worn by Edward, the Black Prince, and beneath this ruby in a circular rim is an oblong sapphire of extraordinary size and beauty. Clusters of drop pearls add to the resplendent effect of the massing of the diamonds, emeralds, rubies, and sapphires.

The exquisite beauty of the jewels of Queen Isabella of Spain has been particularly noted. At the London Exhibition in 1851 two sets of her jewels were shown. One consisted of a diamond necklace, in the form of a ribbon, interlaced with foliage of emeralds. Brilliants were arranged also to form a bouquet of lilies with emerald leaves, encircled with ribbons of brilliants and pendants of pearls. A ribbon of brilliants, interlaced with emeralds, formed a bracelet, and the crown of this set was of the like combination of gems, with aiguillettes of flowers whose stamens were pearls. The second set of jewels was made up entirely of diamonds and sapphires of the finest quality and most artfully matched.

It is scarcely to be expected that any private collections of gems should rival in extent the treasures of sovereigns, whose crown jewels may be the display of centuries of accumulation, but some of the noble families of Europe and other wealthy owners have gems that any monarch in the world might covet, and there are a considerable number of collections ranging in value over a million dollars. In the United States it is estimated that there are at least half a dozen such collections, one of which contains a necklace valued at $320,000.[1] At every leading court reception, or grand ball or opera, the display of jewels may be measured in millions of dollars, and the diffusion of gems is constantly spreading with the extension of wealth.

[1] George F. Kunz.

CHAPTER XVIII

CUTTING AND POLISHING

T has been shown in the opening chapter of this work that fancy has still, and probably must forever have, a free range for its surmise when and how the first diamond crystal was picked from the river-shore wash of the Indo-Gangetic plain. Equally vague and conjectural must be any effort to fix the period when a rough or natural diamond was first artificially ground or polished. It is only certain that some rude polishing, at least, was essential to the revelation of any notable beauty in the diamonds of India; for the surface of these crystals is covered with a grayish white film or incrustation, veiling their refulgence so completely that the rough stones are scarcely more ornamental than common quartz pebbles.

It was in view of this obscuring that the apostle of deportment, the Earl of Chesterfield, wrote to his son: "Manners must adorn knowledge and smooth its way through the world. Like a great rough diamond, it may do very well in a closet by way of curiosity and also for its intrinsic value."[1] A contemporary of this high authority, Dr. Samuel Johnson, was able to controvert this dictum by demonstrating that knowledge can rise from obscurity without any adornment of manners, but polish is indispensable to the revelation of the latent beauties of the rough diamond.

Indian tradition runs back romantically five thousand years to the first gleam of the Koh-i-nûr or "Mountain of Light" in the serpench of a chief who fell in the great battle described

---

[1] Letters of the Earl of Chesterfield, July 1, 1748.

in the epic poem " Mahabharata " ;[1] but nothing more solid
than tradition sustains this tale.   If it were true, it would
demonstrate incontestably a very ancient proficiency in the art
of grinding and polishing a rough Indian diamond, as the figure
of the Koh-i-nûr on page 1 shows, illustrating the appearance of
this famous gem before it was recut by modern lapidary art to

Examination of the Diamonds before and after Cutting.

hold the foremost place in the jewels of the British crown.[2]   The
Italian, Augusto Costellani, is the mouthpiece of another tradition, little firmer than a floating pipe-bubble, that a certain King
Carna of India, who lived some three thousand years before
the Christian era, possessed a diamond whose natural planes or

[1] " Indian Epic Poetry," Sir Monier-Monier Williams, 1863.
[2] " A Popular Treatise on Gems," Dr. Lewis Feuchtwanger, 1867, Plate
VIII, No. 15 and No. 15a.

2 M

facets were polished; but what the good king did with his sparkling treasure, or where it has wandered, is unfortunately left to the drift of fancy.

It has been shown that the earliest known catalogues of gems do not include the diamond, and that the references to it in the Hebrew Scriptures and other writings before the Christian era are far from decisive, in view of the likelihood that the white sapphire was the ancient adamas.[1] The failure to bring to light any diamond in the exhumation of ancient gems is further significant.[2] If it be true that a genuine diamond, bearing the engraved head of the philosopher Posidonius, exists in the collection of the Duke of Bedford, as reported by Streeter,[3] this is a solitary instance, so far as is known, of the application of engraving to this adamantine surface at a date probably prior to the birth of Christ, for Posidonius was a Tyrian Greek, living in the second and first centuries B.C.[4]

It is, however, highly probable that the genuine diamond crystals were discovered in India hundreds, if not thousands, of years before the Christian era, and partially polished, at least, in the primitive method of rubbing or striking the planes of one crystal against the other, or even by laborious friction with gritstone by hand or a grinding wheel.

It is certain that revolving stones or metallic wheels for grinding gems were in use in remote antiquity, perhaps two thousand years or more before the Christian era. From the softer stones, carnelian, onyx, and jasper, the ancient workmen advanced to harder gems, preparing their face first chiefly by a smooth polish for the sculptors of cameos and intaglios. Their mode of

[1] "Precious Stones noted in the Sacred Scriptures," R. Hindmarsh, 1851. "Precious Stones and Gems," Edwin William Streeter, 1880.

[2] The Story of the Nations, "Phœnicia," George Rawlinson, M.A., 1894. "Ancient Mineralogy," N. F. Moore, 1834.

[3] "Precious Stones and Gems," Streeter, p. 46.

[4] The Story of the Nations, "Phœnicia," George Rawlinson, M.A., 1894. "Ancient Mineralogy," N. F. Moore, 1834.

[5] "A Treatise on the Ancient Method of Engraving Precious Stones," Laurentius Natter, London, 1754. "A Treatise on Diamonds and Precious Stones," John Mawe, 1813.

working was very simple, as Feuchtwanger notes.[1] The polishers prepared the stones on a plate by means of the powder of harder stones, either round, oval, flat, or in shield form, according to the designed subjects, and the sculptors cut the engraving with iron tools or diamond splinters mounted in iron.

The Egyptians taught the art of carving to the Phœnicians, Etrurians, and Greeks. The Indians and Persians learned to carve and polish gems perhaps as early as the Egyptians. Representations of the adored beetles or scarabs were the earliest known Egyptian engravings, while the Persians engraved chiefly mythological animals or figures of their priests. Cabalistic devices and Arabic letters on gems formed the doubly precious " talismans," and even without talismanic lettering, marvellous or supernatural origin and powers were attributed by current superstition to all the notable gems.[2] Alexander's seal typified the sovereignty transferred to his vicegerent, Perdiccas. Augustus Cæsar cherished his seal engraved with a sphinx as a token of his divine authority.[2]

In the carving of cameos, precious stones with layers and veins were employed with great skill, bringing out contrasted effects, as where a face is shown in one color and the hair and dress of a figure in different colors. Sometimes certain colors were made typical. Thus Bacchus was carved in amethyst, the color of wine, while Neptune or nymphs of the sea were cut in aquamarine.[3]

Such surface polishing and engraving antedated, however, very far any grinding or faceting of the harder gems, and the intractable diamond especially, for uses of ornament. Pliny writes, " The polished hexahedral Indian diamond thins to a point."[4] As the crystallization of the diamond is much more

[1] " A Popular Treatise on Gems," Feuchtwanger, 1867.

[2] " De duodecim Gemmis in Veste Aäronis," Epiphanius, 1565. " Gemmarum et Lapidum Historia," Boetius, 1647. Theophrastus — " History of Stones and Modern History of Gems," Sir John Hill, 1746. " Precious Stones and Gems," Streeter. " A Popular Treatise on Gems," Feuchtwanger, 1867.

[3] " A Popular Treatise on Gems," Feuchtwanger, 1867.

[4] " Naturalis Historia," Caius Plinius Secundus, 23 A.D.–79 A.D.

commonly octahedral and dodecahedral than cubical, the adamas of Pliny may have been the white sapphire crystal, a hexagonal prism.[1]

Long before the days of Sindbad the sailor,[2] when the true diamond was unquestionably known and prized, and when the lucky adventurer filled his pockets with the choicest crystals, copper had been substituted for lead in revolving wheels used by the most skilful lapidaries for grinding the harder stones; and powdered stone, moistened with oil or water, was sprinkled on the grinding wheel or pressed into furrows on its face. Cutting in the scientific method of the modern art was of comparatively recent development. The grinding or cutting of the Indian stones by native lapidaries was, at first, only a surface polish of natural planes, and later proficiency did not extend beyond an irregular and unsymmetrical fashion, which rarely ventured the risk of cleavage. There are perhaps no known samples indicating with certainty a higher proficiency in the art at the beginning of the Middle Ages than the four large diamonds now to be seen on the buckle of the mantle of the Emperor Charles the Great,[3] which were planed and polished on their natural faces.[4]

There is a particular Oriental cut of diamonds, still followed, which had its origin about the year 1000 A.D. This bears the distinctive name of " Indian " or " Lustre of India." It had four rectangular plates and one upper facet in the form of a parallelogram. The stone was polished highly on all surfaces except the under side, which was left in its natural state. It is thought that the wandering merchants of the East, who travelled

---

[1] "De Gemmis Plinii," Ernst Friedrich Glocker, 1824. " Precious Stones and Gems," Streeter.

[2] Ninth century A.D. "Origin of Tales of Voyages," " Cyclopædia of India," Balfour.

[3] " Charles the Great, King of the Franks and Emperor of the Romans," 742–814 A.D.

[4] " Great Diamonds of the World," Edwin William Streeter, 1882. "Handbook of the Arts of the Middle Ages," Jules Labarte, London, 1855.

by caravan, brought these stones, or a knowledge of their style, from the far Orient to Constantinople, whence they were made known to France, Italy, and Holland.[1]

That such forms of gems were made in Paris and in Venice as early as the thirteenth century is certain. In 1290 A.D. a society of lapidaries was formed at Paris, and at the close of the fourteenth century there were professional diamond cutters of somewhat higher skill in Nuremburg. In 1365 A.D. an inventory of the jewels of Luigi d' Angio was made, which mentions a diamond having eight facets and another shaped like a shield. The facets here spoken of may be only flat sides such as any true octahedral crystal presents.[2]

One of the first, if not the first, of European workmen to attain any distinction as a diamond cutter was named Hermann, living in Paris about 1407 A.D., and it seems to be certain that from his time or the beginning of the fifteenth century the business of polishing and developing the diamond became an established industry in western Europe. Gems in the rough were somehow finding their way from India and Borneo, and were coming into the market not only among kings and the members of the royal households but among noblemen and burghers of great wealth. In 1465 A.D. there were three registered diamond cutters living in the city of Bruges. Perhaps these cutters were associated with Louis de Berquem, a native of that city, who announced in that year a new method of cutting diamonds and established a guild of diamond cutters.

The method which he pursued and the forms which he evolved were deserving the name of a new discovery of which he was truly the inventor. With whatever assurance others may claim to have invented the art of faceting or of cutting diamonds,

[1] "Precious Stones and Gems," Streeter. "A Treatise on Diamonds and Precious Stones," Mawe, 1813. "Treatise on Diamonds and Pearls," David Jeffries, 1750–1751. "On Gems and Precious Stones," Robert Dingley, Phil. Trans. Abi. IX, 345, 1747. "Le Grand Lapidaire," Sir John Mandeville, Paris, 1561. "Les Merveilles des Indes Orientales," etc., Robert de Berguen, 1661. "Voyages en Turquie, en Perse et aux Indes," Tavernier, 1676. [2] *Ibid.*

it is very evident that none before him had done so on any scientific basis of geometrical relations. Berquem was not merely a craftsman; he was an accomplished mathematician, highly versed in optical science, and he had determined the true angles at which the planes of each facet should lie in reference to its crystallization and to its size, in order to make its reflections of light most perfect and its color most complete.

He discovered that in the development of the octahedral form there are certain measurements of relation which must be preserved in the trimming of the diamond for the perfect reflection of all the light which enters the crystal. By this scientific formation he completely changed the basis for estimate of the value of diamonds. Under his treatment the diamond of largest size and weight was not most valuable, but the gem which was transcendent as a light producer or reflector and as a crystal of symmetrical parts. The connoisseur, the artist, and the thrifty merchant alike have vastly profited by the principles evolved by Berquem. He raised the craftsmen of his day from the common plane of gem polishers to the higher position of artists and skilled lapidaries. The successful lapidary of to-day — to whose cutting is intrusted the gems of India, Brazil, and Africa — must be a close student of optics as well as a dexterous stone cutter.

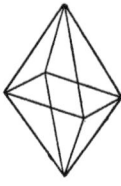

Fig. 1.   Fig. 2.   Fig. 3.

Figs. 1 and 2 above represent the simple octahedral form of diamond crystallization. By the second figure it will appear that if two pyramids of four triangular sides were joined together at their bases, we should have a diamond form with eight triangular surfaces, or an octahedron. Fig. 3 is the same octahedron with its corners either rounded or ground flat as additional facets. The diamond's natural edges are not often so straight and sharp as here represented, but are usually convex, that is, bowing outward ; but when mechanically trimmed to perfect their shape, each line and angle must be unerringly true.

Fig. 4 is a cube of six faces having its corners rounded or flattened, and Fig. 5 is a double cube or dodecahedron, having

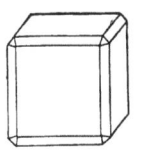

twelve equal rhombic faces. Some diamonds can readily be made to receive these shapes with little loss of substance. Fig. 8 represents a gem shaped as a parallelogram with a facet

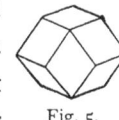

Fig. 4.·   on one upper corner, the lower side showing   Fig. 5.

its natural state. It is called " Indian " or " Lustre of India."

Figs. 6 and 7 represent the oldest and simplest form of gem cutting, called the " table cut." It suits the other precious gems

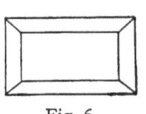

much better than the pure dia-mond. A celebrated " table "

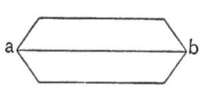

diamond was given by Prince

Fig. 6.   George (afterwards George IV.)   Fig. 7.

to Mrs. Fitzherbert. She had it split along the line from *a* to *b*, and used each half to fit in the face of a locket ; one holding her own portrait, and the other that of her princely lover. The diamond with the portrait of Mrs. Fitzherbert was buried with the old king in the locket which hung on his neck.[1]

Fig. 8.

The first and simplest modification of the " table cut " of a diamond is called the " Old English sin-gle " or the " star single cut." By this arrangement the table cut diamond had its top part planed down about the edges to represent an eight-pointed star whose centre figure was an octagon, or elongated octagon, if the stone was longer than its width. This style of cutting appears in sets of old diamonds for crown jewels or ordinary wear. These sloping triangular faces were ground upon the edge of the upper surface of the stone only, reaching from the flat part, which is then technically called the " table," to the central line which is called the " girdle," and these cut surfaces are called " facets " or small

[1] " Macaulay's Essays," Thomas Babington Macaulay.

faces. Their size and shape are most accurately measured and most exactly ground.

Figs. 9 *a*, 9 *b*, 9 *c* represent, successively, side or girdle, top or table, and back or culet of the next most simple cut of

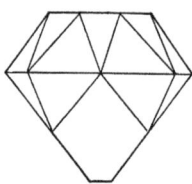
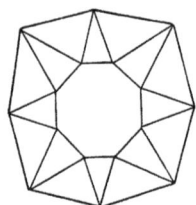
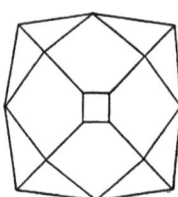

Fig. 9 *a*.   Fig. 9 *b*.   Fig. 9 *c*.

modern date, which is of a scientific cast. It is called the "single cut brilliant," a modification of the simple table cut. Fig. 10 below represents a single cut having sixteen triangular facets on its upper section and twelve facets on the under

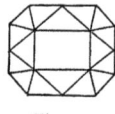

Fig. 10.   Fig. 11.   Fig. 12.

section, plus eight long facets. Figs. 11 and 12 show one of half that number, but both belong to the style here described — the single cut brilliant. Indeed, with very small stones, the single cut has but four faces above and four below. In commercial circles they are called "single sets."

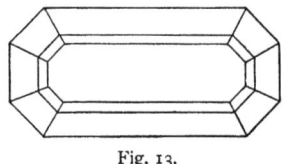
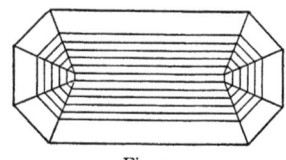

Fig. 13.   Fig. 14.

The two figures above present another modification of the simple table cut of India. It is called the "step cut." In this style the plane above the girdle is only half as thick as that below the girdle. From Fig. 13 it appears that the part above the girdle has been bevelled off at two different angles, making two "steps" besides the table. The other figure represents the part beneath the girdle, which being twice as thick as the upper section is cut with six steps instead of three. In closely studying this step cut and the table cut it was discovered that the diamond crystallized in thin laminæ or plates, and that it might be split into very thin sections resembling plates of mica. By taking advantage of these "lines of cleavage," as they are called, many large diamonds were split into thin leaves and used as faces of small pictures enclosed in lockets. At Queen Victoria's coronation this thin sheet diamond was so common that many distinguished guests were favored with a gift of their own likenesses encased in golden frames and covered with a diamond instead of a glass face.

In the plates below, Fig. 15 and Fig. 16, facets of the "rose cut" pattern are represented. It will be seen that the bottom

Fig. 15.

of the diamond is flat, though not unpolished, while all the facets lie above the girdle. This design, which is called the "Holland," groups twenty-four facets, but a simpler style known as the "Antwerp rose" shows facets

Fig. 16.

ranging from six to sixteen. This rose cut is a very convenient style to adopt for fragments which have been cleft from large stones, or for diamonds which are imperfect in their crystallization on one side. If well proportioned, the depth of the rose must be one-half its breadth at the base.

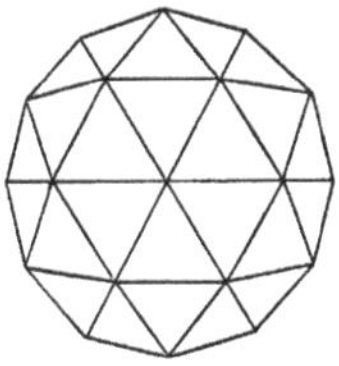

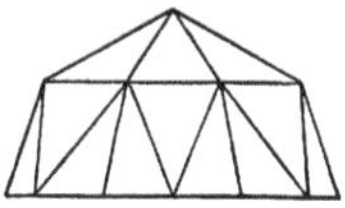

In the rose cut diamond every facet is a triangle and all meet at the central apex, forming a cupola. When the facets on large stones number thirty-two, the dealers call it "fiam minghi" or "half brilliant." A common practice of the trade is to obtain a second "fiam minghi" of the same size, but cut in quartz crystal or even in glass, and glue their bases together with gum mastic, thus forming the "briolet" or "brilliolet," which is palmed off for a pure diamond. Briolets are pear-shaped or oval stones, having neither table, culet, nor edge, but covered with triangular-shaped facets, sometimes pierced at their points of greatest diameter, to be suspended on an axis.

It has been told how the diamond by Berquem's talent was first cut in harmonious and systematic proportion and regular facets at such an angle to its axis and to each other that the fullest play of reflected light is secured from every surface on which it strikes.   His art produced the single cut brilliant, the highest achievement of the lapidary of his day.   Near the close of the seventeenth century a Venetian engraver, named Vincenzo Peruzzi, while experimenting to get rid of obnoxious color in small diamonds, invented the double faceting which is now known as the "brilliant."   It is regarded as the perfection of the lapi-

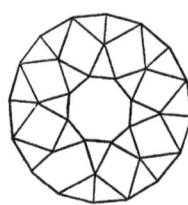

Fig. 17.

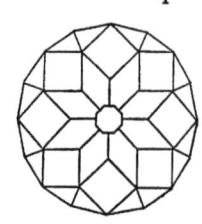

Fig. 18.

dary's art, and is adopted in cutting the most costly gems now put upon the market.   There are thirty-two facets in its upper section, and twenty-four below the girdle.   A diamond cut in this style is shown in Figs. 17 and 18.

The usual double cut brilliant has only fifty-six facets, but, of late years a supposed improvement has been made by adding eight star facets around the culet, which makes a total of sixty-four facets.   The proportions of measurements for the perfect brilliant diamond do not hold for other colored gems whose depth increases or diminishes their color.   The triangular facets on the bezel, which touch the table, are named "star facets," while those which touch the girdle are "skill facets.'"

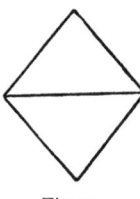

Fig. 19.

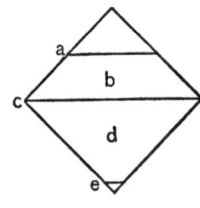

Fig. 20.

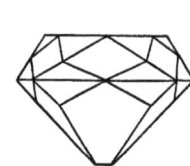

Fig. 21.

In order to show the names of lines and the geometrical relations of diamonds as a lapidary sees them, the above figures may prove helpful.   Fig. 19 shows the side view of an ordinary octahedral or eight-sided diamond.   Fig. 20 shows first the line at *a* cutting off the upper point of the diamond. When this is accomplished, the flat top surface is called the " table." The line at *c*, which is the largest girth of the diamond, is called its " girdle." The space *b*, between the girdle and its " table," is called the " bezel." The line at *e* cuts off the sharp lower point, and its flat surface is the " culet." The space between the culet and the girdle is called the " pavilion."

Cleaving Diamonds

There are three distinct processes in the treatment of diamonds by the lapidary — cleaving, cutting, and polishing. To split the diamond successfully demands a thorough knowledge of its individual character as well as of its generic crystallization and lines of cleavage. The skilled lapidary takes in hand a large rough diamond. If it is an Indian or Brazilian stone, it is coated or partly coated with a hard dull crust. Its corners are perhaps abraded. It may have defects or cracks in its surface, unequal coloring, or black deposits in its interior. He must needs remove the crust, correct the distortion of the crystal, remove or conceal its defects, and decide what is the largest perfect gem which can be cut from the rough stone. He must be able to see the priceless jewel through its shrouding veil, and determine on which surfaces of the stone its prominent corners must rest. Having decided what shape will best befit the stone, he must know whether the rejected portions can be split off safely or whether they must be ground off. Grinding away the rejected portions is probably the safest procedure, but it is the slowest and most expensive. The quickest method is to split off the surplus material. The process will be easy if the proposed fracture is in the direct line of cleavage in that particular stone. If not, his attempt at splitting may ruin a gem of countless price. Shall he make the attempt? He must be both an expert and a man of nerve. If he be so, a single feat of successful polishing may bring him fortune and the reputation of a master, while a single disastrous venture may quite undo him.

The early lapidaries dared not attempt the splitting of a stone to correct its faults or alter its natural form. Every stone was estimated according to the impression it made upon the scales. Hence its facets were only smooth flat surfaces ground upon the rounded exterior, — an unmitigated rose cut of trivial triangles, or a terraced surface of rings and bands. The master of his craft to-day must make his diamonds perfect reflectors of light at

Cleaving.

all hazards.   If any excrescence exists, he must cut it away, or the light which enters a flattened surface may be so entangled that it will never emerge.   When he takes up a cross-grained, defective stone, he will reject it.   Like a true surgeon he will quickly discern how he may remove most safely a defective part, and will proceed boldly with his task.   His first step in the work is to scratch the surface round the part to be split off with another diamond.   Having made the diamond fast in a cement bed composed of brickdust and resin, he applies the edge of a steel knife to the scratched surface, and strikes a quick, hard blow with a slender rod. If he has struck the lines of cleavage, the external scale is at once removed, for the diamond, despite its hardness, is quite easily fractured. Then the split surface must be polished. If no other scale or marked inequality needs removing by splitting, the next operation is that of grinding.

Cutting

As the diamond is the hardest of all known substances, it is evident that much patience and strength are required in reducing

Cutting.

its size or altering its rough figure by grinding. The ordinary file would serve to reduce some other gems, but it will not touch the diamond. Diamond cut diamond is not merely a current phrase, for diamond dust is now invariably used in polishing or grinding this precious stone.

For cutting and polishing purposes the lapidary has a table above which a flat steel wheel revolves horizontally. On the upper surface of this wheel are fine grooves or striæ, cut angling

from its centre to its perimeter.   By means of belts beneath the
table, the grinding wheel is made to turn at a rate from three to
four thousand revolutions a minute.   Diamond dust mixed with
olive oil is applied to the upper face of the wheel, and against
this erosive surface is held the diamond to be ground or cut.

For this object the diamond is set in a fusible solder on the
end of a copper cupel which is held firmly against the surface of
the wheel by a small projecting arm and clamp.   By adjusting
this holder, the lapidary presses the exposed face of the stone on
the revolving wheel until the desired amount of material has
been ground away and the proper angles turned.   Such work in
its finishing stages cannot be intrusted to a tyro or experimenter.
Unusual patience and steadiness of nerve are required for such a
task.   When the facet is finished, the workman wipes the dust
off and tests its smoothness and finish, after which he resets the
diamond, leaving the uncut facet exposed which he intends to
cut next.

Most of the "skill" facets and "underskill" facets are made
by grinding, while the lozenge and larger faces are first shaped,
when possible, by cleaving.   If the stone is thick enough to
form a brilliant, the lapidary first forms the table, and then successively the adjacent facets and lozenges.   The table must be
absolutely flawless and smooth, while all the surrounding facets
in an ideal brilliant must hold the same precise angles and have
their shape correspond to the thousandth of an inch.   After
completing the bezel, the pavilion is next developed.   The
underskill facets of the pavilion must match exactly at the girdle
with those of the bezel, and the girdle when finished should be
as sharp as a knife.   Some lapidaries leave the girdle blunt, but
with a great sacrifice of brilliancy in the gem.   The triangles
and lozenges of the pavilion must, of course, be much larger
than those of the bezel.

There is a still simpler method of cutting diamonds by a
device attributed to Berquem.   Two uncut stones are cemented
into the ends of two sticks resembling penholders in shape.
Then the operator grasps these handles and presses the stones

CUTTING.

against each other with a rubbing motion over a trough. Considerable leverage is obtained for the rubbing by resting the holders against projectors at the sides of the trough. The exposed face of the stones is coated with diamond dust to advance the process. In this laborious way facets may be ground, and the cutting may be completed by repeatedly refixing the stones in the cement. Expert handling is necessary to keep the diamonds from becoming overheated by the constant friction.

Polishing

The third process is that of polishing. The method employed does not differ materially from that adopted in cutting, described above; but as this is the finishing process, all irregularities in faceting must be corrected and the practised eye of the artist must detect and remedy every defect. Each line and angle must be made geometrically correct; each facet and lozenge must be shaped to perfection. The colorless stone must glisten pure as a dewdrop sparkling in the sun, producing the colors of the prismatic spectrum ; the gem of red or blue or green color must flash forth its hue with intense brilliancy.

Such exact and delicate alignments are not the work of a day, though the time required has been greatly shortened by modern methods. The patience of weeks and even of months must be expended in perfecting these tiny crystals. It is said that it was the work of two years to cut the celebrated Pitt diamond, now among the French jewels, and the lapidary received for his skill and labor the sum of £3500 or $17,500. The last cutting of the Koh-i-nûr by Coster of Amsterdam in thirty-eight days was unusually rapid. The ablest workmen in Holland were engaged continuously on it and the wheel was driven by steam power; yet it cost $40,000 to do the work and the diamond lost eighty-four carats in weight.[1]

---

[1] " Great Diamonds of the World," Streeter, 1882.

POLISHING.

Loss in Cutting

No general rule can be stated covering loss which occurs in cutting gems. The waste depends on the character of the stone, — its perfect natural form and crystallization, its purity, and the style of the cut adopted. Perfect octahedrons lose two-fifths of their weight, if cut as brilliants. Rhombohedrons will lose over half of their weight in taking the same form, and stones of other shape will lose as much or more. The following figures will show at what cost of substance some of the natural gems have been perfected in the process of their cutting: The Mogul in its rough outer coat weighed originally $780\frac{1}{2}$ carats; when cut it weighed only $279\frac{9}{16}$ carats, a loss of nearly two-thirds. The Regent weighed 410 carats and was reduced to $136\frac{14}{16}$ carats. The weight of the Koh-i-nûr was originally 793 carats. It was first cut unskilfully by Hortensio Borgio to $186\frac{1}{2}$ carats, and a second cutting reduced it to $102\frac{1}{2}$ carats, — a loss by both processes of the astonishing amount of $690\frac{1}{2}$ carats, or more than six and a half times its present weight. The L'Etoile du Sud shrunk from $254\frac{1}{2}$ carats to $124\frac{9}{16}$ carats in the process of cutting.[1] The average loss of South African diamonds by cutting is from one-half to three-fifths of their gross weight. The $428\frac{1}{2}$ carat diamond found in De Beers mine lost 200 carats in cutting.

It has been demonstrated in cutting that diamonds are of different degrees of hardness and that the same stone may exhibit different degrees on different faces. The Koh-i-nûr is a signal example of this fact. In cutting the facet near a yellow flaw, the section grew noticeably hard, until six hours' grinding at a speed of 2400 revolutions a minute produced only the faintest change. A speed of 3000 revolutions was necessary to cause any perceptible loss of material on that facet. A speed of 4000 revolutions a minute is about the average now in vogue at Amsterdam.[2]

[1] "Famous Diamonds of the World," Streeter, 1882.

[2] Description by Messrs. Veder and Rozelaar, dated 6th March, 1902.

There is another material loss occurring in cutting or in the handling of rough diamonds from a curious infirmity of some of these crystals. The explosion of diamonds sometimes occurs, and the loss is the greater because large stones are more liable to explode or fly into pieces than small ones. This phenomenon is attributed to the heat of the hot solder, or frictional heat of the revolving disk.

The Lapidaries

Early handlers of the diamond were hardly more than polishers, striving to produce an even, glistening surface, and satisfied to retain the natural face of the stone, or to grind away some upper portion of the crust. This clearly appears from the many old, half-polished stones that have been found in treasuries of gems. A signal instance is shown on the royal mantle of Charlemagne, still preserved in the French National Collection. In the clasp of this robe are diamonds whose natural octahedral faces have been simply polished. In ancient church furnishings diamonds have been found with an upper table and four polished borders, and the lower sides cut as four-sided prisms or pyramids. Streeter quotes this inventory of the Duke of Anjou's jewels exhibited in 1360 A.D.: (1) a diamond of a shield shape, from a reliquary; (2) two small diamonds from the same reliquary, with three flat-cut, four-cornered facets on both sides; (3) a small diamond in the form of a round mirror; (4) a thick diamond with four facets; (5) a diamond in the form of a lozenge; (6) an eight-sided, and (7) a six-sided plain diamond.[1] We must allow, of course, for the mistakes and the ignorance of those who may have catalogued rock crystals for diamonds, but granting that some were diamonds, their existence shows what forms were then prevalent and the real development of diamond cutting.

Previous to the success of de Berquem as a lapidary, there were polishers and cutters in Paris and at Nuremburg, as has been noted. A guild was organized in Paris in 1290 A.D., and the table cutters joined in a guild with the stone engravers in

[1] "Precious Stones and Gems," Streeter, 1880.

Nuremburg, which became a primitive Lapidaries' Union. They received and taught the apprentices on the strict condition of their contract to serve for five or six years before undertaking business for themselves. The artist Hermann won for himself an honorable name in France as early as 1407 A.D., and Gutenberg, the originator of the art of printing from block type, learned the lapidary's art at Strasburg in 1434. At a dinner given by the extravagant Charles the Bold, the Duke of Burgundy, to the king of France, he bestowed upon his guests eleven diamonds partially cut and mounted in gold. Undoubtedly these jewels were declared in the height of the fashion by European artists. But it is from de Berquem's day that the profession reckons its firm establishment, and his contemporaries have acknowledged him as the father of his art.

Though many Europeans have become skilful workers at this trade, the most successful lapidaries have been of Hebrew stock. The Jews had, at one time, the monopoly of the trade in diamonds in Portugal, and their especial centre of business was Lisbon. The old "Lisbon cut" of diamonds has never been surpassed for perfection and beauty of workmanship. But unfortunately for Portugal and for the Jews, religious bigotry kindled the fires of persecution against this ancient people, and they were expelled from the kingdom. Hospitable little Holland opened her doors to receive the exiled merchants and lapidaries, and Amsterdam has since become the central mart for the diamond merchant and his comrade, the diamond cutter. Out of thirty-five thousand Jews who reside there, at least a third are engaged in one department or another of the diamond industry.

The settlement of some of de Berquem's pupils at Amsterdam was probably the reason why the exiled Jews selected that city as their home; but others went to Antwerp and some made their homes in Paris. The descendants of these expatriated Jews received especial encouragement and protection from Cardinal Mazarin two hundred years later. Twelve of the largest crown jewels were intrusted to them to be recut on Berquem's

POLISHING.

principle. Their success was so marked that these stones were afterwards known as the " Twelve Mazarins." Unfortunately these rare gems were poorly guarded and all but the tenth had disappeared by 1791. The French cutter Jarlet gained an international reputation in the seventeenth century by cutting one of the notable jewels of the Russian crown weighing 90 carats, but the industry withered in France in spite of its special encouragement by Mazarin and other powerful ministers.

England and Holland had secured almost exclusive trade relations with the East, from whence the diamond supply was obtained. Hence the Hebrews of these countries secured control of the diamond industry, and French lapidaries sought employment in vain. Then the revocation of the Edict of Nantes flooded Holland with French refugees. Of the seventy-five diamond cutters whom Mazarin had so carefully guarded, only five remained in 1775. Inquiry showed that the total rough diamond stock in Paris, just before the outbreak of the Revolution of 1789, was only 3832 carats, and there was little employment obtainable in recutting old stones. During the Revolution and the troubled Napoleonic reign, the industry was fatally paralyzed, and diamonds were sent out of France to Antwerp for cutting.

In the eighteenth century there was a determined push in England to foster the diamond-cutting industry, and some expert workmen, headed by Ralph Potter, made a stout struggle to hold the home trade. The so-called " Old English style " was developed on strict mathematical lines, and gems cut by these artists are still eagerly sought as models of the lapidary's art; but the centralizing drift to Holland was too strong for competition until the discovery of the South African Diamond Fields. In the last twenty years the languishing art has raised its head in England, and become, without doubt, a well-established industry. A hundred and fifty years ago London was accounted the chief centre of business for lapidaries, and it is not beyond expectation that its former preëminence may be reëstablished. Even now it is thought that diamond crystals are cut in London town as well as the work can be done in Amsterdam.

In the United States the late Henry G. Morse of Boston was the pioneer in establishing the lapidary business on a successful footing early in the last (nineteenth) century. He opened his workshop in Boston in 1866, and made several important improvements upon the cumbrous machinery in use in Europe. His business was confined, at first, to recutting and polishing damaged gems; but the influx of South African diamonds brought about speedily an enlargement of his works and the employment of thirty expert hands. At the start only foreign workmen were engaged, but Mr. Morse succeeded in training American women to a height of proficiency as lapidaries which rivalled the best foreign work. Among the fine gems cut and polished in his shop were four weighing fifty carats each, and he later scored a notable success with the cutting of a superb South African diamond weighing 125 carats. The brilliant fashioned from this stone weighed 77 carats, and has been greatly admired by connoisseurs as a specimen of exquisite beauty and purity developed by perfect workmanship. The cutting and finishing of this gem was a work occupying three and a half months.

In spite of this well-designed and ably pushed venture of Mr. Morse, American lapidaries have struggled continuously under serious handicaps. The United States is not a producer of diamonds, and Europe is the established mart for rough stones from India, Africa, and Brazil. Moreover, the business of diamond cutting has been so firmly rooted in Europe that the work naturally gravitates to these older establishments. Foreign lapidaries and dealers enjoy a further advantage in the fact that the banks of England and Holland make loans on uncut stones, knowing that the finished diamond is much enhanced in value, while American bankers do not grant such assistance to American cutters and dealers. Nevertheless, the work of diamond cutting has been so persistently developed that over half of the diamonds imported now enter as rough stones.[1]

[1] George F. Kunz.

A Group of Directors, General Manager, and Secretary of De Beers Consolidated Mines Ltd.

CHAPTER XIX

AN UPLIFTING POWER

HAT a change came over the dismal face of South Africa with the discovery and development of the diamond mines! In a former chapter it has been shown how dragging had been the advance from the few scattered settlements on the coast up to the year when the revelation of diamonds drew the first rush of prospectors to the banks of the Vaal. The yield of this marvellous field was, from the first, of material consequence in the sum of South African products, but it was of far greater importance in the stimulus which it gave to the flagging and stinted enterprises and the sinking hearts of the colonists.

The bits of iron hoop that scraped the diamond-bearing ground were as transforming as magicians' wands. The river

552

wash of the Vaal glittered like the diamond-strewn valley of
Sindbad. No Man's Land had the sparkle of diamond founts.
No part of the world was too remote to be dazzled by the vision
of the novel Golconda, and the black face of the despised karoo
changed in a twinkling to one of transcendent promise.

Then came the rush from every quarter of the globe of
ardent visionaries and fortune-hunters, streaming over the desert

Natives riding Bullocks.

sands of South Africa from every coast port to the Diamond
Fields, while from far inland the tribesmen flocked to the same
glittering beacon. Bitter experience rubbed the glamour from
the eyes of thousands of visionaries, who trooped back disheartened, but the plucky and the lucky held their ground and
thousands came streaming in to take the places of the faint-
hearted. It followed naturally, too, that thousands who would
never have come to South Africa except at the beckoning of the

diamond lure, remained in the country even after the blasting of their hopes of diamond winning. Many were ashamed to

Pioneers Trekking.

run away with the confession of failure; many were too poor to get away, and many were keen to see the profitable openings in other occupations for their work and savings. So every industry in the colonies gained new headway with the influx

A Prospecting Expedition to Mashonaland.

of capital and labor. Supplies of all kinds were needed for the bustling diamond camps and the flow of travel between the mines and the coast.

Trekking on the Veld.

This demand was quite enough to stir the pulse of production in every part of South Africa, and the heartening impulse thus given was sustained and advanced far beyond the stretch of this novel requirement by the rising faith in the possibilities and future of South Africa as a field for investment, which now began to lift the drooping spirits of the colonists and to attract the coöperation of the home country and the leading nations of the world. Hopeful prospectors rambled off farther and farther over the deserts and ranges, or followed the water-courses, testing the sands for diamonds or gold, and picking at every promising ledge in their search for ore. Pioneer

Trekking on the Veld.

graziers trekked to new pasture grounds with their flocks and herds.   Abandoned farms were reoccupied and virgin soil upturned for crops.   Manufactures of various kinds began to spring up and multiply.   Not only Cape Town but little coast ports were thick set with steamers busily discharging cargoes on piers or in lighters and bidding for exports at rates highly stimulating to the products of the Colonies.

The march of development was signally marked in the con-

The first and only Load of Cotton raised in South Africa.

struction of railways to meet the pressing demands for inland communication and transportation, and especially the imperative call of the Diamond Fields.   The progress of mining was greatly handicapped from the start by the heavy cost of dragging supplies in lumbering ox-wagons for hundreds of miles from the coast ports, and the patent impossibility of moving any large plant in this way for mine opening or diamond winning.   The pioneer railway from Cape Town to Wellington barely covered a twelfth of the stretch from the coast to the mines, and the little lines from Salt River to Wynberg, and from Port Elizabeth to Uitenhage were of no service in the advance of transportation.

However, capital was wary and loath to invest in any of the projects for railway building into the heart of South Africa, until the continued working of the Diamond Fields for three years convinced investors that rich diamond deposits were indeed

Zebra.

open, whose continuance in depth might reasonably be anticipated. When the export of diamonds in 1872 amounted to over £1,000,000, the Cape Government authorized the purchase of the sixty miles of railway then in place in the colony and sanctioned the extension of the existing line and the construction of railways from Port Elizabeth and East London. It was a heavy strain to raise capital for the extension of all simultaneously; so the advances were groping and slow. At

Coach leaving Kimberley for the Transvaal Gold Fields before the Day of the Railway.

some points, too, there was even an obstinate fight against any railway extension in their direction.   Little towns that had been centres of distribution for surrounding districts feared that the

Cape Town and Table Mountain from Table Bay.

railroad would divert some trade and shake their preëminence. It was further contended that the whole scheme was chimerical and that the uncertain yield of the South African farms and the wide-ranging pasture lands could not pay interest on the cost of construction and maintenance. One of the most persistent objectors, the town of Worcester, actually sent in five petitions against the extension of the railway line to that point.

A Group of Well-known Kimberley Men.

There was further protracted disputing over the proper gauge for adoption after an extension had been determined. The short line already constructed from Cape Town to Wellington was of the standard English gauge, 4 feet $8\frac{1}{2}$ inches, but the continuance of this gauge was generally opposed. It was urged that the light traffic of the country would not warrant the heavy outlay requisite for the construction of lines of the standard adapted to the requirements of a thickly settled country like Great Britain, and that only narrow-gauge lines were practicable. Some would have pushed this reduction of gauge to 2 feet 6 inches, or even less, in view of the fact that some of the lines might be reduced to the necessity of resorting to tram car and mule service, but the final conclusion was the adoption of a standard gauge of 3 feet 6 inches. It was properly recognized that uniformity of gauge, at any rate, was essential for intercommunication, and whimsical notions of construction were not suffered to break this uniformity. Time has shown the fallacy of these pessimistic predictions as well as the adoption of the 3 feet 6 inches gauge.

There was, however, one essential error in the whole scheme of construction. The pressure of the demand of widely separated points for railway construction was so hard to resist that the Par-

Eland.

liamentary authorization for railway extensions was far in excess of what was feasible at the time in view of the limited capital that could be secured for the prosecution of the scheme. The rivalry of the principal ports was too keen to permit of the drafting of any coöperative plan of extension, for the superior accommodation, even temporarily, accorded to any one port would be challenged by others as injurious favoritism. So, instead of carrying forward a single main line by the most direct or feasible route to the Diamond Fields to meet the most pressing demands for communication, there was for many years only a crawling advance from the

CAPE TOWN AND TABLE MOUNTAIN.

Table Mountain, from the West.

competing coast ports, Cape Town and Port Elizabeth.   Neither of these competing lines were able to pay any adequate return upon the capital invested, and the common aim of reaching the Diamond Fields was blocked and greatly delayed.   Kimberley was distant only 485 miles from the nearest outlet on the coast, Port Elizabeth; but 1600 miles of converging railway lines were actually built before one was extended from De Aar to Kimberley, in November, 1885, then first putting the richly productive diamond mines in railway communication with the coast.

All the lines in operation at this time were single lines, with the exception of the Cape Town-Wynberg line, and the first six miles of the Port Elizabeth-Uitenhage line.   The most difficult engineering in the course of this railway extension was in the crossing of the barrier range of mountains forming the ridge of the karoo plateau.   After repeated surveys an entrance for the

CAPE TOWN.

line from Cape Town was effected through the Hex River Valley with a gradual ascent to Hex River East, where the line begins to climb the mountains by sweeping curves and zigzags, piercing some of the spurs in tunnels, and spanning gulleys with viaducts, until it attains its highest elevation of 3588 feet at Pieter Meintjes Fontein, 77 miles from Worcester. This is a trifle higher than

the summit of Table Mountain, which rises in air 3582 feet above Cape Town. For a stretch of more than 20 miles in the ascent of this ridge, the gradients are one in 40 and one in 45, with curves of five chains radius.

In the year following the extension to Kimberley there was a fortunate impulse to the extension and operation of all the lines by the discovery of the Witwatersrand

His Excellency Lord Milner, Governor of the Cape Colony.

Gold Fields. Then first appeared some substantial prospect of profit for all the competing lines by the addition of another great centre of attraction and production. The junction of the Cape Town and Port Elizabeth line at De Aar, in March, 1884, had largely diverted the flow of freight and passenger traffic between the Diamond Fields and the coast, which, for some years, had been passing principally along the line through Graaff Reinet; but the rise of Johannesburg offset this loss to the Port Elizabeth,

Graaff Reinet, and East London lines.   The linking of the Diamond and Gold Fields by direct railway communication from Kimberley to Johannesburg was apparently of high importance; but this extension has been blocked for years by the action of the Orange Free State in refusing to build the line themselves or to allow either the Cape Government or private corporations to construct it.   Several years were passed in dilly-dallying before sanction was given to the Cape Government for an extension of a connecting link of the Cape Town and Port Elizabeth lines

Donkey Transport.

from Naauw Port across the Orange River at Norvals Pont, thence to Bloemfontein and the Vaal River at Vereeniging, where it connected with the Netherland Company's system of the Transvaal.   It was not until September, 1892, that the first through train from the Cape reached Pretoria, but after the essential link was constructed, the Cape Town and Port Elizabeth lines contrived to secure the greater share of the Transvaal traffic for the next three or four years.

While these two lines were delayed in reaching out for the business of the Gold Fields, a more favored competitor, the Netherlands Railway Company, was actively building an eastern line from Portuguese territory through Middleburg to Pretoria, and shortly afterward running radial lines southeast to Natal and southwest to Klerksdorp. By the extension of these well-designed lines the first through train from Delagoa Bay arrived in Johannesburg in November, 1894, and the first train from Natal in December, 1895. More than two years later, in March, 1898, the dragging extension of the Graaff Reinet line was opened to Rosmead Junction, on the main line to Port Elizabeth, and was then in a position to assist in carrying merchandise from the coast to the Diamond and Gold Fields.

A South African Farm.

While these railway extensions essential to the development of the existing States and Colonies in South Africa were more or less efficiently accomplished, the grand project of Mr. Rhodes for a railway running far north into the heart of Africa was most energetically prosecuted. By the advance of his exploration and colonization plan, to be hereafter described, the range of British territory was extended from Table Bay to the shores of Lakes Tanganyika and Moero.

The line from Kimberley was opened to Vryburg, 774 miles from Cape Town, December 1, 1890.[1] Thus far the conservative government was prevailed upon to proceed, but the profit

[1] Report of the General Manager of Railways, Cape of Good Hope, 1898.

from any further extension seemed so essentially speculative that it is very doubtful if any further advance would have been made, had it not been for the daring enterprise of the Bechuana-land Railway Company, an organization promoted and financed by Mr. Rhodes and his far-sighted associates.   Following hard

upon the heels of the pioneers in Mashonaland and the conquest of Matabeleland the line from Vryburg was opened to Bulawayo in No-vember, 1897.

When the grand importance of this railway advance became clear, even to the doubters, the Brit-

Natives shearing Sheep.

ish Government subsequently guaranteed a loan of £2,000,000 to carry the line 800 miles farther on to Lake Tanganyika.

With the rate of progress attained it was expected that Aber-corn at the foot of Lake Tanganyika would be reached in four years, but the outbreak of the war with the South African States was an unlooked-for clog to this advance.   As soon as the line has reached Lake Tanganyika a further extension of 600 miles to Uganda through the Congo Free State has been guaranteed by an appropriation of the needed funds by vote of the share-holders of the African Transcontinental Railway Company. Besides this main line of advance, the Beira Railway, which was constructed with a gauge of two feet, had been completed and engines were running as far as Salisbury over a stretch of line 375 miles in length before the close of 1900.   The narrow gauge of two feet was soon found to be unworkable, and the line has already been relaid from Beira to Umtali with heavier rails and with the standard South African gauge of 3 feet 6 inches, the remaining stretch from Umtali to Salisbury having been originally laid with the broader gauge.

In spite of the lack of coöperation and capital, and all other impediments and delays in view of the character of the country, the advance of railway systems in South Africa has been phenomenal in the last few years. Including the six or seven short private lines constructed for the advance of mining operations and suburban and other local traffic, there were 2264 miles of railway in Cape Colony at the close of the year 1898. The Transvaal came next with 777 miles, followed by Rhodesia with 604 miles, Natal with 465 miles, and the Free State with 361 miles. Besides this aggregate, 256 miles had been constructed

Dutch Farm in the Karoo.

in Portuguese territory, making a total of 4727 miles of railway actually opened and working in South Africa, and more than half as much more in process of construction, or guaranteed by appropriations.

In the struggle to reach the goal of the Diamond Fields, with the handicap of the lack of capital, it is not surprising that much of the roadways and the rolling stock fell below any high modern standard. The light rails and rickety cars answered the purpose

Beaufort West, a Little Town on the High Veld.

of the day, however, fairly well, and have since been largely
replaced by a plant that will bear wear and tear, but is still not
up to the requirements.   Makeshift bridges were soon supplanted by durable structures, and other engineering works on
the lines of the various systems have also been greatly improved.
The engineers who advanced the pioneer lines deserve, on the
whole, high credit for their energy and talent in piercing or traversing the barriers in their way.   At the time when the first
train reached the Witwatersrand Gold Fields, at the close of
1892, there were somewhat more than 8500 bridges, culverts, and
cuts to be counted on the various lines.   Some of this bridge
construction, especially the bridges across the Orange and Vaal
rivers, was of a high order of excellence.   The Orange River
bridge on the Kimberley line has a length of 1230 feet, with
open spans of 130 feet each between the piers.   The Bethulie
bridge is 1486 feet long, and the Norvals Pont bridge,[1] the
longest of all, has 13 spans of 130 feet, and a total stretch
closely approaching 1700 feet.   The total cost of this fine
bridge was £76,593.   At Fourteen Streams,[2] on the Vaal River,
there is a bridge of ten spans of 133 feet that is fittingly classed
with the chief Orange River bridges.

[1] See photograph, Chapter XX, of this bridge after destruction by the Boers.   [2] *Ibid.*

For rapidity of building railways the palm has heretofore been claimed by America, but the best American records have been challenged repeatedly in the advance of the African Transcontinental Railway, and it is now claimed that the world's record for rapid construction and bridge building has been captured by The Patent Shaft and Axletree Company, of Wednesbury, England. The Boers had effected the isolation of General White and his men in Ladysmith by blowing up the two railway bridges on the Tugela River at Colenso and Frere, and, promptly on learning of the destruction of these bridges, the Natal Government took steps for their rebuilding. The crossing over the Tugela at Colenso was designed in five spans of 105 feet each, and the crossing at Frere of two spans of the same length. The call for the utmost haste in construction was imperative, and tenders were invited, both in England and America. The contract was awarded to The Patent Shaft and Axletree Company upon its undertaking to deliver the first span in six weeks from the day of the contract. The order was given on the 21st of December, 1899, and the first span was finished on the 13th of January, 1900, or in nineteen working days. When the

Norvals Pont Bridge.

order was received, nothing was in stock at the company's works from which the structural steel was rolled, yet at five o'clock on the afternoon of the day of the order 100 tons had been rolled at the company's works, and tested and approved by the engineer of the Natal Government. Each of the spans weighs 105 tons, or a ton to the lineal foot of the bridge. There was about 7500 feet of planing work, and 69,000 rivet holes were drilled in each span; yet on January 19, two of the spans had been built and work begun on three more, while the material for the whole seven for both bridges had been rolled, cut to size, tested, and approved.[1]

Besides government railway building, important private lines have been constructed for the operation of large mining works, and local or suburban traffic. The oldest of these lines was the undertaking of the Cape Copper Company, covering a stretch of 100 miles from Port Nolloth to Ookiep. Some of the grades on this line are very notably steep, exceeding any others where ordinary steam-engine traction is employed. For this service special engines were constructed by Litson & Co. of Leeds, which have been working very successfully since October, 1890. On Klipfontein mountain there is a rise of 1330 feet in $7\frac{1}{2}$ miles, and in several sections the gradient reaches the extreme of 1 foot in 19. This line was built for the development of the Namaqualand copper mines, one of the most profitable undertakings in the Colony. The Cape Copper Company owns most of the paying mines, and has been extracting annually about 30,000 tons of ore, averaging nearly 20 per cent in copper.

The Indwe Railway Company's line is only second to this, with a length of $66\frac{1}{2}$ miles. This line was opened in 1896 to reach the Indwe coal mines, and is operated by the Cape Government as a branch of the Eastern System which it joins at Sterkstroom. It was built by the Indwe Company with the material assistance of De Beers, which subscribed £75,000 to its working capital. The Company owned all its rolling stock, but it was operated under the supervision of the Cape Government Railway Department. This railway has been lately sold to the Cape Government, and is to be extended to Natal.

It is computed that the lines owned and operated by the Cape Government have cost, with their rolling stock, about £20,000,000, representing the investment of about £7200 per mile. The capital invested in the Natal lines was £6,750,000; showing an outlay of £15,000 per mile. The 777 miles of

[1] *The Engineer*, London, England, February, 1900. *The Scientific American*, February 24, 1900.

Netherlands Company cost nearly £9,000,000, or £13,500 per mile. This gives a total of £38,000,000 for the construction of 3500 miles of railway, not including lines owned and operated on private account. With all lines included, it is estimated that there is a total outlay of £56 per head of the white population of the country, which does not average more than 163 to the mile of railway opened.

In 1896 the earnings of the Cape Government railways came to something over £4,000,000, of the Natal railways £1,000,000, and of the Netherlands Company nearly £3,000,000. The net profit after paying interest on capital in the Cape was £1,221,675; in Natal £464,762; and in the Transvaal £1,328,424, making a total of over £3,000,000, not including the Free State share of profit, which for 1896 was £289,553.

Five extensions were authorized by the Volksraad resolution of the Free State in October, 1896. One line

His Excellency Sir Walter Hely Hutchinson, Late Governor of Natal.

through Fauresmith was to serve the diamond mines of Jagersfontein and Koffyfontein and place them in direct communication with the coast ports. In 1898 the Free State decided to build a railway by concession from Bloemfontein to Kimberley, and to extend the Springfontein-Fauresmith line to join the Bloemfontein-Kimberley line at a point near Petrusburg. The Springfontein-Fauresmith line forms a direct route between East London and Kimberley, shortening the present route by 100 miles, making East London 40 miles nearer to Kimberley than Port Elizabeth. The Bloemfontein-Kimberley line will reduce the present distance by rail between Kimberley and Johannesburg and Cape Town.   The war has, for the time being, stopped this work.

The Natal Government is also proceeding with the construction of north and south coast lines: one through Verulam to Zululand, and the other to the Cape border, where it will connect with the extension of the Storkstroom-Indwe line.

Twenty-five years ago only 781 miles of telegraph were open in all of South Africa.   A message of twenty words from Cape Town to East London cost 17s. 6d.   At the outbreak of the late war 19,000 miles of wire were working in Cape Colony, and probably 10,000 miles in other states and colonies.   The march of the telegraph through South Africa will be later detailed.

In addition to the railway and telegraph, several thousand miles of excellent roads have been made, and every river of magnitude has been spanned by substantial bridges.   The great Zwarte Berg Pass, which rises 3400 feet in eleven miles from base to summit, is one of the finest monuments of road construction to be seen in any country.

At every port the shipping accommodations have been extended and improved, and approaches to the coast have been made safer by construction of numerous lighthouses.

The impulse given by the Diamond Fields development for prospecting for mineral deposits of all kinds led to the discovery of the mines of Lydenburg, De Kaap, and the Rand.   In the year preceding the discovery of diamonds Thomas Baines had led a party from Durban to prospect for gold in Matabeleland, and secured a concession from Lobengula in April, 1870, to dig for gold in the district between the Gwelo and Ganyona rivers.   But Baines's party found no largely promising deposits, and without the excitement of the rush to Kimberley there would hardly have been any considerable and determined effort to push prospecting far beyond the Vaal.   Luckily, shortly after the rush to the Diamond Fields in 1871, reef gold was found by prospectors at Eersteling and Marabastad, and, two years later, gold placers were discovered about thirty-three miles east of Lydenburg, at Pilgrim's Rest.

These discoveries were greatly magnified in the fever of speculation excited by the opening of the diamond beds, and companies were formed in Natal and England to develop these gold-fields, while daring adventurers pushed still farther on, to the region north of the Limpopo, seeking the traces of the ancient mining works that were known to exist. Upon the report of the discovery at Lydenburg some fifteen hundred prospectors flocked to this field, and a year or two later gold was found in the Kaap Valley, fifty miles south of Lydenburg. The returns from the placers were hardly tempting enough to hold the gold seekers, and conflicts with the natives, followed by the outbreak of the war with the South African Republic in 1880, were further discouraging to any development in this region. After the war the exactions imposed by the South African Republic upon the prosecution of mining in the Lydenburg district were a check to outside prospecting.

In 1882 an Australian digger, Charles Durnin, found some very rich patches of gold-bearing ground on the Kantoor plateau in the Kaap Valley, and the rush to the Duivels Kantoor and Moodies brought to pass the first considerable undertaking of gold quartz mining in South Africa. Some gold mines showing great richness of ore were soon developed in this district, and the bustling mining town of Barberton marked the centre of a field which was thought to be of marvellous promise. Unfortunately the booming of the district ran to a pitch of insane and fraudulent speculation that was greatly damaging to the reputation of this field of investment, and gold mining undertakings in South Africa would commonly have been reckoned as "bubbles," had it not been for the uncovering, at this juncture, of the astonishing riches of the Rand.

Nearly twenty years before, the famous elephant hunter, H. Hartley, after marking the gold-bearing ground in Matabeleland and the region of the Zambesi, made his home on a farm in the Witwatersrand, unconsciously settling on the face of deposits of gold far more marvellous than any tradition of King Solomon's mines. Hartley died without any vision of the treasures over which he and others were tramping day after day. It was soon noticed by roving prospectors, and by settlers in the district, that there was gold-bearing sand in the beds of the little river and creeks rising in the Witwatersrand, but no noteworthy search for gold was attempted until an Australian mining man, Armfield, a reputed expert, was sent to prospect in this region during the British occupation from 1876 to 1880. He made some tests of quartz ledges on a farm adjoining Paardekraal, but found nothing of value.

The credit of the first important revelation of gold in the district undoubtedly belongs to Mr. Fred Struben, who had given his earnest attention to the gold developments in the Transvaal, and who prospected the Witwatersberg district in 1883, and found traces of gold in creeks and reefs, as well as ancient workings for copper. In the following year his elder brother, Mr. H. W. Struben, purchased two small farms on the northwestern end of Witwatersrand, and both the Strubens continued their prospecting energetically during the year. In the summer of 1884 a gold-bearing vein or reef was discovered and traced for several miles by Fred Struben. Ore shoots and pockets were found which assayed over one thousand ounces of gold and silver to the ton. The rich ledge, named the Confidence Reef, was supposed to be of prime importance; and the Strubens erected a five-stamp battery on the ground to crush the ore of this and neighboring ledges. Several samples of the ore were tested in the stamp mill, but the best ore yielded only eight pennyweight to the ton. Work on the Confidence Reef was greatly disappointing, for the gold-bearing rock was soon proven to be a small deposit.

In August, 1885, I visited a small mine called Kromdraai, situated about twenty miles, in a northwesterly direction, from the present site of Johannesburg, but at a much lower elevation, near the old Pretoria and Kimberley wagon road. A small reef of gold-bearing rock was being mined, and the ore crushed in a little mill in the immediate vicinity. I also spent a few days looking over the Confidence Reef, with the Strubens, who were

MR. RHODES'S HOUSE,
Groote Schuur, near Cape Town.

Entrance — Groote Schuur.

at that time the most enthusiastic and energetic prospectors, as well as the most enlightened and progressive men, in the Transvaal.   Mr. Henry Struben owned large estates near Pretoria.

The Strubens had spent over £11,000 in their mining operations in the Witwatersrand, and their venture seemed a losing one, when, in the spring of 1886, one of their employees, Walker, found the rich reef, now known as the Main Reef Leader, on the farm Langlaagte, about two miles west of the present Johannesburg. In July the first sample of the conglomerate from the reefs on the Langlaagte farm was panned in Kimberley. The showing was so remarkable that Mr. J. B. Robinson, backed by Mr. Alfred Beit, who saw the panning, started on the following day for the Rand. The Kimberley-Pretoria coach road ran through Potchefstroom, and thence northeast, leaving the little pioneer town called " Ferreira's Camp " (now Johannesburg) some fifteen or twenty miles to the east; but Robinson drove in a cart from Potchefstroom to the little sprawling camp that was the first sprout of Johannesburg. Within a day or two after his arrival he bought the Langlaagte farm for £7000. Scoffers, who posed as experts, told him bluntly that "a fool and his money were soon parted "; but he did not take heed of their gibes, and, before the end of the year, bought the whole of the ground comprised in the holdings of the " Robinson " Company for £13,100.

Messrs. Rhodes, Porges, Beit, and other enterprising men of Kimberley shortly followed Robinson in the pioneer work of the Rand. In January, 1897, the development work on the Robinson mine consisted of a hole in the ground, fifty feet deep, which was full of water. Robinson, who had been somewhat unfortunate on the Diamond Fields, went " nap " on the Gold Fields, and the rivalry between him and Rhodes was very keen. One story, with some foundation of fact at least, will show this. While Rhodes was trying to buy a farm from the Dutch owner, and they were parleying in the orchard, Rhodes conversing in English and bad Dutch, and the Boer in Dutch and bad English, Robinson arrived on the ground. He went direct to the farmhouse, and at once opened negotiations for the purchase of the farm with the Boer's vrouw. His familiarity with the " Taal," South African Dutch, was a telling advantage in his competition with Rhodes, and he reckoned shrewdly that the wife would jump at a bargain more quickly than the husband. So he slapped a handful of golden sovereigns on the table, saying smartly, "Those are for you." The old vrouw clutched greedily at the gold and called shrilly to her husband to come to the house. He obeyed the call dutifully, and when he entered the door he found that his wife had already sold the farm to Robinson. Even a henpecked man might have grumbled at such a sale, but when the simple Boer saw the heap of glittering sovereigns on the table,

Groote Schuur, after its Destruction by Fire.

he could not hold out stubbornly against a man who had so kindly presented his vrouw with so great "a mark of respect." While Rhodes stood in the orchard, Robinson got the farm.

In the early rush to the Rand, farms and mines were bought, not so much for any phenomenal richness, as for the fact that they showed more gold distributed over a greater stretch of country than had ever been disclosed in South Africa. The first two or three years were very disappointing, for the total output did not cover the taxes levied upon the mines by the Government. A large percentage of the gold was lost in working the

2 P

ores, for the precious metal was so extremely minute that it floated away with the water, and, at no considerable depth, a portion of the gold was held in the pyrites, and could not be recovered by means of the ordinary process of amalgamation. Some other process was needed that would save the minutely fine gold which became suspended in the water owing to the attachment of globules of air. When the Rand was discovered, no such process had been developed beyond the experimental stage. MacArthur and Forrest, of Glasgow, were experimenting with a solution of cyanide of potassium, which was known to be a solvent of gold. They found that the ores from the Rand readily yielded their gold when treated by this process, which soon came into general use. This was the saving of the Rand, for without such treatment only a few of the richer mines would to-day be paying properties.

A little more than a year after Robinson bought properties on Witwatersrand, the despised "cabbage field" of the Langlaagte farm was floated with a capital of £450,000, and yielded £950,000 in gold in the next five years, with a profit of nearly seventy-five per cent in dividends on the par value of the capital stock. The holdings of the Robinson Company, in the same time, produced over £1,400,000 in gold and paid £570,937 10s. in dividends to shareholders.

By the discovery of the diamond mines in Griqualand West, a product ranging over £80,000,000 in value in less than thirty years had been added to the meagre output of South Africa, and the gold mines of the Witwatersrand began, about fourteen years ago, to swell this great exhibit of the mineral riches of the land by the addition of gold already aggregating over £70,000,000.

The annual flow from the diamond mines has averaged, for years, over £4,000,000 in value and the Rand has greatly outstripped even this rich showing. Prior to the discovery of diamonds, the total tally of South African exports and imports combined was not £6,000,000 in value. In 1898 it was nearly £50,000,000, and, of the total exports, eighty per cent were mineral products.

Pioneers Trekking in Mashonaland.

With this general survey it is now practicable to trace with more clearness the essential and special services rendered in this grand development by Rhodes and his associates in De Beers Consolidated Mines and other organizations. Viewing, as he did, the control of the Diamond Fields very largely as the intermediary step toward the attainment of an aim far grander, the consolidation of the chief diamond mining properties had hardly been effected when Rhodes took action swiftly to extend and intrench the range of British influence north of the Transvaal by obtaining the concession of the mineral rights in Lobengula's kingdom of Matabeleland, through the adroit agency of Messrs. Charles D. Rudd, Rochefort Maguire, and Frank Thompson, in return for an annuity of £1200 and a coveted stock of rifles and ammunition. Lobengula made the grant which gave to Rhodes the needed nucleus for the creation of the grand exploring and developing agency which he pressed for incorporation as the British South Africa Company.

There was a natural hesitancy on the part of the public in

supporting this scheme at the outset, for the Matabele concession seemed to investors at large little more solid than a moonlit cloud bank, and even venturesome speculators shrank from buying shares in a prospecting license in a country held by savage blacks, trained in the school of Chaka to pillage and murder. But this incredulity was anticipated by Rhodes, and a solid backing was given to the enterprise by the subscription of De Beers Consolidated Mines for more than £200,000 of the working capital. This was a demonstration of good faith and practical intent so convincing that the British Government granted a charter formally to the new company in October, 1889, and it has since been popularly known as the Chartered Company. The government was reluctant to extend the working scope of the charter north of the Zambesi, but Rhodes's aim was not pent

Khama, a Noted Chief whose Country lies just South of Rhodesia.

up in Matabeleland, or Mashonaland, and by his forceful representations the British South Africa Company was left unrestricted in its range to the north, as far as it could advance without infringing on other concessions, or entering territory acquired by Germany or other nations of Europe.

There seemed, at first, some likelihood of competition and possible conflict of interests in the race of extension with another adventurous association, that applied for a charter as the African Lakes Company. But the risk was forestalled by Rhodes's foresight and promptness of action. The promoters of the African Lakes Company had spent all the capital they could raise, and were so dangerously near the verge of collapse that they welcomed the helping hands of Rhodes and his friends without much quibbling over the terms exacted. At once

£20,000 were subscribed by the organizers of the British South Africa Company to float the African Lakes Company, and a further subscription of £9000 a year was pledged in return for the right under certain conditions of merging the subsidized company in the British South Africa Company.

Then, with his unhampered charter and its range cleared to the source of the Nile, Rhodes was ready, like Davy Crockett,

Khama's Town.

to go ahead. After consulting with Frank Selous, the famous African hunter, and others familiar with the field, he pitched upon Mashonaland as the first base of operations. Dr. L. S. Jameson was deputed to go to Bulawayo and get Lobengula's express license for this undertaking. The envoy made all possible haste in his mission, and won the king's favor so quickly by his tactful bearing that the entry to Mashonaland was conceded. Rhodes lost no time in taking advantage of this opportunity. A force of five hundred armed men were enlisted under the chartered right to an adequate "police," and two hundred pioneers were hired to make a passable wagon road to Mashonaland. Colonel Pennyfather was placed in command.

Meanwhile, the fickle Lobengula changed his mind when Dr. Jameson was no longer by his side to persuade him, and sent

a message to the expedition, forbidding the road-making. The
messenger of the king met the British at Tuli, but the men
picked by Rhodes were not of a temper to be checked or frightened away, and the road was pushed ahead as fast as possible
through the thick brush and woods of the lowland, where the
peril from attack was most to be dreaded. Dr. Jameson rode
in the van with forty of the best mounted men as an advance

Khama's Hut.

guard. Selous led the pioneers and marked the roadway. Finally, on the 13th of August, 1890, the road-makers came to the
great plateau of Mashonaland, through an easy mountain pass,
and a heavy weight was lifted from the minds of the leaders, for
on this open plateau hostile attacks were no longer to be dreaded,
and a few hundred well armed and mounted men might well defy
a horde of marching Matabeles. It is probable that this daring
advance would not have been made unmolested, if Lobengula's
attention had not been artfully distracted by a feint of entry in
another quarter made by a body of Bechuanaland police on the

southwest border of Matabeleland. Thus was the first grip of civilization secured on the rich territory which now bears fitly the name of " Rhodesia," in lasting commemoration of the grand foresight and enterprise of its redeemer from barbarism, Cecil John Rhodes.

No sooner was this entry effected than Rhodes's untiring energy sought further extensions of British control. By treaty with the native chief, Umtasa, the neighboring Manica was brought under the same protecting power as Mashonaland, and

Matabele Women, Carrying Water.

a footing was gained with the like expedition in the native province of Gazaland. It was obvious that no extended development of the resources of this territory or stable colonization could be effected without railway connection with the Cape, and Rhodes at once undertook the provision of capital for the essential extension of the Transcontinental Railway through Mafeking to Mashonaland. He raised the money required, besides drawing heavily upon his private fortune, at the same time, for the Beira Railway extension. He contributed also four-fifths of the capital of the Transcontinental Telegraph, and, all this

A Group of Visitors at Groote Schuur.

while, bearing in great part the extraordinary expenses, amount-ing to £250,000 annually for the first two years, of the develop-ment of the undertaking of the British South Africa Company in Mashonaland.

Fortunately, by the extraordinary executive ability of Dr. Jameson, who was appointed Administrator for the Chartered Company in 1891, the immense outlay required of the company was reduced to only £30,000 annually. The thriving town of Victoria was founded and the settlement of the country was most energetically pressed in spite of every obstacle. But when the way for the profitable advance of the company's operations seemed to be clearing, its Colony was menaced in 1893 with utter destruction by the attack of the fierce Matabeles.

Lobengula had viewed the entry of the Rhodes expedition into the territory north of his kingdom with rising disgust, accentuated by his failure to stop it, but it was two years before he came to the point of open attack. He had been accustomed, all his life long, to regard the district occupied by his neighbors, the weak and unwarlike Mashonas, as convenient harrying ground for his brutal forays. Marauding troops of freebooters were constantly harassing the poor Mashonas, and oftentimes the king would send his robbing and murdering expeditions to scourge the land, just as he sent his impis to take Ugami, — to despoil and enslave and massacre the Batuwani, — and, across the Zambesi, to raid the Mashukulumbwe or the Barotse.

To the sorely persecuted Mashonas the coming of the English was an assurance of protection which was greatly welcomed,

Great Kafir Kraal, Mochudi, Matabeleland.

but even the presence of the bold white men and the unfolding of the British flag did not stop the marauding. Dr. Jameson protested over and over again to Lobengula, but the king was deaf. Finally, in July, 1893, parties of the Matabeles pushed their ferocious raids contemptuously up to the very bounds of the township of Victoria, and the English could not look on unmoved. Then Dr. Jameson sent a squad of police to warn off the marauders. The Matabele insolently fired on the guard, and the police charged and drove them flying.

This wholly rightful rebuff upset the temper of Lobengula, who was stuffed with barbaric conceit. His impis began pouring over the border, and the infant Colony was threatened with extinction. The menace was met by a heroic response. There were only a handful of police at the time in Mashonaland, but the settlers were men who could defend themselves.

It was judged best to meet the roving assaults by a direct counter attack on Lobengula's stronghold, his capital of Bulawayo. The Chartered Company's funds were drained out; but Rhodes, as ever, rose to the occasion, and raised the money imperatively needed to arm and conduct the little force that was to make the daring venture into the heart of the most savage and warlike province in South Africa. With this backing Dr. Jameson raised a force of about nine hundred men, and, placing himself at their head, as Commander-in-chief, marched on Bulawayo. Just after crossing the Shangani River, his little army was attacked by the Matabele in force, but beat off their assailants. With the encouragement of this success the English pushed on to the Imbesi.

Here they were attacked again in the old Zulu fashion by desperate charges of seven thousand frantic blacks in rank after rank of impis upon the well-prepared English camp. It was a fierce fight, but the issue did not hang long in doubt. The Matabele were as dashing and reckless as the impis that had fallen like breakers of surf on the laagers of the Boers during their "Great Trek." But they were overmatched by the ceaseless belching of machine guns and repeating rifles, mowing them down swath by swath when they charged within close range. At last they broke and fled, and Dr. Jameson's little army marched on to Bulawayo, which was entered without further fighting, for the disheartened Lobengula abandoned his capital.

The British pursued him hotly, for it was highly important to put a decisive end to the war by his capture before the advance of the rainy season. Unfortunately this pursuit was too daringly pressed. Major Wilson, with a little force of less than forty mounted men, nearly plucked Lobengula out of the

GROOTE SCHUUR, BACK OF HOUSE.

midst of his retreating impis.   His impetuous rush reached the cart that carried the king; but the desperate Matabele flung themselves upon this little troop in such masses that its advance was checked.   Messages for help were sent to the main troop in pursuit, under Major Forbes, but succor was cut off by the rising of the Shangani River.   Flight would have saved most of them, but Wilson and his men who were able to ride off scorned to abandon their wounded comrades.   So, hard pressed by the Matabele on all sides, they made a barrier of their horses, living

and dead, and held their ground until their last cartridge was fired.   Then they stood up defiantly and fought hand to hand until the last man was cut down and trampled under foot in the crush of the savage blacks. The troop under Major Forbes was forced to retreat,

Major Wilson's Grave.

and suffered much privation before it was met by a relief party headed by Rhodes, who rode out from Bulawayo.

The loss of Major Wilson and his gallant men was deeply mourned; but the campaign as a whole was a most brilliant success.   Lobengula's power was completely broken, his impis scattered, and he soon afterward died a fugitive.   The royal city of Bulawayo was made the capital of Rhodesia, the province of the Chartered Company, and Dr. Jameson took his seat there as Administrator.   The rich mineral ground near Bulawayo soon attracted a considerable influx and made a rising town, which in less than three years boasted of its banks, clubs, newspapers, electric lighting, and water-works. The brave colonists who made up the force of "Jameson's Volunteers" were disbanded, and began to prospect for gold and pick out farms in the new province.

With the fall of Lobengula, a standing menace to the march of settlement was removed, and the attractions of Rhodesia began to come out in a brighter light. It was the settled purpose of the Chartered Company, from the outset, to do everything that was feasible to encourage investigation and the taking up of farms by honest and thrifty colonists. This was regarded, by Rhodes at least, as transcending in importance even the development of the mineral riches of the country, though the latter was naturally the chief object with most investors. Particular pains has been taken, in directing the colonization, to harmonize relations between the men of different races and nations, and to draw as closely as possible all together in a common bond of union as Africanders.

Considerate and elevating treatment of the natives has also been a notable feature of the determination and policy of the Chartered Company. The relief of Mashonaland from the ferocious forays of the Matabele was a memorable service which will be credited at the outset to this company. It has further given to all within its jurisdiction the fullest protection of English law, and safeguarded all working in service from abusive treatment by their employes, prohibiting the use of the lash, and enforcing other humane regulations. The sale of liquor to the natives is forbidden by stringent laws, and the most discreditable and demoralizing influence in South Africa is barred out of Rhodesia, at least.

To determine the extent of arable and pasture lands, deputations of experienced farmers were appointed to inspect and report by public meetings in the Cape Colony and Orange Free State. Their examination of less than half the area of Mashonaland and Matabeleland reached the conclusion that at least 40,000 square miles were well adapted for colonizing purposes. It may further be noted that highly favorable reports of the agricultural and mineral resources in the vast territories of the British South Africa Company, north of the Zambesi, have been furnished by Joseph Thomson, Alfred Sharpe, and other well-known explorers.   Actual experience and the medical officers' reports have shown that the climate is not unhealthy for any white man who will avoid undue exposure and observe a few simple precautions.[1]

Victoria Falls, Zambesi River.

The advance of immigration and development has been remarkable in view of existing conditions.   There were inevitable hardships and discouragement to check the first rush of gold seekers.   The gold-fields were only slightly explored and lay far from any base of supply.   There was lack of resources and means of communication to develop even the most promising

[1] " Minutes of Progress in Mashonaland," by the Secretary of the British South Africa Company.   " The Ruined Cities of Mashonaland," pp. 405–412.

openings. Yet, in spite of all obstacles, prospecting has been pushed far enough to show the range of gold-bearing ground and ledges for hundreds of miles. Convincing evidence of the mineral richness of the country is given to the extent of the ancient workings that have been traced through Mashonaland far beyond the southern end of Lake Tanganyika. There can be no doubt that an enormous amount of gold has been taken

Victoria Falls.

from this region both by placer washing and quartz mining. There are no other ancient workings, on the face of the Old World at least, of like extent, and this undeniable evidence weighs heavily for the contention that the flow of gold from this source was the main supply, for centuries, of Arabia and Asia Minor. In view of the superiority of modern appliances for mining and the extraction of gold, it would seem, at least, probable that the yield of this territory may, in time, be large.

There are apparently well-verified reports also of the dis-

covery of extensive copper and iron deposits in North Rhodesia
and in the region lying along the western shores of Lake Tanganyika.   The missionaries of the Roman Catholic society

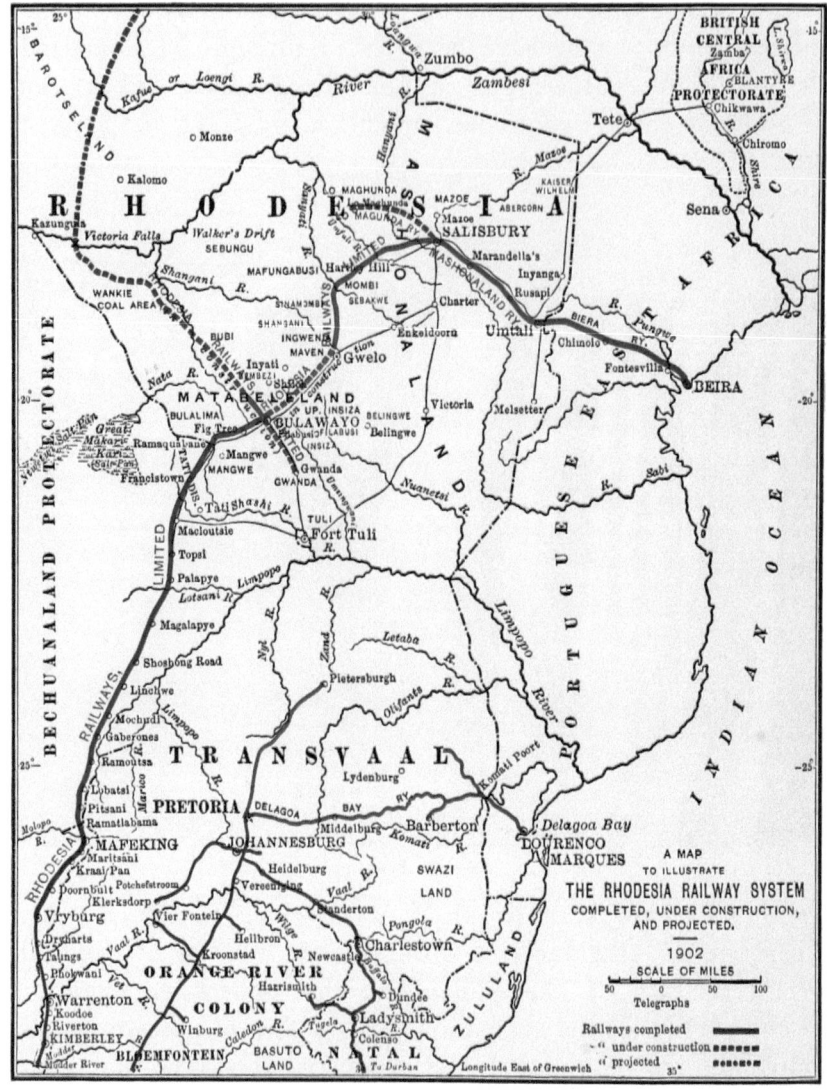

A MAP
TO ILLUSTRATE
THE RHODESIA RAILWAY SYSTEM
COMPLETED, UNDER CONSTRUCTION,
AND PROJECTED.

1902
SCALE OF MILES

Telegraphs

Railways completed
" under construction
" projected

BORNAY & CO., ENGR'S, N.Y.

known as the White Fathers have long been at work on the
shores of Lake Tanganyika, and a report of their explorations
has been published lately in *Petermanns Mitteilungen*.   It is

VICTORIA FALLS,
East Corner.

noted that great quantities of iron ore have been found along the banks of the rivers flowing into Tanganyika, particularly along the Lufuko and Miobosi.

The wide-ranging Marungu district is said also to be exceedingly rich in copper ores ; and the copper areas, better withstanding denudation than the surrounding country, are reported to stand comparatively high above the general level and to be easily recognizable. Agents of the Chartered South Africa Company have also reported the discovery of a rich copper field, estimated to cover 40 square miles, in north Rhodesia. This field lies about 150 miles north of Victoria Falls, near the Congo Free State, and runs over the border. De Beers Company has already taken an active part in the development of the copper mines of Namaqualand, and the new field may prove to be of even greater importance.

The rapid extension of the railway lines of the Bechuanaland Railway Company from

Victoria Falls.

Vryburg to Bulawayo was mainly due to the aid given by Mr. Rhodes and his associates in the Chartered Company. This line reached Bulawayo, a total distance of 1360.4 miles from Cape Town, in 1897, and has since been extended northward about 30 miles. The war has interrupted this work during the past two and a half years. The main line north, it is expected, will reach the enormous coal beds at Wankie, 200 miles north of Bulawayo, in about eighteen months, and will be pushed on

2 Q

ZAMBESI RIVER, ONE MILE BELOW VICTORIA FALLS.

to the Victoria Falls on the Zambesi and thence north as rapidly as possible to tap the rich copper districts.

The telegraph line running ahead of the railway was carried across Rhodesia and reached Lake Tanganyika, from which it is fast extending to Uganda, so that Cape Town and Cairo will soon be in direct overland telegraphic communication. The inspiration of this work of the Transcontinental Telegraph Company was due to Rhodes, and the greater part of the capital needed to extend it was contributed by him personally. The postal service already effected is as remarkable as the telegraph. Even from points hundreds of miles beyond Bulawayo, which, eight years ago, was the heart of savage Matabeleland, the pioneer can send a letter home to England for twopence halfpenny, and the settler on the remote shores of Lake Moero can get by mail a pound package of tea from Liverpool at the cost of a shilling.

Already the Chartered Company has carried the work of exploration and expanding control to Lakes Tanganyika, Moero, and Nyassa, and made treaties with the native chiefs north of Rhodesia, as well as with Lewanika, king of the Barotse, to the west. The range of British influence and civilizing advance now reaches to the heart of Africa from the south, embracing all the country not within German control in the west, and the Portuguese domain in the east. The grand aim of Rhodes has been swiftly advanced in realization even beyond sanguine expectation.

Another undertaking, less far-reaching and impressive in scope, but of evident material importance to the development of the industrial resources of South Africa, was liberally and energetically supported and advanced by De Beers Mines. The cost of importing coal was from the outset, and still continues to be, a crippling handicap upon the advance of the mining and manufacturing industries of the South African Colonies. Persistent searches for coal deposits throughout the country were made, but no coal seams of high quality were uncovered. The best apparent prospect for opening deposits that might compete in the market with imported coals was shown in the Stormberg and neighboring hilly districts lying between Queenstown and the Orange River.

Summer House at Groote Schuur.

Progressive colonists developed the outcrops in the Stormberg district, and in the face of grave discouragements opened seams of importance in the Molteno and Cyphergat mines, but it was impracticable to work these mines with any prospect of profit until railway communication was opened from Stormberg Junction via Steynsburg to Middelburg, connecting the East London and the Midland or Port Elizabeth lines. The possibility of supply from this district was immediately grasped by De Beers Company, both for the sake of an eventual saving in the cost of its fuel, and the public-spirited object of coöperating, so far as was feasible, in the development of a resource of such importance to the colonies.

The Stormberg coal was so mixed with shale that even the shipping coal after sorting held about one-third waste, which clogged the furnaces. But special grates were designed to burn this coal, and by this resort it was practicable to use a supply from this field at the diamond mines. De Beers Company was soon taking by contract practically the entire product of the Stormberg seams at a price of about 20s. per ton at the shipping point.

Not long after the opening of the Stormberg mines, coal seams of much greater width and promise were discovered at Indwe, a point about seventy miles east from Molteno and Cyphergat. Here the prospective returns from energetic development were really very bright, but, to market the coal, the construction of an expensive railway line from Indwe to the East London or Eastern systems was indispensable. In spite of the unwearied and cogent representations of Colonel Schermbrucker and his associates in control of the Indwe field, the Cape Government was reluctant to defray the cost of building this line. The scheme was a dragging one for years, until De Beers Company came forward with a subscription of £75,000 to the shares of the Indwe Railway Collieries and Land Company, organized to extend the necessary railway lines and operate the mines.

"Stoep," Groote Schuur.

In view of this essential backing of capital, coupled with the cogent appeals of Rhodes and his associates, the Cape Government was moved to contribute a grant of £50,000 toward the expense of construction, with an additional allowance of 50,000 acres of land, worth about one pound an acre. Then a line of sixty-six miles was laid at half the rate per mile that was paid for building the lines under Government Administration, and the mines were opened very successfully. It was supposed by the projectors of the scheme at the outset that the main business of the company would be the supply of coal for steamship use at East London; but it was soon demonstrated, upon the completion of the railway, that De Beers Company was the principal customer, consuming about 5500 tons of the average monthly production of 12,000 tons.   This coal supply was delivered to De Beers by agreement for 15*s*. per ton at Sterkstroom, the point of junction of the Indwe and Eastern system lines.   In spite of the inferior quality of the coal, compared with Welsh coal, the South African coal at this price was a good bargain for De Beers, and the very profitable record of the Indwe Company proves that the interests of its shareholders were not sacrificed in making the bargain.   The mines of the Stormberg district are still continuous producers, and supply about 1000 tons monthly to the mines at Kimberley not under control of De Beers Company.

Coal mining in the Orange Free State has not been carried on very energetically on account of the distance of the coal measures from the existing railways.   But the developments in this field are already promising, and the Kroonstad Coal Company, in particular, has opened up a bed of very good coal.   A railway is in course of construction from the main Free State line to the Kroonstad coal fields.   When this line is completed these mines will be in a position to compete with any others, and if the long-promised line is constructed from Branford or Bloemfontein to Kimberley, Kroonstad coal can be delivered at the diamond mines cheaper than any other coal yet discovered. Beyond these undertakings is the opening of the promising coal mines in Natal to which De Beers Company has liberally contributed.   (See Appendix IV.)

Other enterprises, too, of public service are worthy of mention.   De Beers Company is steadily furthering fruit and stock farming, and has constructed storage buildings in various locations in order to prevent a monopoly of the meat supply which was threatening South Africa.   It is constructing, also, one of the largest dynamite factories in the world, near Cape Town, under the able superintendence of Mr. W. R. Quinan.

Of course Rhodes could not foresee the marching steps of this progress in varied lines, but it is none the less certain that

THE TERRACE GARDEN,
Back of Groote Schuur.

the expansion of the undertakings of De Beers Consolidated Mines was the carrying out of his long-cherished aims. It was for this chiefly that De Beers Charter was drawn with so free a hand. Assured control of the great South African diamond mines was the assurance of great wealth, — from Rhodes's point of view, great power that should be greatly used. His aims ranged far beyond any personal ex-alting. His heart was set on the making of Greater Britain by expansion and loyal federal union. In the Dark Continent, beyond the confines of civilization, he saw the open field for British occupation and development, and was unresting till it was grasped. How great this attainment was in actual stretch of territory may best be comprehended,

A Corner in Mr. Rhodes's Library.

as the London *Times* notes, "by any one who will take the trouble to contrast the map of Africa as it appeared in 1881, when Mr. Rhodes first entered public life, with that which is open to his study to-day. At the earlier date, the line of the 28th degree of south latitude bounded our possessions in South Africa; the later map he will find coloured red right up to the shores of Lake Tanganyika — within a few degrees of the Equator."

That this annexation has been, and will be, greatly to the advantage of the territory and its occupants will not be seriously questioned. Its material advance and the security to life and property stand already in bright contrast to its barbaric state — a land which knew only the rudest tillage and was ravaged at the whim of savage chiefs. It is too early yet to think of measuring its resources and probable advances, but enough is known to warrant high confidence in its future, with the assurance of alert grasp of its openings for immigration and capital.

Another View of Mr. Rhodes's Library.

To any eye the gaining of Rhodesia was a long step forward toward the attainment of Rhodes's hope of carrying British dominion from the Cape to Cairo. But the ordinary observer would not mark, as intently as Rhodes did, the force of this acquisition in determining the control of South Africa. Seventeen years ago, in addressing his constituents, at Barkly West, he declared publicly, as a settled conviction: "I came to the conclusion that the key to the (South African) puzzle lay in the possession of the Interior, at that time an unknown quantity. In a humble way I have been mixed up with the politics of the

Interior during the last four years, and such politics, I contend, will be in future most intimately connected with the settlement of the South African Question, for I believe that whatever State possesses Bechuanaland and Matabeleland will ultimately possess South Africa." It was his view, asserted in repeated conversations with Mr. Edward Dicey, that the taking of Rhodesia necessitated the creation of a predominant South African Confederacy, which would be brought to pass by the force of circumstance. In the interest of South Africa and Great Britain Rhodes sought the inclusion of this Confederacy in the British Empire.

It is plainly to be seen that Rhodes's view of the interests of South Africa and the drift of his anticipated confederation were inevitably antagonistic to the attitude and policy of the men controlling the South African Republic. In stretching the arm of Great Britain over Mashonaland and Rhodesia, Rhodes unquestionably blocked the extension of the Transvaal State and the schemes of Krüger. In this brief marking of progress and attainment I would not attempt any measuring of responsibility for the collision that finally resulted in the war just closed. South Africa is now completely under British Imperial control. Whatever view may be taken of the conflict, its practical outcome plainly clears the way for the systematic development of this vast territory under liberal colonial institutions.

Cecil John Rhodes did not live to see the ending of the contest so long maintained by the unyielding temper of the Boers. He died on March 26, 1902, near Cape Town, of the disease of the heart which had long clouded his hope of life.

His visionary political projects ran far beyond any exact defining or determination of method, but, in the main, "the lay of his ideas," to use his own phrasing, is clear. He would urge the union of all English-speaking people to dominate the world, transform barbarism to civilization, do away with poor and hampering government, maintain enduring peace, and promote universal progress.   His last will and testament has proven that the advance of this union was at the core of his heart.   The image of his fancy may never come into being, but he has, at least, done something for an uplifting union in the gathering of young scholars, representing all English-speaking people, in the ancient mother university, to recall their common inheritance and join their hands.

APPENDICES

APPENDIX I

THE MINES BESIEGED

The siege of Kimberley was one of the striking episodes of the late war. As an interruption to the peaceful progress of diamond mining in the South African Fields, it has a place apart from the industrial story. Yet no history of the Diamond Fields would be complete without some

A Boer Commando.

account of its course, and my personal view may be of interest in the possible emphasis of the part taken by De Beers in the maintenance of the defence.   I would mark, too, precisely how the war affected the working of the mines, and tell from my own observation how the call to arms made soldiers of men accustomed to the use of drill, pick, and shovel, and caused our mechanics to turn their hands to the making of ordnance.

Lieut. Col. Kekewich and Staff of Imperial Officers.

For some time previous to the actual outbreak of the war (October 11, 1899), it was apparent to us who were living upon the border of the Orange Free State that both the South African Republic and the Orange Free State were making preparations for war with England, and that the invasion of the Cape Colony was but a matter of a short time. These preparations had been going on for many years until the magazines and arsenals of the Transvaal were filled with the finest munitions of war that the works of Schneider at Creusot or of Krupp at Essen could produce. The Mauser with which the Boers were armed was as good as the small arms of any Continental power, and better than the Lee-Metford which the British brought against them.

In July, 1899, Major Scott-Turner came to Kimberley, and Lieutenant McInnes, Royal Engineers, followed him shortly after. Colonel Trotter, R. A., Chief Staff Officer, also came to stay a short time. He had made a report on the defences of Kimberley as early as 1896, and an accurate military map had been prepared of the town and surroundings. Major O'Meara came later as Intelligence Officer. The Imperial Government sent these officers to

Lieut. Col. Robert George Kekewich, Commandant of Kimberley during the Siege.

prepare for the defence of Kimberley, and on the 13th of September, shortly before the war was declared, there arrived a half regiment of the Loyal North Lancashires (infantry), and a battery of Royal Artillery, consisting of six muzzle-loading seven-pounders of obsolete pattern, and some Maxims.

On the 30th of September the Governor of the Cape Colony gave his consent to the formation of a Town Guard, " solely for local defence in case of attack from without." The radius of the circle in which the Town Guard must confine their operations was eight miles, with the market square as the centre. Lieutenant Colonel Robert George Kekewich was appointed commandant. Lieutenant Colonel Harris, V. D., a director of De Beers, was second in command and was placed in charge of the Town Guard. Major Peakman, an officer of the local

Fort on Tailings Heap, Kimberley Mine Floors.

volunteer force, who had had a considerable amount of experience in the Kafir wars, was appointed Staff Officer. On the 4th of October the local volunteers, five hundred strong, were called out by the Governor, and went into camp.

On the 5th of October the first serious disturbance of the work at the mines occurred. An alarm was sounded at one o'clock in the morning of that day, and all the forces in town, including the men working in the mines, were called out to do military duty, as it was rumored that an attack was contemplated by the Boers, who were massing commandoes in the Orange Free State, only a few miles distant. It had been arranged that the whistles (sirens), commonly called " hooters," at the various engine houses of De Beers Company, should be blown in case an alarm had to be given.   The first alarm caused great consternation throughout the whole town.   Men were running, helter-skelter, in the dark, seeking their various redoubts, the moving guns and ammunition wagons rattled through the streets, and the gardens of the houses were filled with men, women, and children, anxiously awaiting some news as to the cause of the alarm.   The screeching of the hooters was appalling.   These sirens, which in times of peace could " blow the boilers dry " and not disturb the quiet morning slumbers of the dwellers of the Diamond City, had, all in a moment, become a nerve-shattering mechanism.   In later days the roar of the Boer artillery and the bursting of shell all over the town did not so frighten the mass of people.   The horrifying effect was so lasting, that when work at the mines was resumed after the siege, many people

One of the many Redoubts, looking East.

in the town asked me to discontinue the use of the hooters, and, in compliance with their wishes, the old whistles were for a time put into service.

Kimberley, as may be imagined, was quite unprepared for an attack on the 5th of October, as war had not been declared.   The Intelligence Department had received some false reports, and those in charge thought it best to have every man at his post; hence the alarm.   The proven falsity of the reports did not, however, dispel the menace of the situation, and it was considered necessary to make better preparations for the defence of the town.   Our miners were called out to drill during a part of each day.   Our tailing heaps, which formed natural defensive positions, were taken possession of by the military.   Strong forts and redoubts were constructed on the tops of these heaps, and mines of dynamite were laid at their bases.

Sir Edwin Arnold, writing in the *Daily Telegraph*, said: " There is something singularly picturesque and suggestive in the thought of the Diamond City of South Africa being defended by her own waste heaps.

Fort Rhodes, Kenilworth, and its Defence Force.

View from the Conning Tower, looking Northeast.

2 R

Since Syracuse was fortified against Nicias with the columns of her own
white marble temples, and the breaches of Badajoz were filled up with
the empty wine casks, there has been no such curious use made of local

The Waterworks Company's Reservoir as a Fort.

material. Strange, indeed, is the destiny of matter. It may turn out
that the blue clay will prove more valuable to Mr. Rhodes, to the isolated
garrison, and to the little city, than all the diamonds she ever dug up."

Fort Rhodes, on Top of No. 2 Tailings Heap, De Beers Floors.

Other defensive fortifications were made upon the ground lying
between the tailing heaps. The labor necessary to do this work was
drawn from the mines and works. Nearly all the men working in and
about the mines joined the various military organizations, and slept in the

forts and redoubts. Owing to this distraction, work at the mines proceeded very slowly.

Rhodes, accompanied by Dr. Smartt, member of the Legislative Assembly, arrived in Kimberley a few days before the investment. He took

No. 2 Redoubt, near Kimberley Mine.

up his residence at the Sanatorium. Mr. and Hon. Mrs. Maguire
arrived a day or two later and were his guests during the siege. Upon
his arrival at Kimberley, Rhodes realized at once the gravity of the situation, both as regards the defence of the town and the food supply. Orders
for large quantities of provisions were wired to Cape Town, Port Eliza-

A Barrier on the Road leading to Kenilworth.

beth and East London, with the hope that we might be able to add to the
seemingly large stock already on hand — but these supplies never arrived.

The siege of Kimberley commenced on the night of the 14th of October, a little before ten o'clock, when the wires to the south were cut, the

The Sanatorium in Time of Peace.

The Sanatorium during the Siege.

wires to the north having been cut about an hour before. The last train from the south arrived at Kimberley about 11 P.M., bringing several truck-loads of supplies which were at Modder River Station, destined for the Free State Boers.

Colonel Kekewich at once issued a proclamation, declaring the district in a state of siege. The war had actually begun. The various fortifications were made stronger, military organizations were increased in numbers, a mounted force of four companies, known as the Kimberley Light Horse, was formed, and on all sides there was the greatest activity

Mounted Camp, Kimberley.

in making Kimberley a strongly garrisoned town. When all the military organizations were completed, the forces at the disposal of Colonel Kekewich were as follows : —

The Imperial Garrison consisted of the 23d Company of Royal Artillery, one section of the 7th Field Company Royal Engineers, and four companies of the 1st Royal North Lancashire Regiment. There was also a small detachment of the Army Service Corps. The total strength of the regulars was about 600 officers and men. Volunteer companies had been enrolled from the early days of the Fields, and at one time comprised a very considerable force of men, but of late years the community had lost nearly all interest in the volunteer service.

Still the organizations had been kept up, and when the muster roll was taken, shortly before the siege, it showed the following numbers : One battery, Diamond Fields Artillery, consisting of six seven-pounder muzzle-loading guns, with 3 officers and 90 men, in charge of Captain May; the Kimberley Regiment (infantry), under Lieutenant Colonel Finlayson, with 14 officers and 285 men; the Diamond Fields Horse, Major Rodger, 6 officers and 142 men.   The total force of regulars and volunteers was about 1100.

The Town Guard was organized, and the men were drilled in the use of the Lee-Metford rifle.   At the beginning of the siege this force numbered about 1200 men, but both the volunteer corps and the Town

Camp of the Royal North Lancashires.

Guard were soon increased until the total strength of the garrison reached 4500 men.   This included the Cape Mounted Police, number-ing about 360 officers and men, and unmounted police to the number of 175.   The limit of the defence force was gauged by the number of rifles in Kimberley — which had been considerably increased during the previous year by the importation by local merchants of 1000 rifles and six Maxims, together with a considerable amount of ammunition for the use of the rifle clubs.

Our forts and redoubts were in many ways unique and picturesque. The waterworks reservoir was surrounded by a huge fortification, made of grain and coal sacks filled with soil.   The forts on the tailing heaps were made with rows of the trays of trucks which in times of peace

convey the diamond-bearing ground to the floors.   The trays were filled
with tailings, banked up on the outside with the same material, and
coped with sand bags.   Large shelters were made within the forts for

Officers of the Diamond Fields Horse.

the protection of the garrisons.   As tents were not to be obtained,
spacious houses with roofs of corrugated iron and sides of canvas were
constructed as sleeping and eating rooms, and for protection against the
tropical sun and violent thunderstorms.   When the supply of corrugated

The Diamond Fields Horse at Kenilworth, during Siege of Kimberley.

iron gave out in town, for even the enormous stock of De Beers did not
prove equal to the demand, the iron fence which surrounded the race-
course was taken down and carted to the various fortifications.

A Group of the Town Guard.

Barriers were constructed around Kimberley to check any sudden
attack upon the town.   The roads leading from the town were strongly
guarded and barricaded with barbed-wire entanglements, with mining
trucks filled with earth, and with camelthorn trees.   Of late years the
outskirts of Kimberley had begun to assume quite a parklike appearance,
by the growth of young trees from the roots and stumps of those that
had been cut down during the early days of the Fields.   It seemed a
pity that the little natural beauty which these afforded should be destroyed;
but the preservation of the town was of first importance, and all the trees
were cut down and dragged into long lines of fences, where they were
interlaced with barbed wire, making most formidable barriers.   When

Defence Guns and Maxims massed in the Gardens.

the siege was over, these fences disappeared, almost in a day, to supply the inhabitants with firewood, which had been cut down to the scantiest allowance, — a week's supply being barely sufficient to do a day's cooking. The defences were in places supplemented with dynamite mines planned by the Royal Engineers, and carried out by the electrical department of De Beers. On one occasion the officer in charge gave instructions to put down *ten* pounds of dynamite every *thirty* feet, and returning

Headquarters Staff, Cape Mounted Police.

later in the day he asked if his instructions had been carried out, and received the reply, "Yes, sir, we have put down *thirty* pounds of dynamite every *ten* feet."

Premier Mine

Premier mine occupied a unique position during the siege. It was isolated from Kimberley and Beaconsfield, the former town being about four miles, and the latter two miles, distant. There is a large, disused tailing heap near the mine, on the top of which is a small reservoir, into which water from the mine is pumped for distribution to the washing plant and floors. Around this reservoir a fort was built and made almost impregnable. Large shell-proofs were made for storing supplies and ammunition for a local siege, should communication with the Kimberley and Beaconsfield defences be cut off. One of the three searchlights which De Beers Company uses on their "floor" for preventing theft of diamonds by night was placed at this fort. The Boers called these

Kimberley Waterworks Reservoir, with Royal Artillery.

searchlights "Rhodes' eyes." About 150 of De Beers employés and one hundred regulars, with two seven-pound guns and a Maxim, were constantly on duty at this fort.

The pumping plant which supplied Kimberley was down in the open mine. This plant, as well as all the machinery of the mine, was pro-

Premier Mine Fort, Royal Artillery in Action.

tected with sand bags. In heaps about the mine, and in all the buildings on the side of the mine adjoining the Free State, mines were laid, with wires leading from them to the fort. One of the powerful electric searchlights was placed in the fort, and so arranged that it could be lowered out of harm's way during the daytime. Connections were made between the two sets of boilers and the pumping and electric light plants, so that, in case a shell damaged one set, the other could be used. A large

number of hand grenades filled with
dynamite, with fuses and detonators
fixed, were made and kept in the
magazine. An underground hospital
for the wounded was constructed.
In fact, everything necessary was
done to make this fort independent
and secure. There was apparent
need for these precautions, for the
Boers constructed the most formidable fort of any about Kimberley
on a low range of hills about three
miles distant, where they kept two
guns and a pom-pom, which they
fired nearly every day during the
siege, except Sundays.

Premier Mine Searchlight.

Great credit is due to the man-
ager of the mine, Mr. J. M. Jones, and to Captain O'Brien, who was
in charge of the garrison, for the manner in which the defences were
constructed, and to all who occupied the fort during the long, weary four

Canvas House erected for Protection from the Sun and Thunderstorms.

months, for their courage and patience. On several occasions lightning
struck the wires connecting the mines and exploded them. One explo-

sion carried away part of the mine compound, and another wrecked the end of the large stables.   Fortunately no harm came to any of the garrison or to any of the machinery of the mine.   Although it was isolated from Kimberley, the Boers never made an attack upon it nor came within rifle range.

A few days after the commencement of hostilities the Boers took possession of the Kimberley Waterworks Company's plant on the Vaal River, some sixteen miles distant, and cut off the water-supply.   Connections were made between Premier mine pumping system and the Kimberley Waterworks Company's reservoir, and a supply of eight to

Railway Bridge over the Vaal River at Fourteen Streams, destroyed by the Boers.

ten million gallons of water per month was delivered by De Beers, free of cost to that Company, on the understanding that only half rates should be charged to the inhabitants of the town.   The water was perfectly clear, pure, and wholesome.

As the supplies of food in hand seemed ample for any emergency that was thought possible, there were practically no restrictions upon the consumption of supplies during the early part of the siege, except that the amount of meat was fixed at one pound per diem for each adult, and one-quarter of a pound for children under fifteen years of age.   As there were no restrictions as to prices, the speculating part of the community soon took advantage of the situation.   Few had laid in stocks of food, and, as the greater number of people had not the means of making large purchases, they saw starvation staring them in the face.   It

Boer Laager, near Kimberley.

was impossible for many even to purchase their daily requirements at the fabulous prices to which the necessaries of life suddenly rose. Paraffin, which usually sold for 15 shillings a case, jumped to 100 shillings. Naturally the community rebelled against this extortion, and the daily newspaper was full of complaints. As some of them put it, they had taken up arms to defend the very people who were starving their families by putting the prices for the necessaries of life beyond their means. Colonel Kekewich was equal to the occasion, and wisely issued a proclamation fixing the price of all supplies at the same figures as formerly existed.

For the support of people too poor to pay even for the barest necessaries of life, thoughtful provision was made by Rhodes in the institution of a soup kitchen in De Beers convict station. The details of the work were ably carried out under Captain Tyson, Dr. Smartt, and the Hon. Mrs. Maguire, the latter attending to the distribution at Beaconsfield. The soup was excellent, being composed of beef or horsemeat (with now and again a donkey or a few Angora goats thrown in),

Group of Typical Boers.

and a variety of vegetables from Kenilworth, and thickened with Boer meal or mealie meal. Captain Tyson carried pockets full of small bottles, the contents of which would be emptied in the brew, "just to make it a little more appetizing, don't you know." The allowance of meat was a half pound for two days, which could be exchanged for soup.   Long rows of people stood for hours awaiting their turn to be served.

When the siege commenced, De Beers had 8000 tons of coal in stock and also about 2000 tons of wood.   There were about 1500 cases of dynamite belonging to merchants, and De Beers had several hundred cases in stock.   Owing to the dangerous proximity of the magazine to the town, it became necessary to remove nearly all the dyna-

A Group of Mercenaries, fighting with the Boers.

mite to a magazine at Dronfield, about six miles north of Kimberley, from which, for a time, supplies were drawn; but these magazines were subsequently blown up by the Boers.

In order to do as much work as possible while the supply of coal and dynamite lasted, permission was obtained from the officer commanding for the miners to resume work in the mines, on condition that substitutes were found to take their places in the forts.   A company of men was organized at De Beers and Kimberley mines by the assistant general manager, which was known during the siege as the Permanent Guard, and was composed mostly of refugees.   Work was continued at Kimberley mine until the 3d of November, and at De Beers mine until the 4th of December, when it was thought advisable to discontinue work and save the supply of coal for pumping water for the use of the town and preventing the mines from being flooded.   The amount of ground hoisted at Kimberley mine from October 14 to November 3 was 60,396 loads, and at De Beers mine to December 4 was 173,447.   The pumps in both mines were kept going until a few days before the siege was raised, and started again before the water had filled the tunnels in the rock outside the mines proper. While the pumps were stopped a gang of natives were kept busy at each mine picking out pieces of coal from the old ash-heaps to supply the boilers with fuel. Fortunately all damage by flooding to the underground works was prevented.

Communication was kept up between Kimberley and the nearest military post, which was at the Orange River bridge on the Kimberley-De Aar railway, by despatch riders who evaded the Boers and found shelter and remounts at several farms of friendly colonists. The distance was

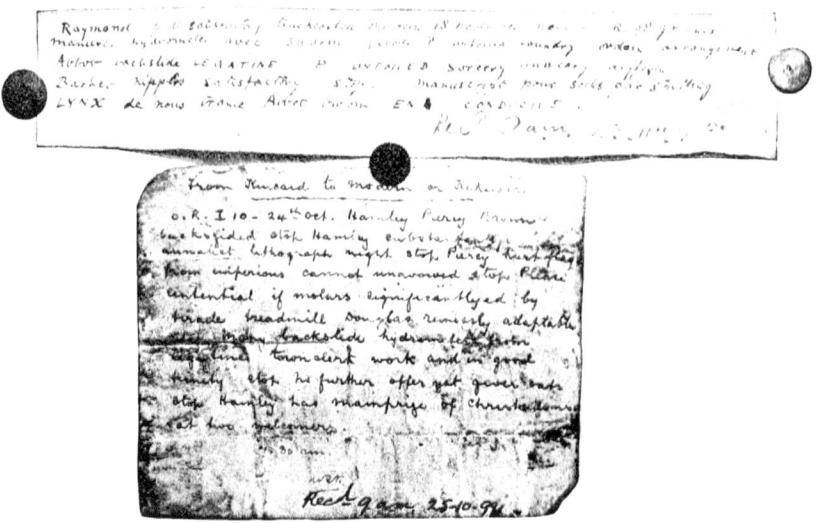

Code Dispatches received during the Siege.

eighty miles. Trooper Brown of the Cape Mounted Police carried the first despatch, and covered the distance in thirteen hours. Great credit is due to these men, who went to and fro at great peril to themselves. Foremost among them were Brown, Cummings, Hambly, and Harding, but there were many others who did good work. The remuneration paid by the military was very small — £5 for the round trip, but in many cases, where private letters were carried, this sum was largely increased by private donations. Later, when the investment of the town was closer, it became very difficult to get through the Boer lines, and despatch riders, carrying private despatches, were paid as high as £100 for a round trip. Many of these men were captured and taken to Bloemfontein as prisoners of war.

How zealously and efficiently Rhodes took part in the preparations for the defence of Kimberley has been particularly noted by Mr. George A.

Typical Boer.

L. Green, editor of the *Diamond Fields Advertiser* in his able and accurate description of the siege. "The need for mounted troops to watch the enemy's movements was early felt. The formation of a new corps, to be called the Kimberley Light Horse, was one of the last things authorized by the High Commissioners before Kimberley was cut off, but the trouble was to find the horses. Mr. Rhodes came to the rescue, and in a few days presented the corps with five hundred admirable mounts; he also did some good work as recruiting sergeant. Largely through his efforts the mounted arm of the defence forces was thus increased to nearly nine hundred men. Major Scott-Turner was appointed with the local rank of lieutenant colonel to command the mounted corps, which now comprised Cape Mounted Police, Diamond Fields Horse, and Kimberley Light Horse.

"It was Mr. Rhodes's pleasant custom to go round asking the question, 'Do you want anything?' Needless to say he rarely met any one who did not want something.

"One evening, while Major Cha-

Major W. H. E. Murray.

mier was dining with Mr. Rhodes, they were discussing the artillery branch of the defence forces, when Mr. Rhodes asked him if he needed anything for his artillery. The Major replied quickly, 'Yes, I want to make my guns mobile. [*Note.* — It is mentioned elsewhere that these guns were small mountain guns without limbers.] I require, to do that, 43 horses, 62 mules, 7 buck wagons, and 4 Scotch carts.' It was a tall order, but Mr. Rhodes made a mental note, without any comment, and three days later Major Chamier found that the whole requisition had been delivered at the artillery camp. All he could

Armored Locomotive.

say, when he saw what had been done in so short a time, was, ' What a wonderful man Mr. Rhodes is.' It was an object lesson to the military officers to see how quickly provisions of this kind could be made by a civilian who was in no way handicapped by official red tape."

From the first threat of the outbreak of hostilities, the resources of De Beers were at the command of the garrison for any needed service. At

Armored Train, constructed at De Beers Workshops.

De Beers workshops several engines and trucks were armored in the manner shown in the accompanying illustrations.

These trains were useful in many ways, and of very great service in keeping the lines of communication open. Those running between Kimberley and De Aar were manned in part by De Beers men. The

2 s

military organization known as Scott's Railway Guards was also mostly made up of De Beers men, with Lieutenant Colonel R. G. Scott, one of the officers of De Beers Company, in charge.

The first encounter with the attacking Boers was on the 24th of October, ten days after the investment of Kimberley. Shortly after the water-supply had been cut off, Lieutenant Colonel Scott-Turner made a reconnoissance in the direction of the pumping station, but took the precaution to follow the line of the railway as far as Macfarlane's farm, which lies eleven miles to the north of Kimberley. His force consisted of detachments of the Kimberley Light Horse, Captain R. G. Scott, V. C.; Cape Mounted Police, Major Elliott; and the Diamond Fields Horse, Major Rodger. The armored train, in charge of Lieutenant Webster,

Railway Bridge at Modder River, both ends of which were blown up by the Boers.

Loyal North Lancashires, supported the troops. On arriving at the farm-houses at Macfarlane's, which stand on a knoll from which the country recedes in all directions, the troops halted and had breakfast. Immediately afterward Lieutenant Colonel Scott-Turner with 180 men proceeded on his mission, but soon after his departure Boers were seen in several directions.

Upon the appearance of the enemy Lieutenant Colonel Scott-Turner took up a strong position with his men. In a short time the Boers sent a few of their number under a flag of truce. Major Elliott of the Cape Mounted Police met them, and was told that if he and his command were on police duty the Boers would not molest them, but if he was there for a fight, they would put a bullet through his head. Major Elliott returned, however, without hindrance. In the meantime the armored train had

Effect of a Nine-pound Shell.

proceeded beyond Macfarlane's, but was soon recalled, as the Boers were evidently trying to cut it off. Later in the morning Boers continued to arrive from the north and east, and came within rifle range of Macfarlane's farm, not knowing that it was occupied by the British. The patrol opened fire on them, and several of them were seen to fall and their riderless horses ran across the veld. The Boers retreated helter-skelter. Shortly afterward five Boers from another com-

Effect of a Nine-pound Shell.

mando came forward, bearing white flags, and were met by Major Elliott, who received the same message as before. The Boers evidently had little knowledge of the proper use of the white flag.

Trophies of the Siege — The Author's Collection.

In pursuing his advance Lieutenant Colonel Scott-Turner fell into an ambuscade, for, owing to the very long grass, which was nearly waist high, he was unable to detect the position of the Boers, who were strongly posted behind the wall of a dry reservoir in numbers greatly exceeding the British force.   Not a shot was fired until the British came within easy rifle range, when they were met with such a fusillade from the magazine Mauser rifles that they sought the nearest cover.   In this repulse the losses on the British side were three killed and nine wounded, and fourteen horses were killed or disabled.   The wounded men were taken up and carried back with the retreating force, but the dead were left behind, to be brought in two days afterward, as the searching party failed to find the bodies, on the first day, in the tall grass.

Lieutenant Colonel Scott-Turner had heliographed to the conning tower to have two mountain guns and two Maxims sent out.   These were despatched at once, and the armored train took out 150 of the Loyal North Lancashires under command of Major Murray.   The Boers were seen to be moving toward Dronfield, a ridge halfway between

Kimberley and Macfarlane's. The armored train proceeded beyond Dronfield, but was ordered back to that place, and the troops left the train near the siding. In the meantime Captain May, with two guns, had reached a position just south of the siding, when the Boers opened fire on him at short range, having allowed his scouts to come close to the place where they were in ambush. Captain May quickly unlimbered his guns under a hot fire, and began to shell the Boers in return. Fortunately for him most of the Boer bullets went over the heads of his men, while he fired his guns with great precision, riddling the gamekeepers' houses, behind which the Boers had taken shelter, and soon driving them to the rocky ridge beyond.

Hearing that the guns and regulars had gone out, I drove to a position north of Kenilworth, where this part of the engagement was in full view. Captain May fired eighty rounds at the Boers, and his men behaved splendidly under a rain of bullets from the enemy, only a thousand yards distant. Out of a total of twenty-six men and eighteen horses, he had seven men wounded, three horses killed and nine wounded. Gunner Payne, who was wounded in the foot early in the fight, continued to lay his gun until the end of the firing; and bugler Dickinson, who was wounded in the right hand, changed the bugle to his left hand and finished his notes.

While this fight was going on Major Murray had taken his men from the train near Dronfield, and had begun to ascend the hill. At his first advance the Boers opened fire. Forming his men in skirmishing order with all possible speed, he led the way up the rocky ridge where the Boers were lying closely under

Boer 9 lb. Shrapnel, ¼ Actual Size, showing Time Fuse.

cover. Fortunately for the Major and his troops, the ascent of the Dronfield ridge on the north was comparatively easy, being over a gently rising country covered with small brush, with here and there a shallow ravine which gave a little shelter to his men. While they were moving forward, three men, not in uniform, rode up to him. At first he took them for Boers, but the Northumberland accent of the first who hailed him was convincing. They were men in charge of De Beers farms, and when the firing began they were looking after the large herd of De Beers cattle.

One of these keepers, Dott, guided the troops up the hill, taking them out of sight of the enemy as much as possible, and shouting " This way, Mr. Officer ! " " This way, Mr. Officer ! "

British and Boer Shells, fired within and around Kimberley and Beaconsfield.

Their scramble up the hill was very plucky. In front lay the Boers hidden in the rocks, and on their left was a magazine containing 1500 cases — 37 ½ tons — of dynamite, which might explode at any moment should a Boer bullet strike it, as it was protected only by a thin sheet of galvanized iron. Two firing parties of twenty-five each went ahead gallantly, with the main force, a hundred strong, following close behind. The men volleyed and ran forward alternately, until they reached the crest of the hill, when they saw Commandant Botha and two or three companions standing near the large Griqualand West triangulation beacon which stood upon the summit. Most of the Boers made their escape by clambering over the precipitous ridge which forms the south and east boundary, but their brave commander, who held his ground to the last, was killed. The mass of Boers reached their horses, which stood among the trees below the ridge, and rode off pell-mell over the ridge in the distance, with shells from Captain May's guns bursting over them.

Site of Cronje's Laager, Magersfontein.

In this engagement only one of the Boers beside Botha was killed, but seven were wounded. Major Murray had two officers and two men wounded. Colonel Scott-Turner and his men returned to Kimberley without meeting with any further opposition.

It was fortunate that this reconnoissance was made, for the following despatch was taken from the body of Commandant Botha : —

"HOOF LAGER, OCTOBER 23, 1899.

"VELDCORNET BOTHA, *Bakinkop, Weledele Heer.*

"In reply to your inquiry about the taking of cattle in the neighborhood of Kenilworth, I am ordered by the Head Commandant Wessels to assure you that he considers it highly desirable that the same should be captured as soon as possible.

"I am, &c.

"J. B. M. HERTZOG."

The success of this engagement was encouraging, but the fast-increasing numbers of the Boer besiegers and the extension of their lines soon put a check on such excursions.   Early in November Commandant Wessels offered to receive all Africander women and children into his own camp, and at the same time offered safe-conduct to all other women and children to the Orange River.   The first part of his despatch was made public, but not the last.   Wessels's despatch contained the following passage, " And whereas it is necessary for me to take possession of the town of Kimberley, therefore I demand of your Honour that upon receipt of this you, as Commanding Officer, shall forthwith hand over the town of Kimberley with all its troops and forts."

Colonel Kekewich, in acknowledging receipt of Commandant Wessels's despatch, wrote, " Your desire being to obtain possession of Kimberley, you are hereby invited to effect the occupation of this town as an operation of war by the employment of the military forces under your command." The invitation was a challenge.

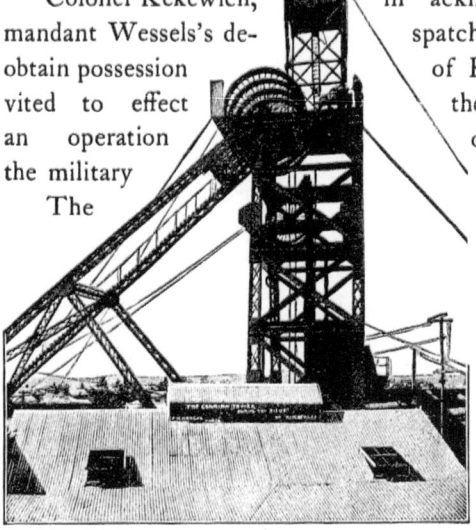

Conning Tower, De Beers No. 1 Shaft.

On the morning of the 6th of November, the Boers fired two shots at Premier mine, and on the following day the first actual bombardment began, from a position about five thousand yards from the mine.   As the compound, containing over two thousand natives, was close to the fort and in the direct line of fire, all these men were taken down into the open mine, where they were protected by an embankment 150 feet high.

On the same day other Boer guns commenced to bombard Kimberley from a ridge nearly five thousand yards distant.   The British guns replied intermittently with a few shots.   Kimberley had no ammunition to waste. The distances were so great that the little popguns in the Kimberley forts frequently " turned turtle," owing to the great elevation at which they had to be fired in order to carry the distance.   The projectiles fell more like meteors out of the sky than shells from modern guns.   For the first few days the Boer shells fell short of the inhabited part of Kimberley.   On the 11th a shell burst in Dutoitspan Road, in front of the Catholic church, and killed an old Kafir woman, which was the only

THE FUNERAL OF COLONEL SCOTT-TURNER AND THE MEN WHO FELL WITH HIM.

casualty from the two hundred shells fired into the town on that day. Seventy shells were fired by the Kimberley artillery during the day. The Kimberley mounted troops also engaged the Boers on the same day near Otto's Kopje mine, and troops under Major Peakman attacked the Boers on Carter's Ridge on their left flank.

Boer Trenches at Magersfontein.

The cessation of active hostilities on Sunday made it a welcome day of rest to all the besieged, and no doubt to the besiegers as well. It gave both sides the opportunity of praying long and hard that their enemies might be confounded. The first bombardment continued for five days, with no further serious casualties on the British side, and the townspeople, appalled at first, began to make light of the danger. More than half the shells fell without exploding, and many children as well as grown people ran up, after each shell struck, to carry off a trophy. These prizes and the fragments and fuses of exploded shells found ready purchasers. The military authorities issued an order forbidding people from collecting these shells and fragments, while a bombardment was going on, owing not only to the risk of death or maiming from the exploding shells, but to the greater danger of the explosion of the dyna-mite mines which were laid around

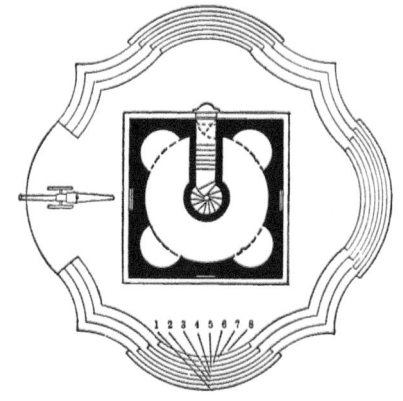

Plan Through Tomb.

the town. The prohibiting order carried this warning, " These mines are at all times ' live,' that is, the fuses and firing arrangements are so arranged that the mines can be fired either automatically or by obser-vation, and they might under certain circumstances be ignited by the enemy's shells." This order should have frightened the average Kimberley urchin, but its apparent effect was to make him all the more eager, for he seemed to think that he had a chance of finding a prize in one of those dynamite mines about which everybody was talking.

The Honoured Dead Memorial.

As the siege dragged along, some of the Imperial officers began to grow impatient. Anticipating the approach of Lord Methuen, they planned a sortie on the 25th of November which was fairly successful; for they took Carter's Ridge, some three miles to the west of Kimberley, and captured thirty-three Boers, including nine wounded. The fighting continued all day, and resulted in a loss to the garrison of six killed and twenty-nine wounded, including Colonel Scott-Turner, and Captains Bowen and Hickson-Mahony. Towards evening the Kimberley troops returned to town, as their ammunition was giving out and it was getting too late to send for more. This was the first fight that many of these men had been in, and their gallantry was greatly creditable, though they were unable to hold the ground they had won. The Boers published their losses as nine killed, seventeen wounded, and *fifteen missing*, instead of thirty-three who were brought into Kimberley.

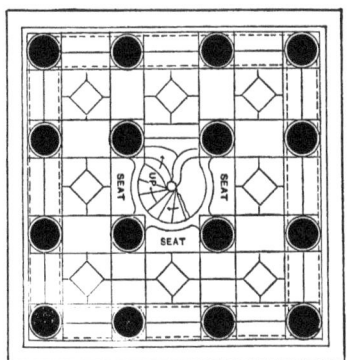

Plan Through Columns.

On November 28th another attempt was made to drive the Boers from Carter's Ridge. Shortly after noon there was great activity in town, and troops were moving in various directions making ready for a sortie. The centre of the advance, commanded by

Colonel Scott-Turner, moved out in the direction of the reservoir and thence along a ridge which gave a little cover.   The first Boer redoubts were quickly taken, and then Colonel Scott-Turner sent for two guns to support him.   He drove the Boers back until they reached their last redoubt, a small fortress dug in the rock, with a coping of sand bags arranged with loopholes.   Colonel Scott-Turner led his last charge and took cover in a small redoubt, only sixty yards from the Boers.   There the Boers had their Armstrong gun.   The Diamond Fields Artillery were obliged to cease firing, owing to the danger of shelling Colonel Scott-Turner and his little body of men.

" Long Cecil" as a Mild Steel Billet.

While this engagement was going on, a small troop of the Diamond Fields Horse attacked the Boer camp in the rear of their redoubt.   This attack was successfully carried out by Captain Shackleton, who dealt the Boers a severe blow.   He captured 149 loaded shells, a considerable quantity of gunpowder, a wagon and span (16 oxen), a Cape cart, and the limber of the gun which Colonel Scott-Turner was trying to take. Among the prizes was a baboon, which proved to be the mascot of the company of Cape Mounted Police stationed at Vryburg, left behind when they evacuated the town somewhat hurriedly.

Meanwhile Scott-Turner and his men were in a most awkward position, lying in a shallow redoubt with its side partly exposed, for the redoubts occupied by the two opposing parties both faced east toward Kimberley, but the one occupied by the Boers was much larger and better built. It was impossible for any of the attacking party to show their heads without receiving a volley from the Boers, and thus one after another of these brave men fell back dead, until finally Scott-Turner took a rifle and was about to fire, when he fell, shot through the head. Major Peakman fought his way with a small force to one of the redoubts, within speaking distance of the survivors. Here he learned that Scott-Turner had been killed, and he at once assumed command as senior officer. He sent a message asking for reënforcements, but, before they

"Long Cecil" in course of Construction.

arrived, darkness had come on, and he decided to withdraw his men to Carter's farmhouse. On the following morning, ambulance wagons were sent out in charge of Captain Robertson under a flag of truce, to collect and bring in the dead. It was then ascertained that Kimberley had lost twenty-two killed and twenty-eight wounded, one of the latter being mortally hurt.

In these encounters, as in all other occasions of their service during the siege, the ambulance corps was notably efficient, and the Kimberley doctors, as a body, did excellent service, both in the field under fire and in the hospitals. Particular mention may fitly be made of Drs. Heberden and Ortlepp, who were attached to the mounted forces, and of Drs. Ashe, Mathias, McKenzie, and Watkins.

The fierceness of this engagement may be judged from Rhodes's statement at a De Beers meeting, held shortly after the siege, " I take this
opportunity of placing it on record that seventy citizen soldiers of Kimberley went to take the position, and out of that number there were only
twenty who were able to creep away alive or unwounded after nightfall."

The 29th of November will long be remembered as the saddest day
during the siege, when the brave men killed in this action were buried
with military and civic honors.

In order to meet the wants of the women and children whose bread-
winners had fallen in battle, a fund was started; to this De Beers gen-

"Long Cecil," just before it was taken out of the Workshops.

erously gave the sum of £10,000, and is now erecting a monument on
one of the most elevated parts of the town, where the heroes who fell in
the defence of Kimberley are to find their last resting-place.

The object of these demonstrations was to detain as many of the be-
sieging force as possible from leaving to join General Cronje at Modder
River, and in this way to assist Lord Methuen in his advance to the
relief of Kimberley.   On December 1st Lord Methuen's first searchlight message reached Kimberley.   This opening of communication was
highly elating and all were eagerly expectant of the news.   Word by
word this message was spelled out, " Please inform the Remount Department, Wynburg, the number marked on the hoof of horse issued to Surgeon O'Gorman of the Kimberley Garrison."

Imagine the disappointment upon receiving this seemingly frivolous

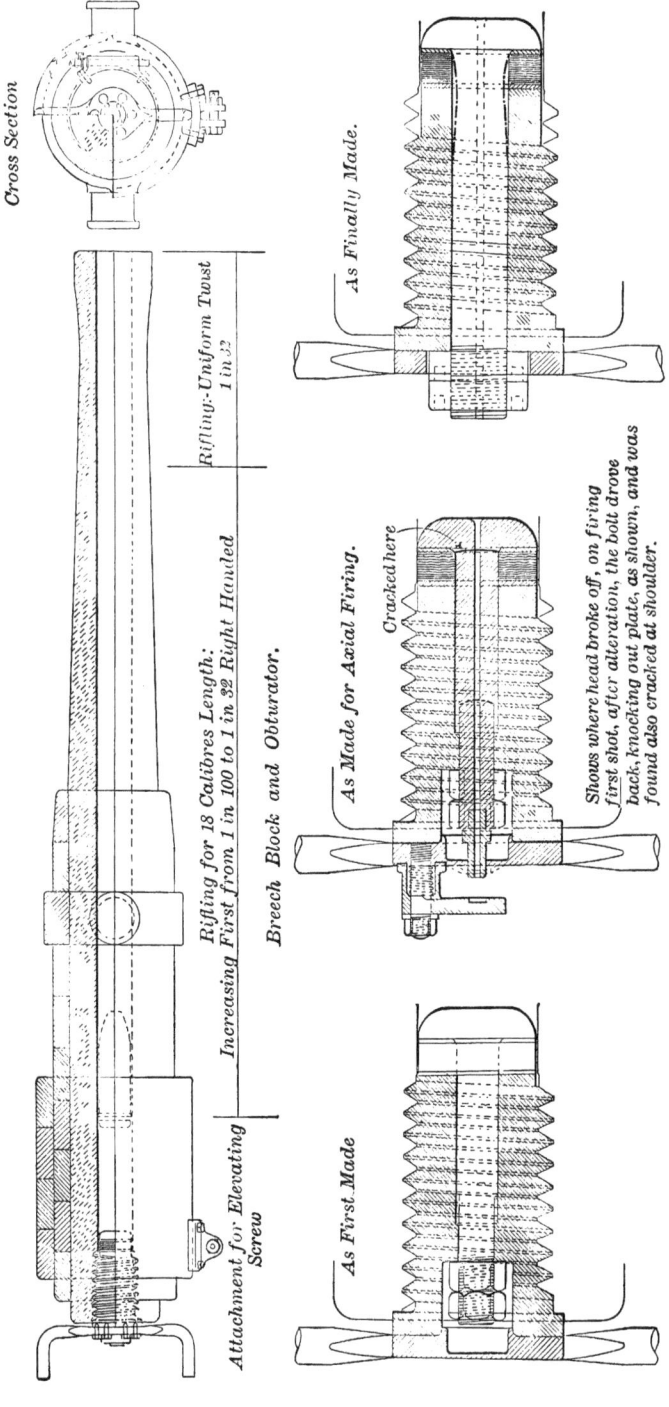

CONSTRUCTION OF "LONG CECIL" GUN.

Cross Section

Rifling:- Uniform Twist
1 in 32

Rifling for 18 Calibres Length:
Increasing First from 1 in 100 to 1 in 32 Right Handed

Breech Block and Obturator.

Attachment for Elevating
Screw

As Finally Made.

As Made for Axial Firing.

Cracked here

Shows where head broke off, on firing
first shot, after alteration, the bolt drove
back, knocking out plate, as shown, and was
found also cracked at shoulder.

As First Made

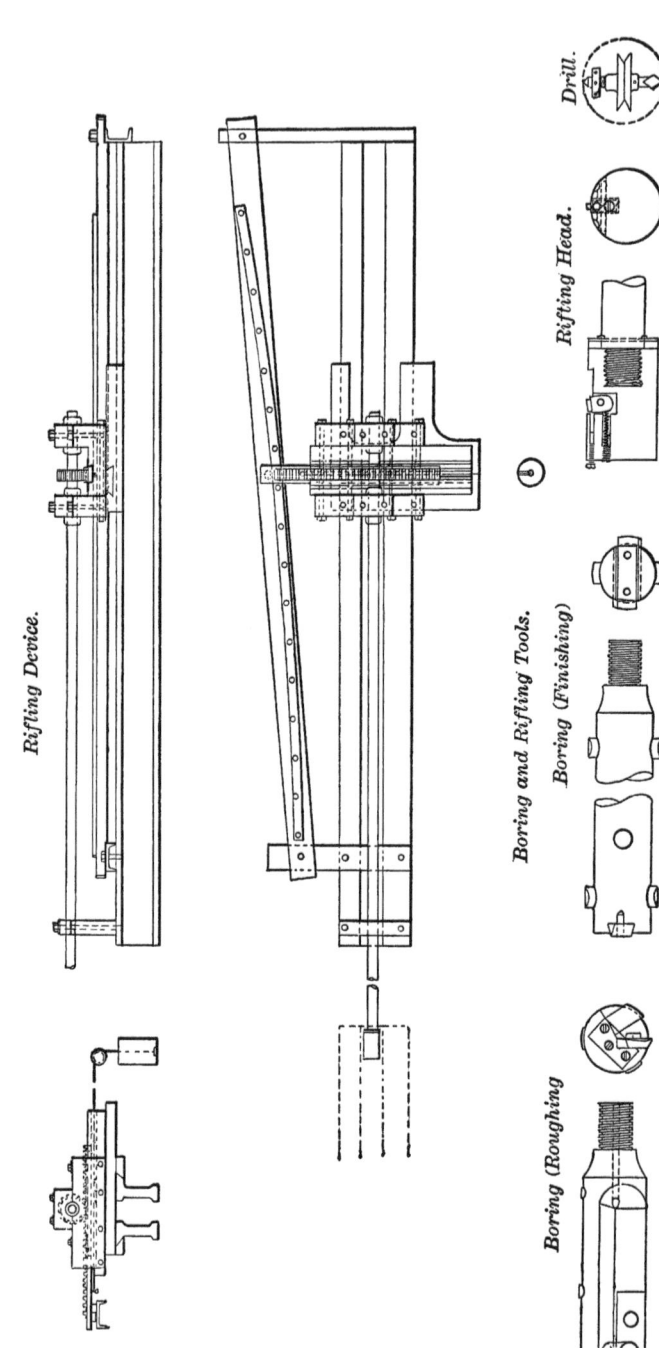

Rifling Device.

Boring and Rifling Tools.

Boring (Finishing)

Boring (Roughing)

Rifling Head.

Drill.

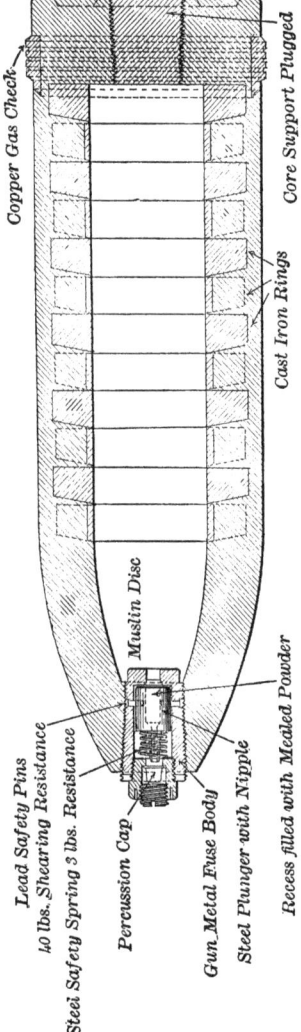

Shell for 4·1 inch B. L. Siege Gun (Long Cecil)

Lead Safety Pins
40 lbs. Shearing Resistance
Steel Safety Spring 3 lbs. Resistance

Copper Gas Check

Core Support Plugged

Cast Iron Rings

Muslin Disc

Percussion Cap

Gun Metal Fuse Body

Steel Plunger with Nipple

Recess filled with Mealed Powder

Shell for 2·5 inch R.M.L. Guns R.A.& D.F.A.

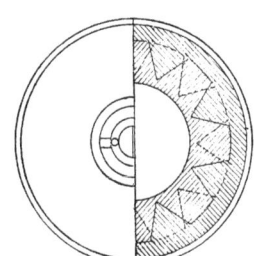

message after the long ten weeks' investment.   It was later reported that this communication was simply a test to ascertain whether the signals were passing between friends or enemies.

On December 11th Lord Methuen met with his first reverse in his march to Kimberley, where he was defeated by Cronje.   I watched this battle from the conning tower, but, as the distance was about sixteen miles, one could see only the bursting of the shells, the big yellow cloud when a Lyddite shell exploded, and the captive balloon giving information as to the position of the Boers.   One could hear the roar of the cannon, which sounded like the breaking of the sea against a cliff.   We waited

" Long Cecil " and the De Beers Men who made it.

anxiously for news of the battle, but for days none came.   The suspense was the more racking from the spread of the report that, as soon as Lord Methuen arrived, there would be an enforced exodus of all the women and children and male non-combatants from Kimberley.   The carrying out of an order to this effect would inevitably have been attended, in my judgment, by great and needless suffering, and the reported determination was rightly resented by all the citizens who had borne so pluckily the strain of the siege.

At length, on December 18th, a week after the battle, we received the first authentic news that Lord Methuen had been defeated at Magersfontein.   This unlooked-for reverse, so blighting to sanguine hopes, cast a deep gloom over the beleaguered town, but there was no lack of heart

"Long Cecil."

in its stubborn defence.   Christmas came, and with it the " Best wishes for Christmas Day and in the coming New Year," from the High Commissioners, and also one from the Queen, " I wish you and all my brave soldiers a happy Christmas; God protect and bless you all."   These messages cheered the garrison and were given a most enthusiastic reception.

On New Year's Day the mayor sent the following message on behalf of the citizens of Kimberley: " The inhabitants of Kimberley humbly beg to send your Majesty New Year's greeting.   The troubles they have

" Long Cecil " firing at the Boers on Carter's Ridge.

passed through, and are still enduring, only tend to intensify their love
and loyalty towards your Majesty's throne and person;" to which the
Queen replied, "Am deeply touched by your kind message and New
Year's greetings. I watch with admiration your determined and gallant
defence, though I regret the unavoidable loss of life incurred."

For some time after the repulse of Lord Methuen, siege life dragged on
from day to day, with nothing very stirring to break the monotony. The
various corps had their "At Homes," when tea would be served, and the
Kimberley Regimental Band would enliven the throngs with martial music.
Every little diversion from the dull routine of camp life was welcome.

2   1

1. "Long Cecil," made in Kimberley.
2. Royal Artillery Gun sent to defend Kimberley.

To provide employment for as many of the inhabitants as possible,
avenues were laid out and macadamized within the municipality of
Kimberley and Beaconsfield, which add much to the convenience and
beauty of the towns. In addition to this street work, Rhodes decided to
make an avenue in commemoration of the siege and to be known as
"Siege Avenue." Years before he had planted rows of grapevines
ranging from 1000 to 2000 feet in length, which were trained upon
trellises, but Siege Avenue was designed to outdo anything in the line of
vine and tree planting that had been done in South Africa. Fourteen
trenches, each over 6000 feet long, were dug. The two centre
trenches were for vines and were 14 feet apart. There were trenches

Casting Shells for Seven-pounders, De Beers Foundry.

The Soup Kitchen.

on either side and at suitable distances for planting orange trees.   The three outside trenches were for ornamental evergreen trees, such as the pepper, eucalyptus, Australian beefwood, and cypress, to serve as a protection to the vines and orange trees from the prevailing winds. Since the siege the vines and trees have been planted, and the wooden trellis has been erected, at a cost of nearly £3000.

When the work of digging the trenches was first started, several hundred natives were employed.   These trenches were about a mile from the nearest fort.   As soon as the Y. A. O. (young artillery officer) in charge saw them, he telephoned about in these words to the O. C. in the conning tower; "A large party of Boers digging trenches just north of Kenilworth. Shall I open fire on them?" The reply came, "Wait and ascertain if they are Boers." Y. A. O. to O. C., "I don't think they are Boers." A min-

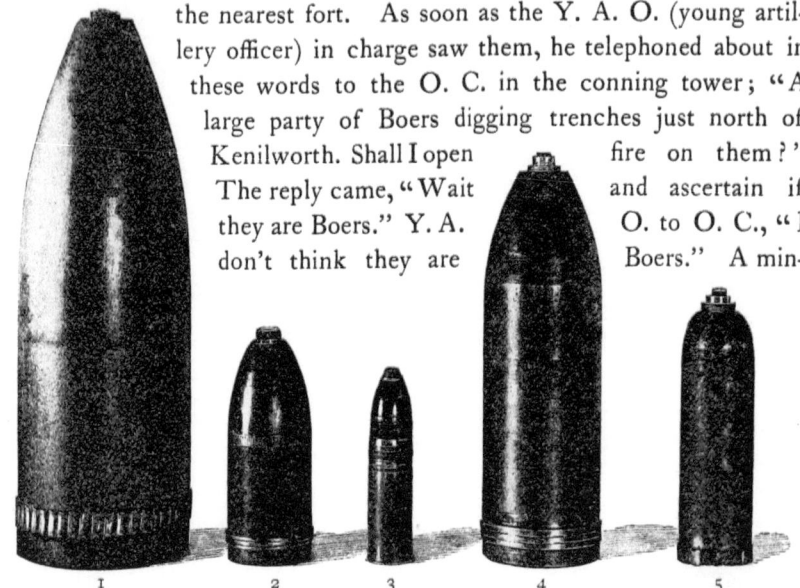

1   2   3   4   5

1. Boer 100-Pounder.   2. Boer 9-Pounder.   3. Boer Pom-Pom.   4. De Beers "Long Cecil,"
28-Pounder.   5. De Beers 7-Pounder.

ute later Y. A. O. to O. C., "They are De Beers workmen digging trenches to plant trees."

The old vines and fruit trees at Kenilworth were of incalculable value to the people of Kimberley, for they bore immense quantities of splendid fruit, which Rhodes sent to the hospitals, to the military camps, and to the citizens generally as far as it would go.

In my own garden there must have been a ton and a half weight of beautiful grapes, which daily reminded one of the old saying, " It is more blessed to give than to receive," as one saw the look of joy on the faces of the women and children as they left the garden.   My mulberry trees were also loaded with fruit, which was eagerly called for.   Some substitute for butter or lard was particularly wanted, for neither of these was

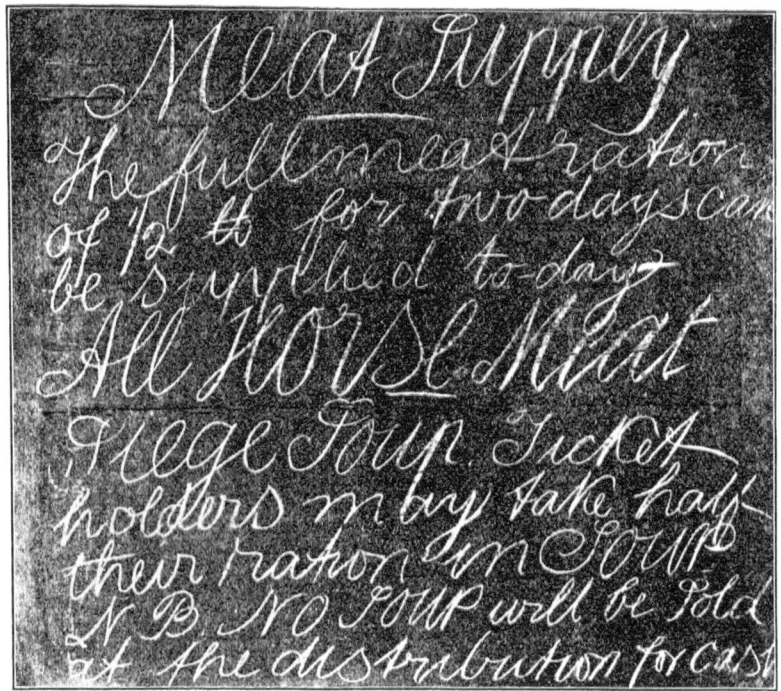

Notice on the Market Buildings during the Siege.

procurable in the town. De Beers again was able to meet this call. In the great warehouses of the company were thousands upon thousands of gallons of lard oil kept in stock — beautifully clear and sweet winter-strained lard oil. Hundreds of people came to the Company's stores daily for this supply. They fried their meat and bread in the oil, and found it much sweeter than most South African butter.

In view of the now obvious certainty of the prolongation of the siege and the call for a gun of greater range and efficiency than any at the command of the garrison, the extraordinary task of the construction of the really formidable piece aptly named " Long Cecil " was undertaken by De Beers Company. It was designed by Mr. George Labram in De Beers workshops. Mr. Edward Goffe, chief draughtsman to the company, describes the making of the gun expertly : —

" Long Cecil " was made from a mild steel billet, $10\frac{1}{2}$ inches diameter and 10 feet long, weighing 2800 pounds, this being turned and rough bored to form the inner tube.

The breech rings were forged from 6 inches $\times$ $2\frac{1}{2}$ inches Lowmoor iron. They were turned and bored, and then shrunk on in place, nine forming the first row shrunk on to the tube direct, and four more the second row over the breech, shrunk on over the first row.   The trunnion ring, carrying the trunnions or bearings, was forged in one piece without weld, from a length of 6 inches square Lowmoor iron, and was shrunk on against a shoulder left on the tube.   The final boring was done after all the rings were shrunk on, the calibre of the gun being 4.1 inches.   The barrel was then ready for rifling.   The rifling is a poly-groove increasing twist, consisting of 32 grooves, each $\frac{1}{4}$ inch wide and $\frac{1}{16}$ inch deep, which, starting with a pitch of 1 in 100 at the breech end,

Inhabitants of Kimberley waiting for their Daily Allowance of Four Ounces of Horse Meat.

and increasing to a pitch of 1 in 32 in a length equal to 18 calibres — 73.8 inches, are uniform at that pitch for the remainder of the length. The curve of increase is the semi-cubical parabola.

The breech block was made of mild steel, screwed to fit the breech with a " V " thread, flattened top and bottom, of $\frac{3}{4}$ inch pitch.   The " De Bange " system of obturation was adopted, the mushroom-headed bolt being of mild steel, annealed in melted tallow, and bored for the friction firing tube.   The pad was made of rings of sheet asbestos soaked in melted tallow.

The carriage was made of steel plates $\frac{1}{4}$ inch thick, cut to shape, and riveted together in pairs, with gun-metal blocks between, for trunnion and axle bearings.   The wheels were taken from a portable engine, bushed with gun metal and bored to fit the axle, whose ends were covered with brass dust-caps.

The shells weighed 29 pounds each, loaded with their bursting charge of one pound of powder, and were fitted with the percussion fuse devised by Mr. George Labram.

The making of the gun was begun on the 26th of December, 1899. It was proved on the 19th of January, 1900, and went into action January 23. From then up to the date of the relief of Kimberly 255 shells were fired from the gun, mostly at ranges approximately 5000 and 6000 yards, the distances of two of the positions of the enemy, which were easily reached with elevations of 12° and 15° re-

Band of the Kimberley Volunteer Regiment playing at the Mounted Camp during the Siege.

spectively, with powder charge of 5 lbs. With the same powder charge another position 8010 yards distant was effectively reached with an elevation of 24° 15'.

The illustrations on previous pages show graphically how "Long Cecil" was made.

The cut on page 643 represents the finished gun ready to go into action. On page 639, the upper figure shows the general construction of the gun barrel; the lower figures on the same plate show the breech block and obturator. The rifling device is given on page 640, and on the same plate the boring and rifling tools are represented.

The upper figure on page 641 gives the details of the construction of the shells and fuse for "Long Cecil," and underneath is shown the shell made for the 2.5-inch popguns with which Kimberley was defended.

The Mafeking "Long Tom," manned by Mercenaries.

The manufacture of these small shells was undertaken in November, and many were thrown into the Boer camps "with C. J. R.'s Compliments" stamped on them.

The powder for charging these shells was fortunately at hand.   The old Central Company of Kimberley mine had a large stock of black powder which was used for blasting in the open mine, as far back as 1888.   When De Beers Consolidated Mines took over this Company, the powder was removed and placed in a magazine on the veld, a mile

"Long Tom," en route to Kimberley.

beyond the Company's washing machines.   Shortly after the opening of
the siege I had stock taken of the contents of outlying magazines and
brought to light *three and a half tons* of good black
powder of various grain.   This discovery was of
much service, for it enabled the garrison to respond
more frequently to the fire of the Boers, and made
the construction and use of " Long Cecil " possible.

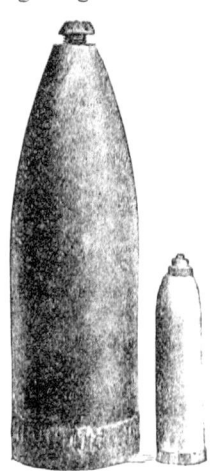

At first, the shells cast in our foundry were not
all perfect, and the bursting of some of them led to
greater care in testing all under hydraulic pressure.
Ring shells made by De Beers are shown on page
641.   Rings with jagged or saw-toothed edges were
first cast; these were stacked one over another
in the mould, and the outer shell cast around them.
When the bursting charge of powder exploded, these
rings were broken into a hundred pieces and thrown
in all directions.

The Boer 100-Pound
Shell and the Imperial
9-Pound Shell.

The Boers evidently resented the firing of " Long
Cecil," for on the 24th of January they kept up a fierce cannonade,
throwing about five hundred shells into Kimberley.   A French officer,
who was at Kampfersdam during a part of the siege, says
that " Long Cecil " did good practice, and with one shell
killed seven Boers, only two less than the Boers killed
with eight thousand shells.   The heavy and continuous
firing which took place on the 25th of January and following days caused many to build " shell-proofs " for the
protection of the women and children.
On the morning of the 8th of February, at about eleven o'clock, I was
in the conning tower, and noticed an
immense volume of smoke belched
forth from a gun on Kampfersdam
tailing heap.   I remarked to those
near me that the Boers had brought a

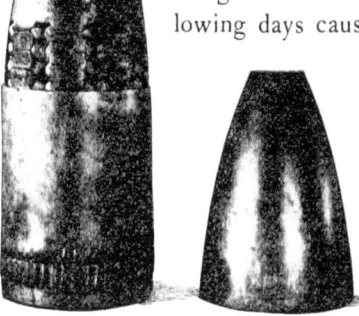

Showing Interior of Boer 100-Pound Ring
Shell and Shrapnel combined.

" Long Tom " against us at last.   In
a few seconds the bang of the gun
was heard, followed a little later by a sound almost indescribable as the
shell came whizzing through the air.   It has been likened not unfitly to
the roar of an express train passing at full speed.   Then a cloud of red

dust was seen where the shell had struck, shortly followed by the crash of the explosion. In the vicinity the air was filled with fragments of the shell or bullets of the shrapnel, which flew on with a singing " ping, ping, ping." Twenty-five of these shells were fired on that day, many of which did not explode. One was brought in and measured, and found to be fifteen centimetres, or about six inches, in diameter.

Premier Studio, showing Effect of 100-Pound Shell.

The " Long Tom " which was brought to Kimberley was a captured piece which had been struck by a shell on the muzzle and broken. This gun was taken to Pretoria or Johannesburg, where the broken part of the muzzle was cut off and a band shrunk on the injured end. The illustration of this gun on page 650, on a railway truck en route from Pretoria to Bloemfontein, shows the method of moving these guns without a limber. The gun was noted for bad shooting. On the afternoon of the 9th of February the Boers turned the gun on the herd of cattle which were being driven in for the night. This shot missed the cattle by half a mile to the left. Three more shots were fired, all falling wide of the target at which they were aimed. The illustrations here given of the effect of these shells are more graphic than words.

On the first day the big gun was fired, the Buffalo Club was struck and sustained considerable damage, and a few private buildings were more or less injured. On the 9th the firing of " Long Tom " commenced at daybreak, and was continued at intervals throughout the day until six P.M., when the last shot was fired. This shot killed George

Effect of a 100-Pounder.

Effect of a 100-Pounder.

Labram, one of the most able men in the service of De Beers Company. He had entered his room in the Grand Hotel only a minute before.   The

Effect of a 100-Pounder.

shell passed through the roof and three brick walls before reaching Labram's room.   During the same day the wife and son (fifteen months old) of Mr. Robert Solomon were struck by the fragments of a shrapnel shell, which burst as it came through the outer wall of the building in which they were temporarily staying.   The child was killed instantly, but the poor mother was taken to the hospital, where she died, thirty-six hours afterward, from her injuries.

During Saturday the firing continued, and buildings in every quarter of the town were struck. The peril of the unprotected people was appalling.   There was the greatest activity in building shelters for the women and children.   The tailing heaps were tunnelled, and the miners erected long rows of tunnel timbers against the débris embankments, and covered them with corrugated iron. Gangs of natives soon protected these galleries with débris several feet deep. Still there were thousands unprovided with any shelter except the thin roofs and walls of their houses, which were absolutely useless against a hundred-pound shell travelling at the rate of

Mr. Compton's Drawing-room barricaded for Shelter from Boer 9-Pounders.

a thousand feet a second.   When firing ceased, about midday, there was a sigh of relief from many hearts, for it was thought that firing would not be resumed until Monday morning.

The funeral of Mr. Labram was timed to leave the hospital at eight o'clock in the evening, as it was thought unsafe to have the funeral by day. He was buried with full military honors, and, as the hour for departure from the hospital approached, the streets were thronged with anxious and sorrowful people.   The troops consisted of regulars, the various volunteer corps, and members of the Town Guard.   My carriage contained Colonel Kekewich, Mr. Rhodes, Mr. Pickering, and myself.   Other carriages followed, and hundreds who were unable to procure conveyances, owing to the scarcity of horses, joined on foot.

Precisely at eight o'clock the procession moved from the hospital, but, before it had gone a hundred yards, the bugler in the conning tower

Shell-proof, constructed by the Public Works Department.

gave the well-known notes which meant that the big Boer gun had been fired.   The band was playing the funeral march at the time, so that few people in the immediate vicinity of the hospital heard the warning notes. Shortly, however, the boom of the cannon was heard, followed by that never-to-be-forgotten hiss of the shell passing through the air.   Traitors in the town had given the Boers information as to the time of the funeral, and doubtless signalled from some elevated place to the besiegers at Kampfersdam the moment the procession started.   There was a sigh of relief as the fearful shell passed over the heads of the multitude, and fell harmless in vacant land behind the hospital.   Colonel Kekewich gave orders for the band to cease playing, and that all carriage lights be put out.   It was a grim and silent funeral.   Shot after shot came thundering over or into the town, as the procession passed through it. At last, as we approached the cemetery, we could see the flash of the gun as it was fired.

Excavations in the Tailing Heaps at Beaconsfield, used as Shelters.

While the last rites were said, the voice of the venerable archdeacon was drowned by the roar of the gun and the hissing of the shells.

When the ceremony was over, every one hastened home to seek whatever cover could be found. Crowds of people were massed for hours behind flimsy walls, which could not protect them, but even this slight pretence of shelter was comforting. The terrible night of the 10th of February, 1900, will never be effaced from the memories of those who passed through it.

Shell-proof, after the Siege.

So great was the strain upon the nerves of the people that it was necessary that some one should come to their help, and as usual that "some one" was Rhodes. Early on Sunday morning he came to my house and said: "You told me, some time ago, that you could put a lot of people down in the mines, and I think the time has now come when we must do it. Will you get your mines ready so that the people can be sent down this evening?"

I supervised the work at De Beers mine, and my son was its director at Kimberley. Tunnels one thousand and twelve hundred feet below the surface were cleaned out — sanitary arrangements were provided, and, early in the afternoon, both mines were ready for occupation. Rhodes had sent a notice about town which is given as an illustration on page 658, and speaks for itself.

Shell-proof at the Convent.

Attention was called to it by the ringing of a bell. Crowds flocked to both shafts during the afternoon and evening; and before midnight nearly three thousand women and children were safely housed, deep down in the subterranean passages of the mines. There was discom-

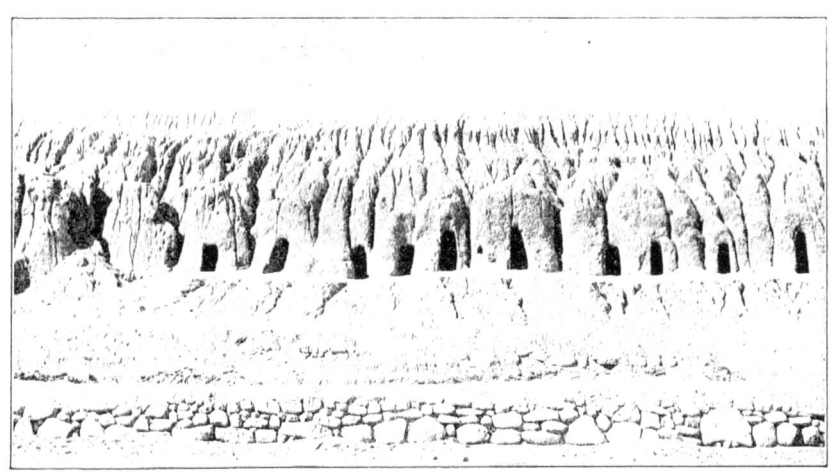

Shell-proof Dugouts.

fort, of course, in this rude lodging, but all were happy in the thought that they were beyond the sound of screeching shells, and out of danger.

I have never seen so much patience and pluck shown by women as was shown by those in the mines. There was no sign of fear in going

2 U

down in the rough mine cages, and when they reached the station, they found to their joy that the tunnel was like a beautiful arcade, brilliantly lighted with electric rays.   Food was served several times a day, and time went so quickly that dates were lost sight of and days and nights became hopelessly mixed.   One lady asked me, " Is this yesterday or to-day or to-morrow ? "   When the glad news was brought to them that Kimberley was relieved, they scarcely believed it, and many preferred to remain in the mines rather than take any chances of hearing " Long Tom "

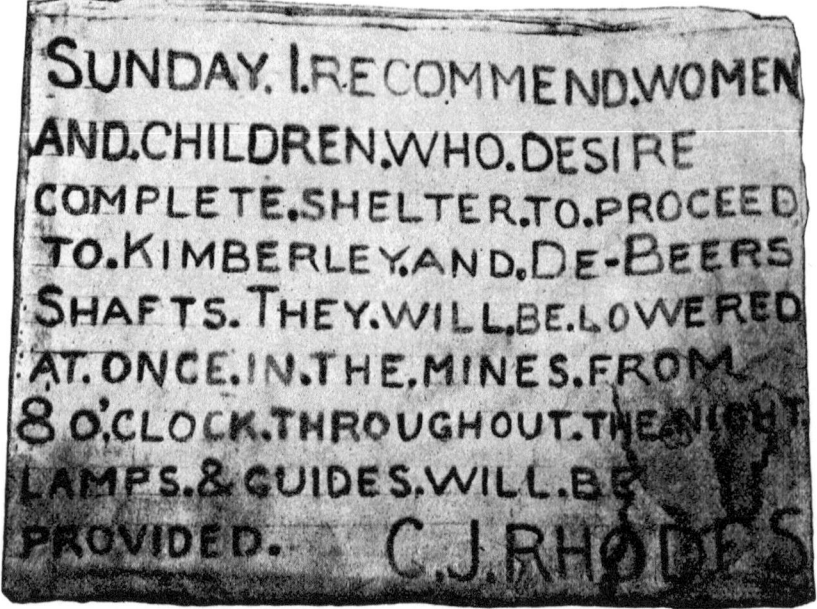

SUNDAY. I.RECOMMEND.WOMEN AND.CHILDREN.WHO.DESIRE COMPLETE.SHELTER.TO.PROCEED TO.KIMBERLEY.AND.DE-BEERS SHAFTS. THEY.WILL.BE.LOWERED AT.ONCE.IN.THE.MINES.FROM 8 O'CLOCK.THROUGHOUT.THE NIGHT. LAMPS.& GUIDES.WILL.BE PROVIDED.   C.J.RHODES

Notice sent round the Town during the Shelling by the Boers' 100-Pound Gun.

give a parting roar, or the awful screech of a flying shell.   On Friday morning all were brought to the surface, thankful for the few days of peace and safety.

The illustration on page 660 shows the people at the shaft waiting to be sent down.   That all were taken down into the mine and brought up again without the least mishap speaks well for those who carried out the details at each mine.

The Boers fired a few shots between eleven and twelve o'clock on the 15th, from " Long Tom."   They knew before we did that a British column was nearing Kimberley, for they had telegraphic communication between all their camps ; and while the column was slowly advancing they were using every effort to remove the big gun, which they did success-

Shelters for Women and Children during firing of 100-Pound Boer Gun.

fully. Over eight thousand shells had been fired by the Boers into Kimberley and its fortifications, with the result that, out of a total population of fifty thousand, only nine were killed, and the majority of these were women and children.

At two P.M. a huge cloud of dust rose in the distant southeast, and shortly afterward one could see mounted troops advancing, and a heliographic message informed the officer commanding Kimberley that it was the relief column. The news spread like wildfire, and from every place which afforded a view, thousands of eager eyes were scanning the veld

Too Late! These two Siege Guns arrived after the Siege.

The United States Consulate.

Women and Children waiting to be lowered down De Beers Mine.

for a glimpse of the troops.   The few public conveyances which were left in Kimberley were quickly taken to convey people to meet the column.

As soon as I received the news, I made an effort to obtain a cab, but found it impossible.   A small spring wagon drawn by a mule and driven by a Kafir passed my door at this time.   Recognizing it as a De Beers fruit and vegetable wagon, I commandeered it, and in company with Captain Bowen was driven to the Sanitorium, which afforded a good view of

Double-decked Cage loaded with People and their Bedding being hoisted from the 1200-foot level of De Beers Mine.

the advancing troops.   With my field glasses I saw the troops slowly advancing, and as they rounded a hill near the farmhouse on De Beers farm, Benaauwdheidsfontein, the Boer gun at Olefantsfontein, southeast of Premier mine, fired a few shots, but the relief column had a battery which soon silenced the Boer gun.   Having telephoned for my light carriage and horses, I soon joined the great crowd which thronged every road leading toward the advancing troops.   It was seven o'clock before they got into camp.   Thus Kimberley was relieved after a long and eventful siege of 124 days.   My old friend, Colonel Rimmington of Rimmington's Scouts, was among one of the first arrivals.   Thinking that I was doing him a good turn, I put him up at my house.   I am sure he enjoyed the bath, but when I went to call him the next morning, at four o'clock, he was gone. Missing the bedclothes, search was made in the garden, and there the poor old tired soldier, wrapped up in clean sheets and blankets, was lying on the ground, sleeping as only a weary soldier can sleep.   He had found the house too stuffy after sleeping so long on the veld.

General French moved at daybreak the morning after his arrival, taking with him about half of his column and four batteries of field guns. He gave battle to the Boers north of Kimberley, and cleared them out of their late haunts.   The Boers left one gun behind, an old Armstrong gun, the limber of which was captured November 25th.   On Saturday morning at daybreak General French left for Paardeberg, taking those of his troops who had rested on Friday, and the others followed the next day.

Boer Gun captured at Dronfield.

It has often been asserted that Rhodes interfered with the military. He did suggest to Lord Methuen that there were more ways into Kimberley than the one over the Magersfontein and Spytfontein kopjes, and mentioned the route over which General French came when he relieved Kimberley. He proposed that small forts be built, every three or four miles, advancing from Modder River and keeping up the base of supplies at that place. His plan was substantially the blockhouse system, which the army later adopted, only that forts, instead of houses, would have been necessary, as the Boers then had cannon.   The only reply to this suggestion was an order to the officers commanding Kimberley to have no communication whatever with Mr. Rhodes on military subjects.

Fortunately for the defence of Kimberley, Rhodes's energies were unflagging, in spite of rebuffs. Throughout the siege no appeal for assistance was ever made to him, nor even a want intimated on the part of the garrison, that he did not do all in his power to meet at once. The formation of the Kimberley Light Horse was due to him. So, too, was the fortification of the village of Kenilworth and the outlying washing machines. The making of the gun, "Long Cecil," was by his order. The employment of thousands of idle hands in street-making in and around Kimberley was at his suggestion, and paid for by De Beers Company, thus assuring a support more welcome than charity. The undertaking of the soup kitchen was his proposal. From the great De Beers dairy milk was supplied to the hospitals, the sick at home, and to a depot where it was distributed under the supervision of a committee.

Surrender of General Cronje. Reproduced by kind permission of Mr. L. S. Amery.

Fruit and vegetables from De Beers gardens were sent to the hospitals, to the camps, and to the poorer families of the town. New gardens were started to enlarge the supply. The ice plant was kept constantly running, and ice furnished to the hospitals, the garrison, and the citizens generally. In everything contributing to the efficiency of the defence and the welfare of the people of Kimberley, Rhodes took the keenest interest, and, whenever possible, a most active part.

A few days before the relief column came in, there was a meeting of a considerable number of the leading citizens of Kimberley, with the object of sending a message to Lord Roberts to inform him of the situation and ascertain whether there was any immediate prospect of relief. Rhodes, the mayor and ex-mayor, a judge of the High Court, several members of Parliament, the author, and other citizens were present, and it was decided to send the following message to Lord Roberts, who was then at

Lord Roberts and General Cronje seated under the Trees on the Modder River at Paardeberg.

Modder River. The military censor at first refused to send it, but the officer commanding finally decided to permit its transmission in an abridged form.

"KIMBERLEY, 10th February. — On behalf of the inhabitants of this town, we respectfully desire to be informed whether there is an intention on your part to make an immediate effort for our relief. Your troops have been for more than two months within a distance of a little over 20 miles from Kimberley, and if the Spytfontein hills are too strong for them, there is an easy approach over a level flat. This town, with a population of over 45,000 people, has been besieged for 120 days, and a large portion of its inhabitants have been enduring great hardships. Scurvy is rampant among the natives; children, owing to lack of proper food, are dying in great numbers, and dysentery and typhoid are very preva-

lent.   The chief food of the whites has been bread and horseflesh for a
long time past, and for the blacks meal and salt only.   These hardships,
we think you will agree, have been borne patiently and without complaint
by the people.   During the past few days the enemy has brought into
action from a position within three miles of us a six-inch gun, throwing a
hundred-pound shell which is setting fire to our buildings, and is daily
causing death among the population.   As you are aware, the military
guns here are totally unable to cope with this new gun.   The only
weapon which gives any help is one locally manufactured.   Under these
circumstances, as representing this community, we feel that we are justified in asking whether you have any immediate intention of instructing
your troops to come to our relief.   We understand that large reënforce-

General Cronje as a Prisoner of War.

ments have recently arrived at Cape Town, and we feel sure that your
men at Modder River have, at the outside, 10,000 Boers opposed to them.
You must be the judge as to what number of British troops would be
required to deal with this body of men, but it is absolutely essential that
immediate relief should be afforded to this place."

The reply received from Lord Roberts was sent to Colonel Kekewich, and was as follows : —

" I beg you will represent to the Mayor and Mr. Rhodes, as strongly
as you possibly can, the disastrous and humiliating effect of surrender
after so long and glorious defence.   Many days cannot possibly elapse
before Kimberley will be relieved, as we commence active operations
to-morrow.   Future military operations depend in a large degree on your
maintaining your position a very short time longer."

Reception of Lord Roberts and Lord Kitchener, Kimberley Town Hall.

What message or messages were sent by the military from Kimberley that conveyed to the mind of Lord Roberts that there was even the remotest chance of the citizens of Kimberley surrendering to the Boers, will probably always remain a military secret. Suffice it to say, however, that such a thought never entered the minds of the men of Kimberley, who would rather have died in their trenches than have surrendered, so long as any scrap of food remained.

The saving of Kimberley from the attack of the Boers was due to the natural strength of the position and its improvised fortifications; to the courage of the citizen soldiers, and the small force of Imperial troops; to the indomitable spirit of Cecil John Rhodes, the chairman of De Beers Company, whose pent-up energies found vent in devising ways and means for adding to the plans of defence; to the forethought of the De Beers men in charge of buying food for man and beast, who laid in supplies far in excess of any expected emergency; and possibly, least of all, to the disinclination of the Boers to attack energetically a fortified town so long invested by commandoes greatly outnumbering the garrison under arms.

APPENDIX II

WINDING ENGINES FOR THE MAIN SHAFT, KIMBERLEY MINE

THIS plant was designed by me for De Beers Consolidated Mines Limited, and built by Messrs. James Simpson & Co., of London, and consists of a pair of inverted vertical tandem compound-condensing engines, driving two reels, which are capable of carrying 1800 feet of flat ropes each. The principal dimensions of these engines are as follows, viz.: —

Diameter of high-pressure cylinders, two	19.5 in.
Diameter of low-pressure cylinders, two	34 "
Stroke of all cylinders	48 "
Diameter of each air pump, two	24 "
Stroke of each air pump	16 "
Diameter of steam cylinder of the reversing engine	7 "
Diameter of oil cataract of the reversing engine	4.75 "
Stroke of reversing engine	18 "
Diameter of each high-pressure steam pipe	6 "
Diameter of each low-pressure steam pipe	10 "
Diameter of each exhaust pipe to the condenser	14 "
Diameter of each high-pressure valve	6 "
Diameter of each low-pressure valve	8 "
Diameter of each high-pressure piston rod	3.5 "
Diameter of each low-pressure piston rod	4.5 "
Diameter of crank-pins	5.5 "
Length of crank-pins	7 "
Diameter of each main bearing	14 "
Length of each main bearing	32.5 "
Diameter of main crank-shaft in the middle	16 "
Smallest diameter of each reel	9 ft. $1\frac{1}{8}$ "
Size of flat ropes used	$3\frac{5}{8}$ in. by $\frac{9}{16}$ "
Capacity of blue ground skips	100 cubic feet.

These engines were intended to hoist six loads, each weighing 1600 lbs., from the 1000-foot level in 45 seconds, including filling, starting, discharging, and stopping; but they do it in from 30 to 35 seconds.

All the steam cylinders are fitted with the Corliss valve gear, having vacuum dash-pots, the cut-off being effected by the same lever that works the throttle valves.

Reversing is effected by ordinary links worked by eccentrics fitted on the tail-shafts; the reversing engine being fitted with a floating lever so that the motion of the piston coincides exactly with the motion of the small hand lever.

The two high pressure cylinders exhaust into the receiver, which is 5 ft. diameter by 18 ft. long, fitted with sixty-eight 2-inch wrought-iron tubes, through which live steam from the high-pressure jackets, but at a reduced pressure, is constantly circulating.   The object of this receiver is to supply the low-pressure cylinders with a considerable volume of dry steam to facilitate a quick starting away.   An 8-inch balanced throttle valve admits steam to the high-pressure cylinders, and a similar valve, 12 inches in diameter, admits steam from the reheater to the low-pressure cylinders.

Each high-pressure cylinder is jacketed with live steam at full boiler pressure, the water of condensation together with a certain amount of steam passing through a Watts pressure regulator, which reduces the pressure in the jackets of the reheater and low-pressure cylinders to about 30 lbs.   The final water of condensation is discharged automatically by a displacement trap into the hot well.

Each air pump of the ordinary marine type is worked off the cross-head.   The condenser, 6 ft. diameter by 16 ft. long, fitted with 125 wrought-iron tubes $3\frac{1}{2}$ in. outside diameter and 16 ft. long, is situated just outside the winding-engine house.   All the water pumped from the mine passes through this condenser on its way to the floors.

A circulating pump on the end of one of the tail-shafts supplies water for jet injection whenever the mine pumps are not supplying sufficient water to condense the steam.

L. I. SEYMOUR,
Mechanical Engineer for D. B. C. M. Ltd.

APPENDIX III

REPORT ON PUMPING PLANT FOR KIMBERLEY MINE

THE new plant consists of a vertical triple-expansion condensing engine, having cylinders $15\frac{1}{2}$ in., $23\frac{1}{4}$ in., and 37 in. diameter respectively, with a stroke of 36 in.

The high and intermediate pressure cylinders are arranged tandem, over one crank, the low pressure working on the other, which is placed at the opposite end of the crank-shaft and at an angle of 90° with the other.

A double acting air pump is driven by a rocking lever from one crosshead and a feed pump in the same manner from the other engine.

A cast steel spur-wheel, 3 ft. 9 in. pitch diameter, is keyed on the engine shaft, and drives a second shaft 27 in. diameter by gearing with a spur-wheel 30 ft. pitch diameter made of cast iron, with teeth 6 in. pitch by 30 in. face. The gears were made by Fraser & Chalmers, of Chicago, U.S.A., the crank-shafts by Sir J. Whitworth, of Manchester, and the rest of the work, including the pumps, by Messrs. J. Simpson & Co. Ltd., of London. A cast-steel crank is keyed on the second motion shaft, and drives the T bob by a pitman with 35 ft. centres.

On the nose of the bob is hung the spear rod 1250 ft. long, of hard pine, 14 in. square for the first 500 ft., 12 in. square for the second, and 10 in. square for the remainder.

The total weight of the rod, including strapping plates and poles, is 61 tons, which will be partially balanced by a counterweight on the top bob, and partly by a second bob placed at the 1200-ft. level.

Attached to the spear rod at the 250 ft., the 500 ft., the 750 ft., the 1000 ft., and the 1200 ft. levels are cast-iron plungers 14 in. diameter, having a stroke of 10 ft., each of which forces the water to the next station above through a riveted steel pipe, 14 in. diameter, with joints riveted together.

The foundations for the driving machinery are made of concrete, with the proportion of cement to stone of 1 : 9 on the average.

L. I. SEYMOUR, Mechanical Engineer.

APPENDIX IV

THE relative values of South African coals are shown in the following table, exhibiting tests made with the Beeley boilers at De Beers mine : —

ENGLISH COAL

1890.		Pounds of feed-water evaporated per lb. of coal from and at 212° Fahr.
April 29.	Nixon's Steam Navigation coal . . .	11.67
July 16.	Nixon's Steam Navigation coal (1st test) . .	10.22
July 16.	Nixon's Steam Navigation coal (2d test) . .	11.11
Aug. 4.	Nixon's Steam Navigation coal . . .	10.40

SOUTH AFRICAN COAL

Nov. 7.	Vaal Drift Mine, Transvaal	4.515
1891.		
Feb. 12.	Newcastle, Natal	9.520
March 24.	Indwe, Colonial	7.090
March 25.	Lewis and Marks, Transvaal	6.734
June 29.	Newcastle, Natal	8.520
July 20.	Kroonstad, Free State	7.084

The relative cost and service of Welsh and Indwe coal delivered at the mines are approximately as follows : —

	£	s.	d.
A ton of 2000 pounds Welsh coal cost . . .	7	0	0
A ton of Indwe coal	1	19	0

Welsh steam coal will evaporate about eleven pounds of water per pound of coal from and at 212° Fahr., and Indwe coal about seven pounds. Indwe coal is, therefore, worth about $60\frac{2}{3}\%$ of Welsh coal, and costs about £3 4s. for the same evaporating value contained in a ton of Welsh coal costing more than double this sum.

APPENDIX V

TABLE OF STATISTICS OF DE BEERS CONSOLIDATED MINES LIMITED,

Since its Formation, 1st April, 1888.

	Year Ending	Number of Loads of Blue hoisted.	Number of Loads of Blue washed.	Number of Carats of Diamonds found.	Amount realized by Sale of Diamonds.	Number of Carats per Load of Blue.	Amount realized per Carat sold.	Amount realized per Load.	Cost of Production per Load.	Number of Loads of Blue on Floors at close of Year, exclusive of Lumps.	Dividends Paid. Amount.	Dividends Paid. Equal to
					£ s. d.		s. d.	s. d.	s. d.		£ s. d.	
De Beers and Kimberley Mines	March 31, 1889 prior to Consolidation	944,706	712,263	914,121	901,818 0 5	1.283	19 8¾	25 3¾	9 10½	476,403	188,329 10 0	5 per cent
	March 31, 1890	2,192,226	1,251,245	1,450,605	2,330,179 16 3	1.15	32 6¾	37 2¾	8 10½	1,576,821	789,682 0 0	20 per cent
	March 31, 1891	1,978,153	2,209,588	2,020,515	2,974,670 9 0	.99	29 6	29 3¾	8 8	1,525,386	789,791 0 0	20 per cent
	*June 30, 1892	3,338,553	3,239,134	3,035,481	3,931,542 11 1	.92	25 6	23 5	7 4.3	1,624,805	1,382,134 5 0	35 per cent
	June 30, 1893	3,090,183	2,108,626	2,229,805	3,239,389 8 6	1.05	29 0.6	30 6	6 11.6	2,606,362	987,238 15 0	25 per cent
	June 30, 1894	2,999,431	2,577,460	2,308,463½	2,820,172 3 9	.89	24 5.2	21 10.6	6 6.8	3,028,333	987,238 15 0	25 per cent
	June 30, 1895	2,525,717	2,854,817	2,435,541½	3,105,957 15 8	.85	25 6	21 8	6 10.8	2,699,233	987,238 15 0	25 per cent
	June 30, 1896	2,698,109	2,597,026	2,363,437¾	3,165,382 1 4	.91	26 9.4	24 4.5	7 0.1	2,800,316	1,579,582 0 0	40 per cent
	June 30, 1897	2,515,889	3,011,288	2,769,422¾	3,722,099 3 3	.92	26 10.6	24 8.6	7 4.3	2,304,917	1,579,582 0 0	40 per cent
Premier Mine	June 30, 1897	271,777	—	—	—	—	—	—	—	271,777	—	—
De Beers and Kimberley	June 30, 1898	3,332,688	3,259,692	2,603,250	3,451,214 15 3	.80	26 6.2	21 2.1	6 7.4	2,377,913	1,579,582 0 0	40 per cent
Premier Mine	June 30, 1898	1,146,984	691,722	189,356¼	196,659 18 8	.27	20 9.3	5 8.2	2 7.1	727,039		
De Beers and Kimberley	June 30, 1899	3,504,899	3,311,773	2,345,466	3,471,060 12 1	.71	29 7.2	20 11.5	6 7.7	2,937,784	1,579,582 0 0	40 per cent
Premier Mine	June 30, 1899	2,032,771	1,662,778	496,762¼	567,360 11 7	.30	22 10.1	6 9.8	2 3.3	1,097,032		
De Beers and Kimberley	June 30, 1900	1,673,664	1,522,108	1,000,964	1,794,222 9 11	†.67	35 10.2	23 6.9	7 6.2	2,722,595	No dividend owing to the war and siege of Kimberley.	
Premier Mine	June 30, 1900	980,210	736,929	220,762½	276,191 6 6	.30	25 0.2	7 5.9	2 7.5	1,340,313		
De Beers and Kimberley	June 30, 1901	2,120,397	2,616,873	‡2,000,495¾	‡3,959,383 0 11	.76	39 7	30 3.1	8 5	2,226,119	1,579,582 0 0	40 per cent
Premier Mine Bultfontein	June 30, 1901	1,571,631 148,086	1,517,981	447,399¾	610,831 4 10	.295	27 3.7	8 0.6	3 0.9	1,393,963		

* These Figures are for a period of Fifteen Months. † Exclusive of "West End" test.
‡ Exclusive of Tailings Diamonds, amounting to 50,147½ carats, valued at £58,484 12s. 6d.

INDEX

INDEX

For EU product safety concerns, contact us at Calle de José Abascal, 56–1°,
28003 Madrid, Spain or eugpsr@cambridge.org.

www.ingramcontent.com/pod-product-compliance
Ingram Content Group UK Ltd.
Pitfield, Milton Keynes, MK11 3LW, UK
UKHW040618240426
470322UK00010B/188